Simplified Engineering
for Architects and Builders

||||||||||||||||||||||||||||||||||||||||||||||||||||||||||||||||||||||||||||

Simplified Engineering for Architects and Builders

||||||||||||||||||||||||||||||||||||||||||||||||||||||||||||||||||||||||||||||||||||||||

Harry Parker, M.S.

Emeritus Professor of Architectural Construction
University of Pennsylvania

FIFTH EDITION

prepared by

Harold D. Hauf, M.S.

Fellow American Institute of Architects
Fellow American Society of Civil Engineers

A Wiley-Interscience Publication

JOHN WILEY & SONS

New York • Chichester • Brisbane • Toronto

Library of Congress Cataloging in Publication Data:

Parker, Harry, 1887–
  Simplified engineering for architects and builders.

  "A Wiley-Interscience publication."
  1. Structural engineering. I. Hauf, Harold Dana, ed. II. Title.

TA633.P3 1975   690   74-18068
ISBN 0-471-66201-1

Printed in the United States of America

10

Preface to the Fifth Edition

||||||||||||||||||||||||||||||||||||||||||||||||||||||||||

Although structural design procedures tend to become increasingly complex with the effort to achieve more efficient use of materials, the need continues for a simpler approach to the design of structural members in buildings of usual types of construction, spans, and story heights where many of the more sophisticated refinements permitted by modern codes are not warranted. Many persons employed in the offices of architects, designers, and builders, as well as those in training for such positions, require a working knowledge of structural design that will enable them to handle day-to-day problems that arise, and that will provide a background for working effectively with structural engineers on more complicated projects. The wide acceptance of *Simplified Engineering for Architects and Builders* over more than three decades supports the soundness of this point of view.

The purpose in publishing this fifth edition is to maintain the book's level of usefulness by updating the text to conform with current practice. Since publication of the fourth edition in 1967, new industry-recommended specifications and design standards have been promulgated for steel, wood, and reinforced concrete construction. In general, the text is now consistent with the American Institute of Steel Construction's 1969 *Specification for the Design, Fabrication & Erection of Structural Steel for Buildings;* the 1973 Edition of the *National Design Specification for Stress-Grade Lumber and Its Fastenings*, issued by the National Forest Products Association; and the American Concrete Institute's *Building Code Requirements for Reinforced Concrete* (ACI 318-71), insofar as provisions of the 1971 Code may be applied to the working stress design method.

v

The five sections of the book are Principles of Structural Mechanics, Steel Construction, Wood Construction, Reinforced Concrete, and Roof Trusses. As in earlier editions, the first section is presented for those who have not had technical preparation in mechanics, as well as for those who may wish to review this material. All sections have been written without employing advanced mathematics; a knowledge of high school algebra and arithmetic is adequate preparation for following the discussions and design procedures. One of the important features of the book is the large number of illustrative design examples worked out step by step. These are followed by similar problems to be solved by the reader. As an aid to persons studying outside the classroom, answers to many of these problems have been provided.

Many tables of engineering data have been included so that additional reference books are not required to understand the illustrative examples, nor to solve the exercise problems. However, the size and scope of this volume limit the number of such tables that can be presented, and the reader desiring more extensive data should consult the handbooks and other technical literature of the various industry associations; these are identified in the sections relating to steel, wood, and reinforced concrete construction. Grateful acknowledgment is made to these associations for granting permission to reproduce selected tabular data and other technical information from their publications.

The subject of structural design is extensive and in an elementary book of this size only a limited number of items can be discussed. As in earlier editions, the design procedures are based on elastic theory, commonly known as allowable stress design or working stress design. However, brief chapters on plastic design in steel and ultimate strength design in reinforced concrete are presented in this edition to provide an introduction to the theory of these design methods. The scope of the book, its purpose and limitations were stated by Professor Parker in the preface to the first edition; this is reprinted on the following pages.

Harold D. Hauf

Phoenix, Arizona

Preface to the First Edition

||||||||||||||||||||||||||||||||||||||||||||||||||||||

To the average young architectural draftsman or builder, the problem of selecting the proper structural member for given conditions appears to be a difficult task. Most of the numerous books on engineering which are available assume that the reader has previously acquired a knowledge of fundamental principles, and thus are almost useless to the beginner. It is true that some engineering problems are exceedingly difficult, but it is also true that many of the problems that occur so frequently are surprisingly simple in their solution. With this in mind, and with a consciousness of the seeming difficulties in solving structural problems, this book has been written.

In order to understand the discussions of engineering problems, it is essential that the student have a thorough knowledge of the various terms which are employed. In addition, basic principles of forces in equilibrium must be understood. The first section of this book, "Principles of Mechanics," is presented for those who have not had such technical preparation, as well as for those who wish a brief review of the subject. Following this section are structural problems involving the most commonly used building materials, wood, steel, reinforced concrete, and roof-trusses. A major portion of the book is devoted to numerous problems and their solution, the purpose of which is to explain practical procedure in the design of structural members. Similar examples are given to be solved by the student. Although handbooks published by the manufacturers are necessities to the more advanced student, a great number of appropriate tables are presented herewith so that sufficient data are directly at hand to those using this book.

Care has been taken to avoid the use of advanced mathematics, a knowledge of arithmetic and high-school algebra being all that is required to follow the discussions presented. The usual formulas employed in the solution of structural problems are given with explanations of the terms involved and their application, but only the most elementary of these formulas are derived. These derivations are given to show how simple they are and how the underlying principle involved is used in building up a formula that has a practical application.

No attempt has been made to introduce new methods of calculation, nor have all the various methods been included. It has been the desire of the author to present to those having little or no knowledge of the subject simple solutions of everyday structural problems. Whereas thorough technical training is to be desired, it is hoped that this presentation of fundamentals will provide valuable working knowledge and, perhaps, open the doors to more advanced study.

Harry Parker

Philadelphia, Pa.
March 1938

Suggestions

||||||||||||||||||||||||||||||||||||||||||||||||||||||||||||

Those unfamiliar with the terms and basic principles of structural mechanics should study Section I thoroughly before attempting the succeeding sections. Others who have had previous work in mechanics may wish to review this material, since it is fundamental to the technical discussions in all other sections. The following suggestions are offered as aids to study:

1. Take up each item in the sequence presented and be certain that each is thoroughly understood before continuing with the next.

2. Since each problem to be solved is prepared to illustrate some basic principle or procedure, read it carefully and make sure you understand exactly what is wanted before starting to solve it.

3. Whenever possible make a sketch showing the conditions and record the data given. Such diagrams frequently show at a glance the problem to be solved and the procedure necessary for its solution.

4. Make a habit of checking your answers to problems. Confidence in the accuracy of one's computations is best gained by self-checking. However, in order to provide an occasional outside check, answers to some of the exercise problems are given at the end of the book. Such problems are indicated by an asterisk (*) following the problem number where it occurs in the text.

5. If you do not own a slide rule or desk calculator, get one at the first opportunity. The ability to use these computational aids is readily acquired, and whichever one you obtain will shortly become indispensable.

6. In solving problems, form the habit of writing the denomination of each quantity. The solution of an equation will be a number. It may be so many pounds, or is it pounds per square inch? Are

the units foot-pounds or inch-pounds? Adding the names of the quantities signifies an exact knowledge of the quantity and frequently prevents subsequent errors. Abbreviations are commonly used for this purpose, and those employed in this book are identified below for convenient reference.

Abbreviation	Quantity
cu ft	cubic foot
cu in.	cubic inch
ft	foot
ft-lb	foot-pound
in.	inch
in-lb	inch-pounds
kip	1000 pounds
kip-ft	kip-feet
kip-in.	kip-inches
ksf	kips per square foot
ksi	kips per square inch
lin ft	linear foot
lb	pounds
lb per cu ft	pounds per cubic foot
lb per lin ft	pounds per linear foot
psf	pounds per square foot
psi	pounds per square inch
sq ft	square foot
sq in.	square inch

The same abbreviation is used for both singular and plural, ft indicating either foot or feet. Also, it is common practice to omit the period generally used after abbreviations except in the case of inches (in.). Symbols are often used on drawings and diagrams instead of letter abbreviations. Among these symbols are:

$$\# = \text{lb} \qquad '\# = \text{ft-lb} \qquad k' = \text{kip-ft}$$

$$\#/' = \text{lb per lin ft} \qquad ''\# = \text{in-lb} \qquad k'' = \text{kip-in.}$$

The following "shorthand" symbols are also frequently used:

Symbol	Reading
>	is greater than
<	is less than
≧	equal to or greater than
≦	equal to or less than

Contents

‖‖‖

Section I **Principles of Structural Mechanics**

1	**Forces and Stresses**	**3**
1-1	Introduction	3
1-2	Forces	3
1-3	Direct Stress	4
1-4	Kinds of Stress	5
1-5	Compression	6
1-6	Shear	6
1-7	Bending	8
1-8	Deformation	9
1-9	Hooke's Law	9
1-10	Elastic Limit and Yield Point	9
1-11	Ultimate Strength	10
1-12	Factor of Safety	11
1-13	Modulus of Elasticity	11
1-14	Allowable Unit Stresses	13
1-15	Use of Direct Stress Formula	15

2	**Moments and Reactions**	**18**
2-1	Moment of a Force	18
2-2	Laws of Equilibrium	20
2-3	Moments of Forces on a Beam	21
2-4	Types of Beams	23

2-5   Kinds of Loads   24
2-6   Reactions   25
2-7   Distributed Loads   27
2-8   Overhanging Beams   29

3   Shear and Bending Moment   **34**

3-1   Introduction   34
3-2   Vertical Shear   35
3-3   Shear Diagrams   37
3-4   Bending Moment   40
3-5   Bending Moment Diagrams   41
3-6   Relation between Shear and Bending Moment   42
3-7   Negative Bending Moment: Overhanging Beams   47
3-8   Cantilever Beams   53
3-9   Bending Moment Formulas   57
3-10   Concentrated Load at Center of Span   57
3-11   Simple Beam with Uniform Load   58
3-12   Typical Loadings: Shear and Moment Formulas   59

4   Theory of Bending and Properties of Sections   **61**

4-1   Resisting Moment   61
4-2   The Flexure Formula   62
4-3   Properties of Sections: Structural Shapes   64
4-4   Centroids   65
4-5   Moment of Inertia   79
4-6   Section Modulus   82
4-7   Application of the Flexure Formula   83
4-8   Transferring Moments of Inertia   85
4-9   Radius of Gyration   87

Section II   **Steel Construction**

5   Steel Beams   **91**

5-1   Structural Steel   91
5-2   Structural Shapes   92

5-3 Designations of Rolled Steel Shapes   94
5-4 Nomenclature   95
5-5 Allowable Stresses for Structural Steel   98
5-6 Materials and Stresses for Connectors   98
5-7 Factors in Beam Design   100
5-8 Compact and Noncompact Sections   100
5-9 Lateral Support of Beams   101
5-10 Flexure or Bending   102
5-11 Shear   105
5-12 Deflection   107
5-13 Deflection Coefficients   109
5-14 A Convenient Deflection Formula   111
5-15 Safe Load Table for W and S Shapes   113
5-16 Safe Load Table for Channels   118
5-17 Equivalent Tabular Loads   118
5-18 Laterally Unsupported Beams   119
5-19 Beams with Light Loads   124
5-20 Beam Safe Load Table Based on Deflection   125
5-21 Provision for Beam Weight   128
5-22 Floor Framing   131
5-23 Bearing Plates for Beams   138
5-24 Crippling of Beam Webs   141
5-25 Open Web Steel Joists   142
5-26 Plate Girders   147
5-27 Structural Steel Design Methods   148

6 Steel Columns   149

6-1 Introduction   149
6-2 Column Sections   150
6-3 Slenderness Ratio   152
6-4 Effective Column Length   152
6-5 Column Formulas   153
6-6 Allowable Column Loads   156
6-7 Design of Steel Columns   159
6-8 Double-Angle Struts   162
6-9 Steel Pipe and Structural Tubing Columns   166

6-10 Bending Factors for Columns 168
6-11 Trial Section for Eccentrically Loaded Columns 171
6-12 Reduction in Column Live Loads 172
6-13 Column Base Plates 173
6-14 Grillage Foundations 177

7 Bolted and Riveted Connections **179**

7-1 General 179
7-2 Riveting 179
7-3 Gage Lines, Pitch, and Edge Distance 180
7-4 Structural Action in Riveted Joints 181
7-5 Allowable Stresses and Working Values for Rivets 183
7-6 Unfinished Bolts 187
7-7 High-Strength Bolts 187
7-8 Net Sections: Angles in Tension 189
7-9 Beam Framing Connections 192
7-10 Free-End and Moment Connections 193

8 Welded Connections **194**

8-1 General 194
8-2 Electric Arc Welding 195
8-3 Types of Welding Joints 195
8-4 Stresses in Welds 196
8-5 Design of Welded Joints 200
8-6 Beams with Continuous Action 202
8-7 Plug and Slot Welds 204
8-8 Miscellaneous Welded Connections 205

9 Plastic Design Theory **208**

9-1 Introduction 208
9-2 Stress–Strain Diagram 209
9-3 Plastic Moment, Plastic Hinge 209

9-4 Plastic Section Modulus   211
9-5 Shape Factor   213
9-6 Load Factor   214
9-7 Scope of Plastic Design   215

Section III Wood Construction

10 Wood Beams   219

10-1   Structural Lumber   219
10-2   Nominal and Dressed Sizes   220
10-3   Allowable Stresses for Structural Lumber   220
10-4   Design for Bending   227
10-5   Horizontal Shear   229
10-6   Deflection   231
10-7   Beam Design Procedure   233
10-8   Bearing on Supports   238
10-9   Joist Floors: Span Tables   239
10-10 Design of Joists   246
10-11 Plank Floors   250
10-12 Rafter Roofs: Span Tables   255
10-13 Glued Laminated Beams   261

11 Wood Columns   262

11-1 Column Types   262
11-2 Slenderness Ratio   262
11-3 Column Formulas   263
11-4 Allowable Loads on Solid Columns   264
11-5 Design of Solid Columns   265
11-6 Spaced Columns   271
11-7 Glued Laminated Columns   274

12 Timber Connectors   275

12-1 General   275
12-2 Split Ring Connectors   275

12-3   Strength of Connector Joints   277
12-4   Timber Species Groups   277
12-5   Connector Loads   277
12-6   End Distance   281
12-7   Allowable Connector Loads   282
12-8   Edge Distance   283

Section IV   **Reinforced Concrete**

13   Stresses in Reinforced Concrete   **289**

13-1   Introduction   289
13-2   Design Methods   289
13-3   Strength of Concrete   290
13-4   Water–Cement Ratio   291
13-5   Cement   293
13-6   Air-Entrained Concrete   294
13-7   Steel Reinforcement   294
13-8   Notation Used in Reinforced Concrete   296
13-9   Modulus of Elasticity   297
13-10   Flexural Design Formulas for Rectangular Beams   297
13-11   Diagonal Tension: Web Reinforcement   301
13-12   Notation and Formulas for Web Reinforcement   302
13-13   Web Reinforcement for Uniform Loads   303
13-14   Design of Beams with Web Reinforcement   306
13-15   Bond Stress   310
13-16   Development Length of Reinforcement   313

14   Reinforced Concrete Beams   **318**

14-1   Typical Beams   318
14-2   Length of Span   320
14-3   Bending Moments   320
14-4   Design of Rectangular Reinforced Concrete Beams   322
14-5   Alternate Design: Straight Bars Only   328
14-6   Web Reinforcement for Girders   328

14-7   T-Beams   332
14-8   Design of a T-Beam   333
14-9   Compressive Stress in T-Beam Flange   383
14-10 Compression Reinforcement in T-Beam Stem   340

15   Reinforced Concrete Floor Systems   341

15-1   Introduction   341
15-2   One-Way Solid Slabs   342
15-3   Shrinkage and Temperature Reinforcement   343
15-4   Tensile Reinforcement in Slabs   344
15-5   Design of a One-Way Solid Slab   345
15-6   Ribbed Slabs: Concrete Joists   349
15-7   Design of a One-Way Joist Slab   350

16   Reinforced Concrete Columns   354

16-1   Introduction   354
16-2   Tied Columns   355
16-3   Design of a Tied Column   356
16-4   Spiral Columns   357
16-5   Design of a Spiral Column   359

17   Column Footings   361

17-1   Introduction   361
17-2   Independent Column Footings   362
17-3   Design of an Independent Column Footing   364

18   Ultimate Strength Design   368

18-1   Introduction   368
18-2   Loads and Load Factors   369
18-3   Capacity Reduction Factors   369

18-4   Bending Stresses in Rectangular Beams   370
18-5   Design of Rectangular Beams for Bending   372
18-6   Scope of Ultimate Strength Design   373

Section V   **Roof Trusses**

19   Stresses in Trusses   **377**

19-1   Introduction   377
19-2   Force Polygon   378
19-3   Stress Diagrams   379
19-4   Wind Load Stress Diagram   381
19-5   Roof Loads   382
19-6   Equivalent Vertical Loading   385

20   Design of a Steel Truss   **387**

20-1   General   387
20-2   Loads and Stresses   387
20-3   Compression Members   390
20-4   Tension Members   392
20-5   Riveted Joints   393
20-6   Welded Joints   395
20-7   Shop Drawings   395

Answers to Selected Problems   398

Index   403

I

PRINCIPLES OF STRUCTURAL MECHANICS

1

Forces and Stresses

IIIIIIIIIIIIIIIIIIIIIIIIIIIIIIIIIIIIIIIIIIIIIIIIIIIII

1-1 Introduction

Mechanics is the science that treats of the action of forces on material bodies, and *statics* is that branch of mechanics which treats of bodies held in equilibrium by the balanced external forces acting on them.

Strength of materials considers the behavior of material bodies in resisting the action of external forces, the stresses developed within the bodies, and the deformations that result from the external forces.

Taken together, these two subjects constitute the field of *structural mechanics*, and it is the purpose of the chapters in this section to present the key principles of structural mechanics that form the basis of structural design.

1-2 Forces

A force is that which tends to change the state of rest or motion of a body. It may be considered as pushing or pulling a body at a definite point and in a definite direction. Such a force tends to give motion to a body at rest, but this tendency may be neutralized by the action of another force or forces. A force is completely determined when its magnitude, direction, line of action, and point of application are known. In building construction we are concerned primarily with forces in equilibrium, that is, with bodies at rest. For example, a steel column supports a given load which, owing to gravity, is downward. The column transfers the load to the footing below. The resultant upward pressure on the footing equals the load in magnitude; its direction is upward, and it is called the *reaction*. The two forces are opposite in direction, have the same line of action, and are equal in magnitude. The result is equilibrium, that is, no motion. The units of force are pounds, kilograms, tons, etc. In engineering practice the word *kip*, meaning 1000 pounds, is widely used. Thus 30,000 lb may also be written 30 kips.

1-3 Direct Stress

A stress in a body is an internal resistance to an external force. The hanger bar shown in Fig. 1-1a supports a suspended load P, acting vertically along the axis of the bar. The load tends to stretch the bar and is called a *tensile force*. The bar resists the tendency to elongate by developing an internal *tensile stress* equal to the tensile force. The tensile stress produced under this condition of axial loading is called *direct stress*.

A characteristic of direct stress is that the internal resistance may be assumed to be evenly distributed over the cross-sectional area of the body under stress. Thus if P in Fig. 1-1a is 30,000 lb and the area of the hanger bar is 2 square inches, each square inch of the bar cross section is stressed to $30,000 \div 2 = 15,000$ pounds per square inch (psi) or, if P is stated kips, the stress in the bar is expressed as $30 \div 2 = 15$ kips per square inch (ksi). This tensile stress per unit

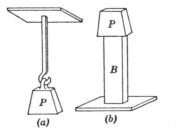

(a)   (b)   FIGURE 1-1

of area is called a *unit stress* to distinguish it from the total stress of 30,000 lb. By calling the load or external force *P*, the area of cross section *A*, and the unit stress *f*, this fundamental relationship governing direct stress may be stated

$$f = \frac{P}{A} \quad \text{or} \quad P = fA \quad \text{or} \quad A = \frac{P}{f}$$

When using this equation, remember the two assumptions on which it is based: the loading is axial, and the stresses are evenly (uniformly) distributed over the cross section. Note also that if any two of the quantities are known, the third may be found.

Example.   Suppose a wrought iron rod with a diameter of $1\frac{1}{2}$ in. is used as a hanger in an arrangement similar to that shown in Fig. 1-1*a*. If the *allowable* unit tensile stress for wrought iron is 12,000 psi, determine the load that the rod will safely support.
Solution: (1) To find the area of the rod we square the radius and multiply by 3.1416,

$$A = \pi r^2 = 3.1416 \times .75^2 = 1.767 \text{ sq in.}$$

(2) Since the allowable unit tensile stress is 12,000 psi, the load the rod will carry is found by using the second form of the direct stress equation given above,

$$P = fA = 12,000 \times 1.767 = 21,120 \text{ lb}$$

1-4   Kinds of Stress

The three basic kinds of stress that concern us are *tension, compression*, and *shear*. As we observed in Art. 1-3, tensile forces tend to stretch a structural member. Compressive forces tend to shorten members, and shearing forces tend to make parts of a structure slide past each other. The stress that accompanies the tendency of a force to twist a member is called *torsion*, but it is generally of less concern to us in building construction than are the other three.

In addition to occurring under conditions of direct stress, tension and compression are also developed in structural members subjected to bending or flexure (Fig. 1-2*e*). This is explained in detail later.

1-5 Compression

The load P on the short square block B shown in Fig. 1-1b exerts an axial force on the block that tends to shorten it. This is called a compressive force and is resisted by an internal compressive stress equal to P. The *unit compressive stress* is given by the direct stress formula $f = P/A$. However, this relationship holds for short compression members only.[1]

Example. A short timber post with nominal cross-sectional dimensions of 8 × 8 in. supports an axial load of 50,000 lb, as indicated in Fig. 1-1b. Find the unit compressive stress developed in the post.

Solution: (1) The nominal size of the post is 8 × 8 in. but the "standard dressed size" or actual size is $7\frac{1}{2} \times 7\frac{1}{2}$ in.; therefore the area of the cross section is 56.25 sq in. (see Table 4-6).

(2) By data, the load to be carried is 50,000 lb. The unit compressive stress is found by substituting the known values in the first form of the direct stress equation,

$$f = \frac{P}{A} = \frac{50,000}{56.25} = 890 \text{ psi}$$

1-6 Shear

A shearing stress occurs when two forces act on a body in opposite directions but not in the same plane. This condition is illustrated in Fig. 1-2a which shows two plates held together by a rivet. Under the action of the forces P, the plates tend to shear the rivet at their plane of contact as indicated in Fig. 1-2b. Another illustration of shear is given in Fig. 1-2c, where a load W rests on a beam which in turn is supported on walls at its ends. It is evident from the sketch that the beam might fail by dropping between the walls, i.e. by shearing along vertical planes at C and D (Fig. 1-2d). This type of shear is discussed in Chapter 3.

Whereas the direct tensile and compressive stresses discussed in Arts. 1-3 and 1-5 act at right angles to the cross sections of the

[1] As the ratio of length to least width of compression members increases, other factors enter the problem; these are considered under the design of steel, wood, and reinforced concrete columns in Sections II, III, and IV, respectively.

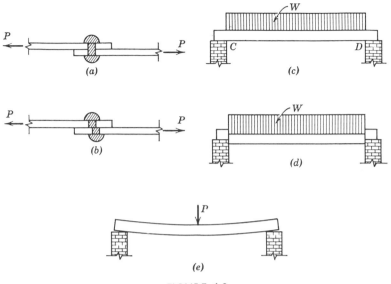

FIGURE 1-2

members considered, shearing stress acts transversely or *parallel* to the cross section. With respect to the rivet in Figs. 1-2*a* and *b*, the unit shearing stress f_v is given by the formula $f_v = P/A$, where P is the shearing force and A is the cross-sectional area of the rivet. It will be noted that this expression is similar to the direct stress formula $f = P/A$. However, it must be understood clearly that the *physical situations represented by the two formulas are quite different.*

Example.   The forces P in the plates illustrated in Fig. 1-2*a* are each 5000 lb and the rivet has a diameter of $\frac{3}{4}$ in. What is the unit shearing stress?

Solution: (1) The area of the $\frac{3}{4}$-in. rivet may be computed from the formula $A = \pi r^2$ or found directly from Table 7-3 which gives a value of 0.4418 sq in.

(2) Since $P = 5000$ lb, the unit shearing stress is

$$f_v = \frac{P}{A} = \frac{5000}{0.4418} = 11{,}320 \text{ psi}$$

1-7 Bending

Figure 1-2*e* illustrates a simple beam with a concentrated load *P* at the center of the span. This is an example of bending or flexure. The fibers[2] in the upper part of the beam are in compression and those in the lower part are in tension. These stresses are not uniformly distributed over the cross section of the beam and cannot be computed by the direct stress formula. The expression used to compute the value of the bending stress in either tension or compression is known as the *beam formula* or the *flexure formula*, and is considered in Chapter 4.

Problem 1-7-A*. A wrought iron bar receives a tensile force of 40 kips. If the allowable unit tensile stress of wrought iron is 12 ksi, what is the required cross-sectional area of the bar?

Problem 1-7-B. What axial load may be placed on a short timber post whose actual cross-sectional dimensions are $9\frac{1}{2} \times 9\frac{1}{2}$ in. if the allowable unit compressive stress is 1100 psi?

Problem 1-7-C*. What should be the diameter of the rivet shown in Fig. 1-2*a* if the shearing force is 9000 lb and the allowable unit shearing stress is 15,000 psi?

Problem 1-7-D. The allowable bearing capacity of a foundation bed is 4 tons per sq ft. What should be the length of a side of a square footing if the load on the column and footing is 240 kips?

Problem 1-7-E. If a steel bolt with a diameter of $1\frac{1}{4}$ in. is used for the fastener shown in Fig. 1-2*a*, find the shearing force that can be transmitted across the joint if the allowable unit shearing stress in the bolt is 15 ksi.

Problem 1-7-F*. A short, hollow, cast iron column is circular in cross-section, the outside diameter being 10 in. and the thickness of the shell $\frac{3}{4}$ in. If the allowable unit compressive stress of cast iron is 9 ksi, what load will the column support?

Problem 1-7-G. Determine the minimum cross-sectional area of a steel bar required to support a tensile force of 50 kips, if the allowable unit tensile stress is 20 ksi.

---

[2] Although steel and concrete are not fibrous materials in the sense that wood is, the concept of infinitely small fibers is useful in the study of stress relationships within any material.

Problem 1-7-H.   A short square timber post supports a load of 115 kips. If the allowable unit compressive stress is 1000 psi, what nominal size square timber should be used? See Table 4-6.

1-8   Deformation

Whenever a force acts on a body, there is an accompanying change in shape or size of the body. In structural mechanics this is called *deformation*. Regardless of the magnitude of the force, some deformation is always present, although often it is so small that it is difficult to measure even with the most sensitive instruments. In the design of structures it is often necessary that we know what the deformation in certain members will be. A floor joist, for instance, may be large enough to support a given load safely but may *deflect* (the term for deformation that occurs with bending) to such an extent that the plaster ceiling below will crack, or the floor may feel excessively springy to persons walking on it. For the usual cases we can readily determine what the deformation will be. This is considered in more detail later.

1-9   Hooke's Law

As a result of experiments with clock springs, Robert Hooke, a mathematician and physicist working in the seventeenth century, developed the theory that "deformations are directly proportional to stresses." In other words, if a force produces a cerain deformation, twice the force will produce twice the amount of deformation. This law of physics is of utmost importance in structural engineering although, as we shall find, Hooke's Law holds true only up to a certain limit.

1-10   Elastic Limit and Yield Point

Suppose that we place a bar of structural steel with a cross-sectional area of 1 sq in. in a machine for making tension tests. We measure its length accurately and then apply a tensile force of 5000 lb which, of course, produces a unit tensile stress of 5000 psi in the bar. We measure the length again and find that the bar has lengthened a definite amount, which we will call X inches. On applying 5000 lb more, we note that the amount of lengthening is $2 \times X$, or twice the amount noted after the first 5000 lb. If the test is continued, we will find that for each 5000 lb increment the length of the bar will increase the same amount noted when the initial 5000 lb was applied; that is, the deformations (elongation) are directly proportional to the stresses. So far Hooke's Law has held true, but after we have applied about 36,000 lb, the length increases more than X in. for each additional 5000 lb. This unit stress of about 36,000 psi (which varies with different grades of steel) is called the *elastic limit*. It may be defined as the unit stress beyond which the deformation increases in a faster ratio than the applied loads.

Another phenomenon may be noted in this connection. If we make the test again, we will discover that when any applied load which produces a unit stress *less* than the elastic limit is removed, the bar will return to its original length. If a load producing a unit stress *greater* than the elastic limit is removed, we will find that the bar has permanently increased its length. This permanent deformation is called the *permanent set*. This fact permits another way of defining the elastic limit: it is that unit stress beyond which the material does not return to its original length when the load is removed.

If our test is continued beyond the elastic limit, we quickly reach a point where the deformation increases without any increase in the load. The unit stress at which this deformation occurs is called the *yield point*; it has a value only slightly higher than the elastic limit. Since the yield point, or yield stress as it is sometimes called, can be determined more accurately by test than the elastic limit, it is a particularly important unit stress. Nonductile materials such as wood and cast iron have poorly defined elastic limits and no yield point.

1-11 Ultimate Strength

After passing the yield point, the steel bar of the test described in the preceding article again develops resistance to the increasing load. When the load reaches a sufficient magnitude, rupture occurs. The unit stress in the bar just before it breaks is called the *ultimate strength*. For the grade of steel assumed in our test, the ultimate strength occurs at about 70,000 psi.

Structural members are designed so that stresses under normal service conditions will not exceed the elastic limit, even though there is considerable reserve strength between this value and the ultimate strength. This procedure is followed because deformations produced by stresses above the elastic limit are permanent and hence change the shape of the structure.

1-12   Factor of Safety

The degree of uncertainty that exists, both with respect to actual loading of a structure and uniformity in the quality of materials, requires that some reserve strength be built into the design. This degree of reserve strength is the *factor of safety*. Although there is no general agreement on the definition of this term, the following discussion will serve to fix the concept in mind.

Consider a structural steel that has an ultimate tensile strength of 70,000 psi, a yield point stress of 36,000 psi, and an allowable unit tensile stress of 22,000 psi. If the factor of safety is defined as the ratio between the ultimate strength and the allowable stress, its value is 70,000 ÷ 22,000 or 3.18. On the other hand, if it is defined as the ratio of the yield point to the allowable stress, its value is 36,000 ÷ 22,000 or 1.64. This is a considerable variation, and since failure of a structural member begins when it is stressed beyond the elastic limit, the higher value may be misleading. Consequently, the term *factor of safety* is not employed extensively today. Building codes generally specify the allowable unit stresses that are to be used in design for the grades of structural steel to be employed.

If one should be required to pass judgment on the safety of a structure, the problem resolves itself into considering each structural element, finding its actual unit stress under the existing loading conditions, and comparing this stress with the allowable stress prescribed by the local building regulations. This procedure is called *investigation*.

1-13   Modulus of Elasticity

We have seen that, within the elastic limit of a material, deformations are directly proportional to the stresses. Now we shall compute the magnitude of these deformations by use of a number (ratio), called the *modulus of elasticity*, that indicates the degree of *stiffness* of a material.

A material is said to be stiff if its deformation is relatively small when the unit stress is high. As an example, a steel rod 1 sq in. in cross-sectional area and 10 ft long will elongate about 0.008 in. under a tensile load of 2000 lb. But a piece of wood of the same dimensions and with the same tensile load will stretch about 0.24 in., or nearly $\frac{1}{4}$ in. We say that the steel is stiffer than the wood because, for the same unit stress, the deformation is not so great.

Modulus of elasticity is defined as *the unit stress divided by the unit deformation.* It is represented by the letter E, expressed in pounds per square inch, and has the same value in compression and tension for most structural materials. Letting f represent the unit stress and s the unit deformation we have, by definition,

$$E = \frac{f}{s}$$

From Art. 1-3 we remember that $f = P/A$. It is obvious that, if l represents the length of the member and e the total deformation, s, the deformation per unit of length, must equal the total deformation divided by the length or $s = e/l$. Now, if we substitute these values in the equation determined by definition,

$$E = \frac{f}{s} = \frac{P/A}{e/l} = \frac{P}{A} \times \frac{l}{e}$$

This can also be written

$$e = \frac{Pl}{AE}$$

in which e = total deformation in inches,

$\quad P$ = force in pounds,

$\quad l$ = length in inches,

$\quad A$ = cross-sectional area in square inches,

$\quad E$ = modulus of elasticity in pounds per square inch.

Note that E is expressed in the same units as f (pounds per square inch) because in the equation $E = f/s$, s is an abstract number. For structural steel $E = 29,000,000$ psi and for wood, depending on the species and grade, it varies from something less than 1,000,000 psi to about 1,900,000 psi. The modulus of elasticity of concrete ranges from about 2,000,000 psi to 5,000,000 psi and over depending on the compressive strength. The important thing to remember is that the foregoing formula is valid only when the unit stress lies within the elastic limit of the material.

Example. A 2-in. diameter steel bar 10 ft long is subjected to a tensile force of 60,000 lb. How much will it elongate under the load? *Solution:* (1) The area of a 2-in. diameter bar is 3.1416 sq in.

(2) Checking to determine whether the stress in the bar is within the elastic limit,

$$f = \frac{P}{A} = \frac{60,000}{3.1416} = 19,100 \text{ psi}$$

This is within the elastic limit of structural steel so the formula for finding the deformation is applicable.

(3) From data, $P = 60,000$ lb, $l = 120$ (the length in inches), $A = 3.1416$, and $E = 29,000,000$. Substituting these values, the total lengthening of the bar is

$$e = \frac{Pl}{AE} = \frac{60,000 \times 120}{3.1416 \times 29,000,000} = 0.079 \text{ in.}$$

1-14 Allowable Unit Stresses

In the examples and problems dealing with the direct stress equation, we have differentiated between the unit stress developed in a member sustaining a given load ($f = P/A$) and the *allowable unit stress* when determining the size of a member required to carry a given load ($A = P/f$). The latter form of the equation is, of course, the one used in design.

From the discussion in Arts. 1-8 through 1-12, we can see that the allowable unit stresses should be set within the elastic limit of the structural material being used. The procedures for establishing allowable unit stresses in tension, compression, shear, and bending are different for different materials, and are prescribed in specifications promulgated by the American Society for Testing and Materials.

In general, allowable stresses for structural steel are expressed as fractions of the yield stress, those for wood involve an adjustment of clear wood strength as modified by lumber grading rules and conditions of use, and allowable stresses for concrete are given as fractions of the specified compressive strength of concrete. Tables 5-2, 10-1, and 13-1 give allowable stresses for steel, wood, and reinforced concrete construction, respectively, as recommended by the industry associations concerned.[3] When scanning these tables, it will be noted that they contain several terms that have not been introduced in this book thus far. These will be identified in subsequent sections dealing with the design of members to which they apply. However, in order to provide information for convenient reference when solving the problems at the end of this chapter, selected data from the more complete allowable stress tables are presented in Table 1-1.

In actual design work, the building code governing the construction of buildings in the particular locality must be consulted for specific requirements. Many municipal codes are revised infrequently

TABLE 1-1. Selected Allowable Stresses Taken from Tables 5-2, 10-1, 13-1 (values in pounds per square inch)

Structural Steel		Concrete	
Yield stress	36,000	f'_c (specified compressive strength)	3,000
Tension	22,000	Compression (bearing)	891
Shear (rivets)	15,000	Shear	60
E	29,000,000	E	3,100,000

Structural Lumber	
Douglas Fir, Select Structural grade	(posts and timbers)
Compression parallel to grain	1,150
E (modulus of elasticity)	1,600,000
Southern Pine, No. 1 Dense SR grade	(5 in. and thicker)
Compression parallel to grain	1,050
E (modulus of elasticity)	1,600,000

[3] These are the American Institute of Steel Construction, the National Forest Products Association, and the American Concrete Institute.

and consequently may not be in agreement with current editions of the industry-recommended allowable stresses. Unless otherwise noted, the allowable stresses used in this book are those given in the three tables referenced above.

1-15   Use of Direct Stress Formula

Except for shear, the stresses we have discussed so far have been direct or axial stresses. This, we recall, means they are assumed to be uniformly distributed over the cross section. The examples and problems presented fall under three general types: first, the design of structural members ($A = P/f$); second, determination of safe loads ($P = fA$); third, the investigation of members for safety ($f = P/A$). The following examples will serve to fix in mind each of these types.

Example 1.   *Design* (determine the size of) a short square post of Southern Pine, No. 1 Dense SR grade, to carry an axial load of 30,000 lb.

Solution: (1) Referring to Table 1-1, we find that the allowable unit compressive stress for this timber parallel to the grain is 1050 psi.

(2) The required area of the post is

$$A = \frac{P}{f} = \frac{30,000}{1050} = 28.6 \text{ sq in.}$$

(3) Referring to Table 4-6, an area of 30.25 sq in. is provided by standard dressed size $5\frac{1}{2} \times 5\frac{1}{2}$ in. Therefore we select a nominal 6 × 6-in. post to carry the load.

Example 2.   Determine the axial *safe load* on a short concrete pier 2 ft square.

Solution: (1) The area of the pier is 4 sq ft or 576 sq in.

(2) Table 1-1 gives the allowable unit compressive stress for concrete as 891 psi.

(3) Therefore the safe load on the pier is

$$P = fA = 891 \times 576 = 513,000 \text{ lb}$$

Example 3.   A running track in a gymnasium is hung from the roof trusses by steel rods, each of which supports a tensile load of 11.2

kips. The rods have a diameter of $\frac{7}{8}$ in. with the ends "upset," that is made larger by forging. This upset is necessary if the full or *gross area* of the rod (0.601 sq in.) is to be utilized; the area of a $\frac{7}{8}$-in. rod at the root of the thread that receives the nuts of the end fastenings is only 0.419 sq in. (the net area). *Investigate* this design to determine whether it is safe.

Solution: (1) Since the gross area of the hanger rod is effective, the unit stress developed is

$$f = \frac{P}{A} = \frac{11.2}{0.601} = 18.6 \text{ ksi}$$

(2) Table 1-1, gives the allowable unit tensile stress for steel as 22 ksi, which is greater than the 18.6 ksi developed by the loading. Therefore the design is safe.

Shearing Stress Formula. The foregoing manipulations of the direct stress formula can, of course, be carried out also with the shearing stress formula $f_v = P/A$. However, as pointed out in Art. 1-6, it must be borne in mind that the shearing stress acts transversely to the cross section—not at right angles to it. Furthermore, while the shearing stress equation applies directly to the situation illustrated by Figs. 1-2a and b, it requires modification for application to beams (Figs. 1-2c and d). The latter situation will be considered in more detail later.

Review Problems

Problem 1-15-A*.   What force must be applied to a 1-in. square steel bar 2 ft 0 in. long to produce an elongation of 0.016 in.?

Problem 1-15-B.   How much will a nominal 8 × 8 in. Douglas Fir post, 12 ft in length, shorten under an axial load of 45 kips?

Problem 1-15-C*.   A routine quality control test is made on a structural steel bar 1 in. square and 16 in. long. The data developed during the test show that the bar elongated 0.0111 in. when subjected to a tensile force of 20.5 kips. Compute the modulus of elasticity of the steel.

Problem 1-15-D.   A $\frac{1}{2}$-in. diameter steel cable 100 ft long supports a load of 2 tons. How much will it elongate?

Problem 1-15-E. What should be the minimum area of a steel rod to support a tensile load of 26 kips?

Problem 1-15-F. A short square post of Douglas Fir, Select Structural grade, is to support an axial load of 61 kips. What should its nominal dimensions be?

Problem 1-15-G. A steel rod has a diameter of $1\frac{1}{4}$ in. What safe tensile load will it support if its ends are upset?

Problem 1-15-H. What safe load will a short 12 × 12 in. Southern Pine post support if the grade is No. 1 Dense SR?

Problem 1-15-I. A short post of Douglas Fir, Select Structural grade, with nominal dimensions of 6 × 8 in. supports an axial load of 50 kips. Investigate this construction to determine whether it is safe.

Problem 1-15-J. A short concrete pier 1 ft 6 in. square supports an axial load of 150 kips. Is the construction safe?

Problem 1-15-K. The shearing load on a $\frac{7}{8}$-in. diameter rivet in a lap joint (Fig. 1-2a) is 8.5 kips. Is this a safe condition?

2

Moments and Reactions

||||||||||||||||||||||||||||||||||||||||||||||||||||||||||||||

2-1 Moment of a Force

The term *moment of a force* is commonly used in engineering problems; it is of utmost importance that you understand exactly what the term means. It is fairly easy to visualize a length of 3 ft, an area of 16 sq in., or a force of 100 lb. A moment, however, is less readily comprehended; it is a force multiplied by a distance. *A moment is the tendency of a force to cause rotation about a given point or axis.* The magnitude of the moment of a force about a given point is the magnitude of the force (pounds, kips, etc.) multiplied by the distance (feet, inches, etc.) to the point. The point is called the *center of moments* and the distance, which is called the *lever arm* or *moment arm*, is measured by a line drawn through the center of moments *perpendicular to the line of action* of the force. Moments are expressed in compound units such as foot-pounds and inch-pounds, or kip-feet and kip-inches. Summarizing,

moment of force = magnitude of force × moment arm

Consider the horizontal force of 200 lb shown in Fig. 2-1. If point *A* is the center of moments, the lever arm of the force is 5 ft 0 in. Then the moment of the 200-lb force with respect to point *A* is

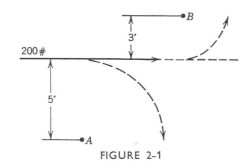

FIGURE 2-1

$200 \times 5 = 1000$ ft-lb. In this illustration the force tends to cause a *clockwise* rotation (shown by the dotted arrow) about point *A* and is called a positive moment. If point *B* is the center of moments, the moment arm of the force is 3 ft 0 in. Therefore, the moment of the 200-lb force about point *B* is $200 \times 3 = 600$ ft-lb. With respect to point *B*, the force tends to cause *counterclockwise* rotation; it is called a negative moment. It is important to remember that we can never consider the moment of a force without having in mind the particular point or axis about which it tends to cause rotation.

Figure 2-2 represents two forces acting on a bar which is supported at point *A*. The moment of force P_1 about point *A* is $100 \times 8 = 800$ ft-lb and it is clockwise or positive. The moment of force P_2 about point *A* is $200 \times 4 = 800$ ft-lb. The two moment values are the same but P_2 tends to produce a counterclockwise or negative moment about point *A*. In other words, the positive and negative moments are equal in magnitude and are in equilibrium; i.e. there is no motion. Another way of stating this is to say that the sum of the positive and negative moments about point *A* is zero, or

$$(P_1 \times 8) - (P_2 \times 4) = 0$$

Stated more generally, *if a system of forces is in equilibrium, the*

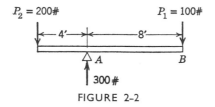

FIGURE 2-2

algebraic sum of the moments is zero. This is one of the laws of equilibrium.

In Fig. 2-2 point *A* was taken as the center of moments, but the fundamental law holds for any point that might be selected. For example, taking point *B* as the center of moments, the moment of the upward supporting force of 300 lb acting at *A* is clockwise (positive) and that of P_2 is counterclockwise (negative). Then

$$(300 \times 8) - (200 \times 12) = 2400 \text{ ft-lb} - 2400 \text{ ft-lb} = 0$$

Note that the moment of force P_1 about point *B* is $100 \times 0 = 0$; it is therefore omitted in writing the equation. The reader should satisfy himself that the sum of the moments is zero also when the center of moments is taken at the left end of the bar under the point of application of P_2.

2-2   Laws of Equilibrium

When a body is acted on by a number of forces, each force tends to move the body. If the forces are of such magnitude and position that their combined effect produces no motion of the body, the forces are said to be in *equilibrium*. The three fundamental laws of equilibrium are:

1. The algebraic sum of all the vertical forces equals zero.
2. The algebraic sum of all the horizontal forces equals zero.
3. The algebraic sum of the moments of all the forces about any point equals zero.

These laws, sometimes called the conditions for equilibrium, may be expressed as follows:[1]

$$\sum V = 0 \qquad \sum H = 0 \qquad \sum M = 0$$

The law of moments, $\sum M = 0$, was discussed in the preceding article.

We shall defer consideration of $\sum H = 0$ for the time being. Our immediate concern is with vertical loads acting on beams where the

---

[1] The symbol $\sum$ indicates a summation, i.e., an algebraic addition of all similar terms involved in the problem.

expression $\sum V = 0$ is another way of saying that *the sum of the downward forces equals the sum of the upward forces.* Thus the bar of Fig. 2-2 satisfies $\sum V = 0$ because the upward supporting force of 300 lb equals the sum of P_1 and P_2, the downward forces.

2-3 Moments of Forces on a Beam

Figure 2-3a shows two downward forces of 100 lb and 200 lb acting on a beam. The beam has a length of 8 ft between the supports; the supporting forces, which are called *reactions*, are 175 lb and 125 lb. The four forces are in equilibrium and therefore the two laws, $\sum V = 0$ and $\sum M = 0$, apply. Let us see if this is true.

First, because the forces are in equilibrium, the sum of the downward forces must equal the sum of the upward forces. The sum of the downward forces, the loads, is $100 + 200 = 300$ lb; and the sum of the upward forces, the reactions, is $175 + 125 = 300$ lb. We can write $100 + 200 = 175 + 125$; this is a true statement.

Second, because the forces are in equilibrium, the sum of the moments of the forces tending to cause clockwise rotation (positive moments) must equal the sum of the moments of the forces tending to produce counterclockwise rotation (negative moments) about any center of moments. Let us first write an equation of moments about

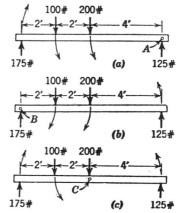

FIGURE 2-3

point A at the right-hand support. The force tending to cause clockwise rotation (shown by the curved arrow) about this point is 175 lb; its moment is $175 \times 8 = 1400$ ft-lb. The forces tending to cause counterclockwise rotation *about the same point* are 100 lb and 200 lb and their moments are (100×6) and (200×4) ft-lb. Therefore we can write

$$(175 \times 8) = (100 \times 6) + (200 \times 4)$$

$$1400 = 600 + 800$$

$$1400 \text{ ft-lb} = 1400 \text{ ft-lb}$$

which is true.

The upward force of 125 lb is omitted from the above equation because its lever arm about point A is zero feet, and consequently its moment is zero. Thus we see that a force passing through the center of moments does not cause rotation about that point.

Let us try again. This time we select point B at the left support as the center of moments. See Fig. 2-3b. By the same reasoning we can write

$$(100 \times 2) + (200 \times 4) = (125 \times 8)$$

$$200 + 800 = 1000$$

$$1000 \text{ ft-lb} = 1000 \text{ ft-lb}$$

Again the law holds. In this case the force 175 lb has a lever arm of zero feet about the center of moments and its moment is zero.

Suppose we select any point, such as point C in Fig. 2-3c, as the center of moments; then

$$(175 \times 4) = (100 \times 2) + (125 \times 4)$$

$$700 = 200 + 500$$

$$700 \text{ ft-lb} = 700 \text{ ft-lb}$$

We have seen that the law of moments holds in each case. It is of great importance that we understand this principle thoroughly before going on. Remember that the loads and reactions are usually in units of pounds or kips and that the moments are compound

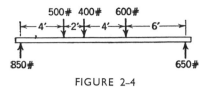

FIGURE 2-4

quantities, usually foot-pounds or kip-feet, the result of multiplying a force by a distance.[2]

Problem 2-3-A. Figure 2-4 represents a beam in equilibrium with three loads and two reactions. Select five different centers of moments and write the equation of moments for each, showing that the sum of the clockwise moments equals the sum of the counterclockwise moments.

2-4 Types of Beams

A beam is a structural member that resists transverse loads. The supports for beams are usually at or near the ends, and the supporting upward forces are called reactions. As noted in Art. 1-7, the loads acting on a beam tend to *bend* rather than shorten or lengthen it. *Girder* is the name given to a beam that supports smaller beams; all girders are beams insofar as their structural action is concerned. There are, in general, five types of beams which are identified by the number, kind, and position of the supports. Figure 2-5 shows diagrammatically the different types and also the shape each beam tends to assume as it bends (deforms) under the loading.[3]

A *simple beam* rests on a support at each end, the ends of the beam being free to rotate (Fig. 2-5a).

A *cantilever beam* is supported at one end only. A beam embedded in a wall and projecting beyond the face of the wall is a typical example (Fig. 2-5b).

[2] When loads are given in kips there is no intrinsic reason why moments could not be stated as "foot-kips," which would be consistent with "foot-pounds." However, foot-pounds and kip-feet (or inch-pounds and kip-inches) are the terms commonly used in practice for the compound units in which moments are expressed.

[3] In ordinary steel or reinforced concrete beams these deformations are not usually visible to the eye but, as noted in Art. 1-8, some deformation is always present.

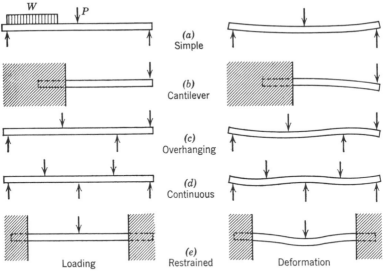

FIGURE 2-5. Types of beams.

An *overhanging beam* is a beam whose end or ends project beyond its supports. Figure 2-5c indicates a beam overhanging one support only.

A *continuous beam* rests on more than two supports (Fig. 2-5d). Continuous beams are commonly used in reinforced concrete and welded steel construction.

A *restrained beam* has one or both ends restrained or *fixed* against rotation (Fig. 2-5e).

2-5 Kinds of Loads

The two types of loads that commonly occur on beams are called *concentrated* and *distributed*. A concentrated load is assumed to act at a definite point, such as a column resting on a beam. A distributed load is one that acts over a considerable length of the beam. A concrete floor slab supported by a beam is an example of a distributed load. If the distributed load exerts a force of equal magnitude for each unit of length of the beam, it is known as a *uniformly distributed load*. Obviously, a distributed load need not extend over the entire length of the beam. Figure 2-5a illustrates a conventional method of representing the two types of loads: W is the total uniformly distributed load and P is a concentrated load.

2-6   Reactions

We have already defined reactions as the upward forces acting at the supports which hold in equilibrium the downward forces or loads. The left and right reactions are usually called R_1 and R_2, respectively.

If we have a beam 18 ft in length with a concentrated load of 9000 lb located 9 ft from the supports, it is readily seen that each upward force at the supports will be equal and will be one half the load in magnitude, or 4500 lb. But consider, for instance, the 9000-lb load placed 10 ft from one end, as shown in Fig. 2-6. What will the upward supporting forces be? Certainly they will not be equal.

Now this is where the principle of moments applies. See Fig. 2-6. Let us write an equation of moments, taking the center of moments about the right-hand support R_2:

$$18R_1 = 9000 \times 8$$

$$R_1 = \frac{72,000}{18}$$

$$R_1 = 4000 \text{ lb}$$

Because we know that the sum of the loads is equal to the sum of the reactions, we can easily compute R_2, for

$$R_1 + R_2 = 9000$$

$$4000 + R_2 = 9000$$

$$R_2 = 5000 \text{ lb}$$

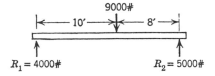

FIGURE 2-6

To check this value of R_2 write an equation of moments about the left-hand support R_1:

$$18R_2 = 9000 \times 10$$

$$R_2 = \frac{90,000}{18}$$

$$R_2 = 5000 \text{ lb}$$

Example.   A simple beam 20 ft in length has three concentrated loads, as indicated in Fig. 2-7. Compute the magnitude of the reactions.

Solution: (1) With the right-hand support as the center of moments, write the equation of moments. Then

$$20R_1 = (2000 \times 16) + (8000 \times 10) + (4000 \times 8)$$
$$20R_1 = 144,000$$
$$R_1 = 7200 \text{ lb}$$

(2) The sum of the reactions equals the sum of the loads:

$$R_1 + R_2 = 2000 + 8000 + 4000$$
$$7200 + R_2 = 14,000$$
$$R_2 = 6800 \text{ lb}$$

(3) To check R_2 write the equation of moments about R_1:

$$20R_2 = (2000 \times 4) + (8000 \times 10) + (4000 \times 12)$$
$$20R_2 = 136,000$$
$$R_2 = 6800 \text{ lb}$$

Problems 2-6-A-B*-C-D-E*-F-G*-H.   Eight simple beams with concentrated loads are shown in Fig. 2-8. Compute the reactions for each beam and check the results in each case.

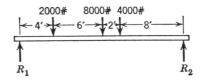

FIGURE 2-7

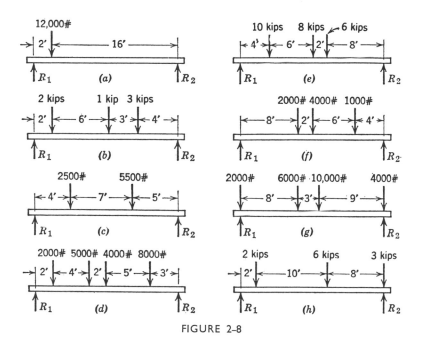

FIGURE 2-8

2-7 Distributed Loads

So far we have considered only concentrated loads in computing the magnitudes of reactions. The method of dealing with distributed loads is quite similar but there is one key point to remember: *a distributed load on a beam produces the same reactions as a concentrated load of the same magnitude acting through the center of gravity of the distributed load.* If we bear in mind that the center of gravity of a uniformly distributed load lies at the middle of its length, the problem becomes a very simple one.

Example 1. A simple beam 16 ft long carries a concentrated load of 8000 lb and a uniformly distributed load of 14,000 lb arranged as shown in Fig. 2-9a. Compute the reactions.

Solution: (1) Note that the uniformly distributed load extends over a length of 10 ft. First, let us write an equation of moments about R_2, considering that the uniformly distributed load of 14,000 lb acts at

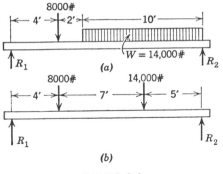

FIGURE 2-9

the middle of its length, or 5 ft from R_2. See Fig. 2-9b. Then

$$16R_1 = (8000 \times 12) + (14{,}000 \times 5)$$
$$16R_1 = 166{,}000$$
$$R_1 = 10{,}375 \text{ lb}$$

(2) To find R_2 take the center of moments at R_1. The lever arm of the 14,000-lb load is 11 ft because we consider that the uniformly distributed load acts at its midpoint, or 11 ft from R_1. Then

$$16R_2 = (8000 \times 4) + (14{,}000 \times 11)$$
$$16R_2 = 186{,}000$$
$$R_2 = 11{,}625 \text{ lb}$$

(3) To check these results, the sum of the loads should equal the sum of the reactions, or

$$8000 + 14{,}000 = 10{,}375 + 11{,}625$$
$$22{,}000 \text{ lb} = 22{,}000 \text{ lb}$$

In this example the values of the reactions have been carried to the nearest pound in order to clarify the explanation. This is not common practice. Three, or at most four, significant figures (slide rule accuracy) are sufficient in structural work. On this basis R_1 and R_2 of this example would have values of 10,400 lb and 11,600 lb, respectively. In general, the answers to examples and problems in

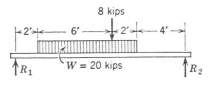

8 kips

W = 20 kips

FIGURE 2-10

this book are the result of slide rule computations except in cases where extension of the numerical answer seemed desirable as an aid in interpreting the result.

Example 2. A beam has a uniformly distributed load of 20 kips extending over a length of 20 ft and a concentrated load of 8 kips, arranged as shown in Fig. 2-10. Compute the reactions.
Solution: (1) The center of gravity of the 20 kip load is 6 ft from R_1 and 8 ft from R_2. Writing an equation of moments about R_2,

$$14R_1 = (20 \times 8) + (8 \times 6)$$

$$14R_1 = 160 + 48$$

$$R_1 = 14.86 \text{ kips}$$

(2) Taking R_1 as the center of moments,

$$14R_2 = (20 \times 6) + (8 \times 8)$$

$$14R_2 = 120 + 64$$

$$R_2 = 13.14 \text{ kips}$$

(3) Checking results,

$$20 + 8 = 14.86 + 13.14$$

$$28 \text{ kips} = 28 \text{ kips}$$

Problems 2-7-A-B*-C-D*-E-F-G-H*-I-J. Compute the reactions for the 10 beams shown in Fig. 2-11 and check the results in each case.

2-8 Overhanging Beams

The method of computing reactions for overhanging beams is the same as that employed in the preceding examples. Select one of the

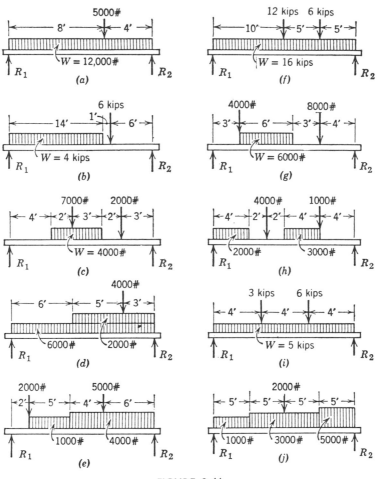

FIGURE 2-11

reactions as the center of moments. On one side of the equation place the sum of the moments tending to cause clockwise rotation and on the other side place the sum of the moments tending to cause rotation in the opposite direction. When writing a moment equation, bear two points in mind: (a) *be consistent and take the same center of moments for each force*, and (b) *consider uniformly distributed loads to act at their midpoints.*

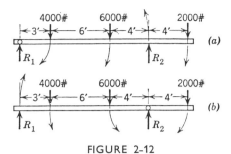

FIGURE 2-12

Example 1. Compute the reactions of the overhanging beam shown in Fig. 2-12a and check the results.

Solution: (1) Select R_1 as the center of moments. The forces tending to cause clockwise rotation *about this point* are the 4000-lb, 6000-lb, and 2000-lb loads, and the only force tending to cause counterclockwise rotation is R_2. Note the direction of the arrows. Therefore

$$13R_2 = (4000 \times 3) + (6000 \times 9) + (2000 \times 17)$$

$$13R_2 = 12,000 + 54,000 + 34,000$$

$$R_2 = 7692 \text{ lb}$$

(2) To find R_1 take the center of moments at R_2. The forces tending to cause clockwise rotation *about this point* are R_1 and the 2000-lb load; those that tend to cause counterclockwise rotation *about the same point* are the 4000-lb and 6000-lb loads. Note the direction of the arrows in Fig. 2-12b. Then

$$13R_1 + (2000 \times 4) = (4000 \times 10) + (6000 \times 4)$$

$$13R_1 = 40,000 + 24,000 - 8000$$

$$13R_1 = 56,000$$

$$R_1 = 4308 \text{ lb}$$

(3) To check these results, the sum of the loads should equal the sum of the reactions; therefore

$$4000 + 6000 + 2000 = 4308 + 7692$$

$$12,000 \text{ lb} = 12,000 \text{ lb}$$

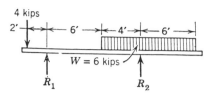

FIGURE 2-13

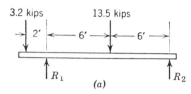

(a)

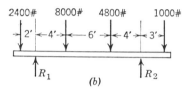

(b)

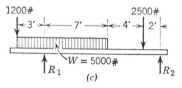

(c)

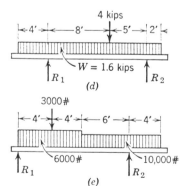

(d)

(e)

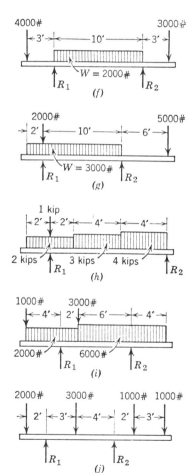

(f)

(g)

(h)

(i)

(j)

FIGURE 2-14

Example 2.   The overhanging beam shown in Fig. 2-13 supports a concentrated load of 4 kips and a uniformly distributed load of 6 kips, arranged as indicated. Compute the reactions and check the results.

Solution: (1) Note that this beam overhangs both of its reactions. The uniformly distributed load extends over a length of 10 ft, and its midpoint lies 1 ft to the right of R_2. Taking R_1 as the center of moments,

$$10R_2 + (4 \times 2) = (6 \times 11)$$

$$10R_2 = 66 - 8$$

$$R_2 = 5.8 \text{ kips}$$

(2) Taking R_2 as the center of moments

$$10R_1 + (6 \times 1) = (4 \times 12)$$

$$10R_1 = 48 - 6$$

$$R_1 = 4.2 \text{ kips}$$

(3) Checking results,

$$10 \text{ kips} = 10 \text{ kips}$$

Note: The importance of your ability to compute the magnitudes of the reactions of beams cannot be stressed too highly. The principle involved occurs time after time in engineering problems; therefore it is necessary to understand thoroughly the foregoing discussion before proceeding. In addition to working the problems below, make up some of your own to test your knowledge. As the above examples show, it is a simple matter to check your results, for after you have computed the reactions by the principle of moments, the sum of the reactions must equal the sum of the loads.

Problems 2-8-A-B-C*-D*-E-F-G-H-I*-J.   Compute the reactions and check the results for the beams shown in Fig. 2-14.

3

Shear and Bending Moment

||||||||||||||||||||||||||||||||||||||||||||||||||||||||||

3-1 Introduction

Figure 3-1a represents a simple beam with a uniformly distributed load W over its entire length. Examination of an actual beam so loaded probably would not reveal any effects of the loading on the beam. However, there are three distinct major tendencies for the beam to fail. Figures 3-1b, c, and d illustrate the three phenomena, with deformations greatly exaggerated.

First, there is a tendency for the beam to fail by dropping between the supports (Fig. 3-1b). This is called *vertical shear*. Second, the beam may fail by bending (Fig. 3-1c). Third, there is a tendency for the fibers of the beam to slide past each other in a horizontal direction (Fig. 3-1d). The name given to this action is *horizontal shear*, and it will be discussed further under steel, wood, and reinforced concrete construction. Naturally, a beam properly designed does not fail in any of the ways just mentioned, but these tendencies to fail are always present and must be considered in structural design.

The forces that prevent failure are supplied by the resisting stresses developed within the beam. Our problem in design is to select beams with dimensions that will provide adequate material to develop these resisting stresses. In this chapter we shall be concerned with methods

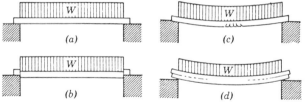

FIGURE 3-1

of measuring the magnitude of shearing and bending forces caused by the beam loading or, as sometimes stated, by the external forces on the beam.

3-2   Vertical Shear

We can define vertical shear as the tendency for one part of a beam to move vertically with respect to an adjacent part. *The magnitude of the shear at any section in the length of a beam is equal to the algebraic sum of the vertical forces on either side of the section.* Vertical shear is usually represented by the letter V. In computing its values in our examples and problems, we consider the forces to the left of the section, but keeping in mind that the same result will be obtained if we work with the forces on the right. We may say then that *the vertical shear at any section of a beam is equal to the reactions minus the loads to the left of the section.* Fix this definition firmly in mind. Now, if we wish to find the magnitude of the vertical shear at any section in the length of a beam, we simply repeat the foregoing statement and write an equation accordingly. It follows from this procedure that the maximum value of the shear for simple beams is equal to the greater reaction.

If the loads and reactions are in units of pounds or kips, the magnitude of the vertical shear will be in units of pounds or kips also. Form the habit of writing the denomination of the units after the numerical values; it will prevent many errors.

Example 1.   Figure 3-2a illustrates a simple beam with two concentrated loads of 600 lb and 1000 lb. Our problem is to find the value of the vertical shear at various points along the length of the beam. Although the weight of the beam constitutes a uniformly distributed load, it is neglected in this example.

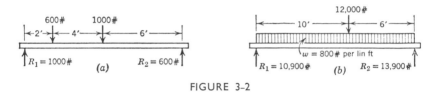

FIGURE 3-2

Solution: (1) The reactions are computed by the principle of moments previously described, and we find that they are $R_1 = 1000$ lb and $R_2 = 600$ lb.

(2) Consider first the value of the vertical shear, V, at an infinitely short distance to the right of R_1. Applying the rule that the shear is equal to the reactions minus the loads to the left of the section, we write $V = R_1 - 0$, or $V = 1000$ lb. The zero represents the value of the loads to the left of the section, which, of course, is zero. Now take a section 1 ft to the right of R_1; again $V_{(x=1)} = R_1 - 0$ or $V_{(x=1)} = 1000$ lb. The subscript $(x = 1)$ indicates the position of the section at which the shear is taken, the distance of the section from R_1. We find that the shear is still 1000 lb and has the same magnitude up to the 600-lb load. The next section to consider is a very short distance to the right of the load 600 lb. Then $V_{(x=2+)} = 1000 - 600 = 400$ lb. Because there are no loads intervening, the shear continues to be the same magnitude up to the load 1000 lb. At a section a short distance to the right of the 1000-lb load, $V_{(x=6+)} = 1000 - (600 + 1000) = -600$ lb. This magnitude continues up to the right-hand reaction R_2.

The preceding example dealt only with concentrated loads. Let us see if we can apply the same procedure to a beam having a uniformly distributed load in addition to a concentrated load.

Example 2. The beam shown in Fig. 3-2b supports a concentrated load of 12,000 lb located 6 ft from R_2 and a uniformly distributed load of 800 pounds per linear foot (lb per lin ft) over its entire length. Compute the value of the vertical shear at various sections along the span.

Solution: (1) Note that the uniform load is given in lb per lin ft (frequently abbreviated to lb per ft). The symbol for uniform load per foot is w, the capital letter W being used to represent the *total* uniformly distributed load. In this instance $w = 800$ lb and $W =$

$800 \times 16 = 12,900$ lb. Writing the equation of moments about R_2 as the center,

$$16R_1 = (800 \times 16 \times 8) + (12,000 \times 6)$$

$$16R_1 = 102,400 + 72,000$$

$$R_1 = \frac{174,400}{16} = 10,900 \text{ lb}$$

In a similar manner R_2 is found to be 13,900 lb. In the quantity $(800 \times 16 \times 8)$ the load (800×16) lb has a lever arm of 8 ft, the distance of its center of gravity to the reaction.

(2) Following the rule used in the preceding example, write the value of V at various sections along the beam:

$$V_{(x=0)} \quad = 10,900 - 0 = 10,900 \text{ lb}$$

$$V_{(x=1)} \quad = 10,900 - 800 = 10,100 \text{ lb}$$

$$V_{(x=5)} \quad = 10,900 - (800 \times 5) = 6900 \text{ lb}$$

$$V_{(x=10-)} = 10,900 - (800 \times 10) = 2900 \text{ lb}$$

$$V_{(x=10+)} = 10,900 - [(800 \times 10) + 12,000] = -9100 \text{ lb}$$

$$V_{(x=16)} \quad = 10,900 - [(800 \times 16) + 12,000] = -13,900 \text{ lb}$$

Note that the value of the vertical shear at the supports has the same magnitude as the reactions.

3-3   Shear Diagrams

In the two preceding examples we computed the value of the shear at several sections along the length of the beams. In order to visualize the results we have obtained, we may make diagrams to plot these values. They are called *shear diagrams* and are constructed as explained below.

To make such a diagram, first draw the beam to scale and locate the loads. This has been done in Figs. 3-3a and b by repeating the load diagrams of Figs. 3-2a and b, respectively. Beneath the beam draw a horizontal base line representing zero shear, such as line A in Figs. 3-3a and b. Above and below this line, plot at any convenient scale the values of the shear at the various sections; the positive or plus values are placed above the line and the negative or minus

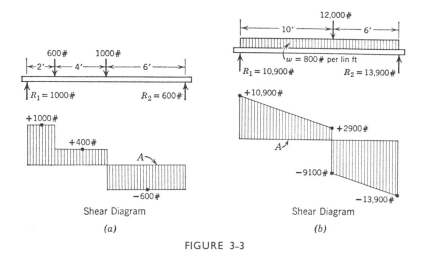

FIGURE 3-3

values below. In Fig. 3-3a, for instance, the value of the shear at R_1 is $+1000$ lb. The shear continues to have the same value up to the load of 600 lb, at which point it drops to $+400$ lb. The same value continues up to the next load, 1000 lb, where it drops to -600 lb and continues to the right-hand reaction. Obviously, to draw a shear diagram it is necessary to compute the values at significant points only. Having made the diagram, we may readily find the value of the shear at any section of the beam by scaling the vertical distance in the diagram. The shear diagram for the beam in Fig. 3-3b is made in the same manner.

There are two important facts to note concerning the vertical shear. The first is the maximum value. We see that the diagrams in each case confirm our earlier observation that the maximum shear is at the reaction having the greater value, and its magnitude is equal to that of the greater reaction. In Fig. 3-3a the maximum shear is 1000 lb, and in Fig. 3-3b it is 13,900 lb. We disregard the positive or negative signs in reading the maximum values of the shear, for the diagrams are merely conventional methods of representing the absolute numerical values.

The other important fact is to note the point at which the shear changes from a plus to a minus quantity. We say, "the point at which the shear passes through zero." In Fig. 3-3a it is under the 1000-lb load, 6 ft from R_1. In Fig. 3-3b the shear passes through zero, 10 ft from R_1. The reason for finding this point is that *the greatest tendency for the beam to fail by bending is at the point at which the shear passes through zero.*

Problems 3-3-A-B*-C*-D-E-F-G-*H-I-J. For the beams indicated in Fig. 3-4, draw the shear diagrams and note in each instance the value of the maximum shear and the point at which the shear passes

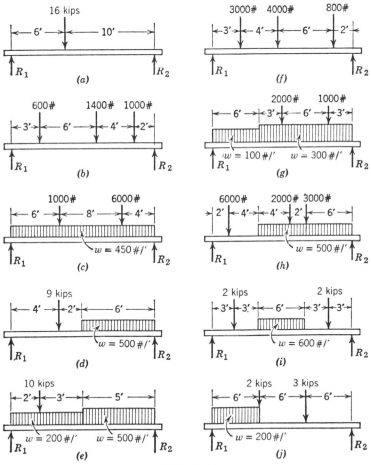

FIGURE 3-4

through zero. Notice that the distributed loads are designated by w in Fig. 3-4, which means so many pounds per foot ($\#/'$ on the diagrams).

3-4   Bending Moment

The forces that tend to cause bending in a beam are the reactions and the loads. Consider the section X–X, 6 ft from R_1. See Fig. 3-5. The force R_1, or 2000 lb, tends to cause a clockwise rotation about this point. Because the force is 2000 lb and the lever arm is 6 ft, the moment of the force is $2000 \times 6 = 12{,}000$ ft-lb. This same value may be found by considering the forces to the right of the section X–X. Let us see. There are two forces to the right of section X–X; R_2, which is 6000 lb, and the load 8000 lb, with lever arms of 10 and 6 ft, respectively. The moment of the reaction is $6000 \times 10 = 60{,}000$ ft-lb, and its direction is counterclockwise with respect to the section X–X. The moment of the force 8000 lb is $8000 \times 6 = 48{,}000$ ft-lb, and its direction is clockwise. Subtracting, $60{,}000$ ft-lb $- 48{,}000$ ft-lb $= 12{,}000$ ft-lb, the resultant moment tending to cause counterclockwise rotation about the section X–X. This is the same magnitude as the moment of the forces on the left which tends to cause a clockwise rotation.

Thus it makes no difference whether we consider the forces to the right of the section or the left; the magnitude of the moment is the same. It is called the *bending moment* because it is the moment of the forces that cause bending stresses in the beam. Its magnitude varies throughout the length of the beam. For instance, at 4 ft from R_1 it is only 2000×4, or 8000 ft-lb. *The bending moment is the algebraic sum of the moments of the forces on either side of the section.* For simplicity, let us take the forces on the left; then we may say *the bending moment at any section of a beam is equal to the moments of the reactions minus the moments of the loads to the left of the section.*

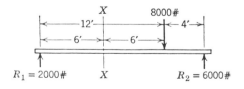

FIGURE 3-5

Because the bending moment is the result of multiplying forces by distances, the denominations are foot-pounds or kip-feet.

Almost everyone confuses shear and bending moment at first. Remember that the shear is the result of subtracting loads from reactions, the units being pounds or kips; the bending moment is the result of subtracting *moments* of loads from *moments* of reactions, with units of foot-pounds or kip-feet.

3-5 Bending Moment Diagrams

The construction of bending moment diagrams follows the procedure used for shear diagrams. The beam span is drawn to scale showing the locations of the loads. Below this, and usually below the shear diagram, a horizontal base line is drawn representing zero bending moment. Then the bending moments are computed at various sections along the beam span and the values are plotted vertically to any convenient scale. In simple beams all bending moments are positive and therefore are plotted above the base line. In overhanging or continuous beams we shall find negative moments, and these are plotted below the base line.

Example. The load diagram in Fig. 3-6 shows a simple beam with two concentrated loads. Draw the shear and bending moment diagrams.

Solution: (1) R_1 and R_2 are first computed and are found to be

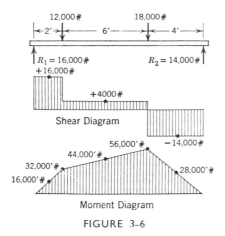

Moment Diagram

FIGURE 3-6

16,000 lb and 14,000 lb, respectively. These values are recorded on the load diagram.

(2) The shear diagram is drawn as described in Art. 3-3. Note that in this instance it is necessary to compute the shear at only one section (between the concentrated loads) because there is no distributed load, and we know that the shear at the reactions is equal in magnitude to the reactions.

(3) Because the value of the bending moment at any section of the beam is equal to the moments of the reactions minus the moments of the loads to the left of the section, the moment at R_1 must be zero, for there are no forces to the left. We might also say that R_1 has a lever arm of zero and $R_1 \times 0 = 0$. Other values in the length of the beam are computed as follows. The subscripts ($x = 1$), etc., show the distance from R_1 at which the bending moment is computed.

$$M_{(x=1)} = (16{,}000 \times 1) = 16{,}000 \text{ ft-lb}$$

$$M_{(x=2)} = (16{,}000 \times 2) = 32{,}000 \text{ ft-lb}$$

$$M_{(x=5)} = (16{,}000 \times 5) - (12{,}000 \times 3) = 44{,}000 \text{ ft-lb}$$

$$M_{(x=8)} = (16{,}000 \times 8) - (12{,}000 \times 6) = 56{,}000 \text{ ft-lb}$$

$$M_{(x=10)} = (16{,}000 \times 10) - [(12{,}000 \times 8) + (18{,}000 \times 2)]$$
$$= 28{,}000 \text{ ft-lb}$$

$$M_{(x=12)} = (16{,}000 \times 12) - [(12{,}000 \times 10) + (18{,}000 \times 4)] = 0$$

The result of plotting these values is shown in the bending moment diagram of Fig. 3-6. More moments were computed than were actually necessary. We known that the bending moments at the supports of simple beams are zero, and in this instance only the bending moments directly under the loads were needed.

3-6   Relation between Shear and Bending Moments

In simple beams the shear diagram passes through zero at some point between the supports. As stated earlier, an important principle in this respect is that *the bending moment has a maximum magnitude wherever the shear passes through zero.* In Fig. 3-6 the shear passes through zero under the 18,000-lb load, that is, at ($x = 8$). Note that

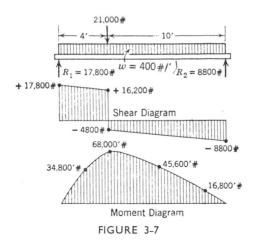

Moment Diagram

FIGURE 3-7

the bending moment has its greatest value at this same point, 56,000 ft-lb. In order to design beams, we must known the value of the maximum bending moment. Frequently we draw only enough of the shear diagram to find the section at which the shear passes through zero and then compute the bending moment at this point.

Example 1.   Draw the shear and bending moment diagrams for the beam shown in Fig. 3-7 which carries a uniformly distributed load of 400 lb per lin ft and a concentrated load of 12,000 lb located 4 ft from R_1.

Solution: (1) Computing the reactions,

$$14R_1 = (21,000 \times 10) + (400 \times 14 \times 7)$$

$$R_1 = 17,800 \text{ lb}$$

$$\text{total load} = 21,000 + (400 \times 14) = 26,600 \text{ lb}$$

Therefore

$$R_1 + R_2 = 26,600$$

$$17,800 + R_2 = 26,600$$

$$R_2 = 26,600 - 17,800 = 8800 \text{ lb}$$

(2) Computing the value of the shear at essential points,

$$V_{(\text{at } R_1)} = 17,800 \text{ lb}$$

$$V_{(x=4-)} = 17,800 - (400 \times 4) = 16,200 \text{ lb}$$

$$V_{(x=4+)} = 17,800 - [(400 \times 4) + 21,000] = -4800 \text{ lb}$$

$$V_{(\text{at } R_2)} = -8800 \text{ lb}$$

Note that the shear passes through zero under the 21,000-lb load; therefore at this point we shall expect to find that the bending moment is a maximum.

(3) Computing the value of the bending moment at selected points,

$$M_{(x=0)} = 0$$

$$M_{(x=2)} = (17,800 \times 2) - (400 \times 2 \times 1) = 34,800 \text{ ft-lb}$$

In the foregoing equation the reaction to the left of the section is 17,800 lb and its lever arm is 2 ft. The load to the left of the section is (400×2) lb and its lever arm is 1 ft, the distance from the center of the load, (400×2) lb, to the center of moments.

$$M_{(x=4)} = (17,800 \times 4) - (400 \times 4 \times 2) = 68,000 \text{ ft-lb}$$

$$M_{(x=8)} = (17,800 \times 8) - [(400 \times 8 \times 4) + (21,000 \times 4)]$$
$$= 45,600 \text{ ft-lb}$$

$$M_{(x=12)} = (17,800 \times 12) - [(400 \times 12 \times 6) + (21,000 \times 8)]$$
$$= 16,800 \text{ ft-lb}$$

$$M_{(x=14)} = (17,800 \times 14) - [(400 \times 14 \times 7) + (21,000 \times 10)] = 0$$

From the two preceding examples (Figs. 3-6 and 3-7) it will be observed that the shear diagram for the parts of the beam on which no loads occur is represented by horizontal lines. For the parts of the beam on which a uniformly distributed load occurs the shear diagram consists of straight inclined lines. The bending moment diagram is represented by straight inclined lines when only concentrated loads occur and by a curved line if the load is distributed.

Occasionally, when a beam has both concentrated and uniformly distributed loads, the shear does not pass through zero under one of the concentrated loads. This frequently occurs when the distributed load is relatively large compared with the concentrated loads. Since it is necessary in designing beams to find the maximum bending moment, we must know the point at which it occurs. This, of course, is the point where the shear passes through zero, and its location is readily determined by the procedure illustrated in the following example.

Example 2.   The load diagram in Fig. 3-8 shows a beam with a concentrated load of 7000 lb applied 4 ft from the left reaction, and a uniformly distributed load of 800 lb per lin ft extending over the full span. Compute the maximum bending moment on the beam.
Solution: (1) Compute the values of the reactions. These are found to be $R_1 = 10,600$ lb and $R_2 = 7600$ lb, and are recorded on the load diagram.

(2) Constructing the shear diagram in Fig. 3-8, we see that the shear passes through zero at some point between the concentrated load of 7000 lb and the right reaction. Call this distance x ft from R_1. Now, we know that the value of the shear at this section is zero; therefore we write an expression for the shear for this point, using the terms of the reaction and loads, and equate the quantity to zero. This equation contains the distance x:

$$V_{(at\ x)} = 10,6000 - [7000 + (800x)] = 0$$

$$800x = 3600$$

$$x = 4.5\ \text{ft}$$

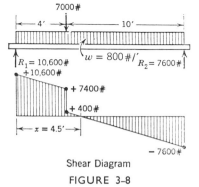

Shear Diagram

FIGURE 3-8

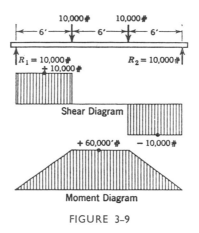

Shear Diagram

Moment Diagram

FIGURE 3-9

(3) The value of the bending moment at this section is given by the expression

$$M_{(x=4.5)} = (10,600 \times 4.5) - \left[(7000 \times 0.5) + \left(800 \times 4.5 \times \frac{4.5}{2}\right)\right]$$

$$M = 36,100 \text{ ft-lb}$$

Under certain conditions the value of the shear is zero for the entire distance between concentrated loads. This occurs when a simple beam has two equal loads at equal distances from the supports. The value of the bending moment is the same at any section between the loads.

Example 3.   The load diagram of Fig. 3-9 shows a simple beam with a span of 18ft, supporting concentrated loads of 18,000 lb located 6 ft from each reaction. Compute the maximum bending moment and draw the moment diagram.

Solution: (1) Because the beam is symmetrically loaded, each reaction is equal to half the total load, or $\frac{1}{2} \times 20,000 = 10,000$ lb.

(2) The shear diagram is readily constructed as shown. Note that the shear has a zero value at all points between the loads. According to the rule, the bending moment must be a maximum over this portion of the beam.

(3) Compute M at various sections along the span and plot the diagram:

$$M_{(x=6)} = 10{,}000 \times 6 = 60{,}000 \text{ ft-lb}$$

$$M_{(x=9)} = (10{,}000 \times 9) - (10{,}000 \times 3) = 60{,}000 \text{ ft-lb}$$

When such a problem occurs, we know, therefore, that the maximum bending moment is equal to one of the loads multiplied by its distance to the nearest reaction. In this example only two concentrated loads are considered. Actually, the weight of the beam constitutes a uniformly distributed load; hence the shear passes through zero at the center of the span. Because the weight of the beam is often quite small when compared with the concentrated loads, it is sometimes neglected in the computations. In this chapter and the preceding one, consideration of beam weight has been omitted in order to focus on the separate effects of concentrated and uniform loads. It will be considered, however, in the sections on steel, wood, and reinforced concrete design.

Problems 3-6-A*-B*-C-D*-E-F-G-H-I-J*.   Draw the shear and bending moment diagrams for the beams shown in Fig. 3-10. In each case, note the magnitude of the maximum shear and the maximum bending moment.

3-7 Negative Bending Moment: Overhanging Beams

When a simple beam bends, it has a tendency to assume the shape shown in Fig. 3-11a. In this case the fibers in the upper part of the beam are in compression. For this condition we say the bending moment is positive (+). Another way to describe a positive bending moment is to say that it is positive when the curve assumed by the bent beam is concave upward. When a beam projects beyond a support (Figs. 3-11b and c), this portion of the beam has tensile stresses in its upper part. The bending moment for this condition is called negative (−); the beam is bent concave downward. If we construct moment diagrams, following the method previously described, the positive and negative moments are shown graphically.

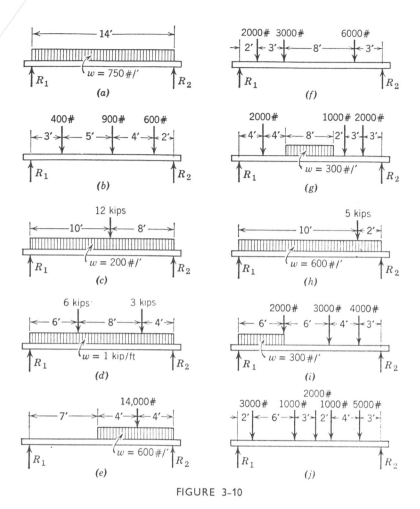

FIGURE 3-10

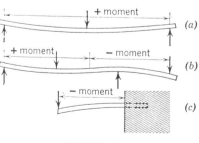

FIGURE 3-11

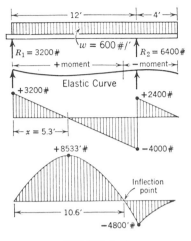

FIGURE 3-12

Example 1. Draw the shear and bending moment diagrams for the overhanging beam shown in Fig. 3-12.

Solution: (1) Computing the reactions,

$$12R_2 = 600 \times 16 \times 8 \qquad R_2 = 6400 \text{ lb}$$

$$12R_1 = 600 \times 16 \times 4 \qquad R_1 = 3200 \text{ lb}$$

(2) Computing the values of the shear,

$$V_{(\text{at } R_1)} = +3200 \text{ lb}$$

$$V_{(x=12-)} = 3200 - (600 \times 12) = -4000 \text{ lb}$$

$$V_{(x=12+)} = (3200 + 6400) - (600 \times 12) = +2400 \text{ lb}$$

$$V_{(x=16)} = (3200 + 6400) - (600 \times 16) = 0$$

To find the point at which the shear passes through zero between the supports (see Example 2, Art. 3-6),

$$3200 - 600x = 0$$

$$x = 5.3 \text{ ft}$$

(3) Computing the values of the bending moment,

$$M_{(at\ R_1)} = 0$$

$$M_{(x=5.3)} = (3200 \times 5.3) - \left(600 \times 5.3 \times \frac{5.3}{2}\right) = 8533 \text{ ft-lb}$$

$$M_{(x=12)} = (3200 \times 12) - (600 \times 12 \times 6) = -4800 \text{ ft-lb}$$

$$M_{(x=16)} = 0$$

To draw the bending moment diagram accurately, the magnitudes at other points may be computed; it will be a curved line.

It is seen in plotting the shear values that there are two points at which the shear passes through zero: at $x = 5.3$ ft and at $x = 12$ ft. The bending moment diagram shows that maximum values are found at each point, one being a positive moment and the other a negative moment. When we design beams, we are concerned only with the maximum value, regardless of whether it is positive or negative. The shear diagram does not indicate which of the two points gives the greater value of the bending moment, and often it is necessary to compute the value at each point at which the shear passes through zero to determine the greater numerically. For this beam it is 8533 ft-lb.

Inflection Point.   The bending moment diagram in Fig. 3-12 indicates a point between the supports at which the value of $M = 0$. This is called the *inflection point*; it is the point at which the curvature reverses, as it changes from concave to convex. It is important to know the position of the inflection point in the study of reinforced concrete beams, for this is the position at which the tensile steel reinforcement is bent upward.

In this problem call x the distance from the left support to the point at which $M = 0$. Then, writing an expression for the value of the bending moment and equating it to zero,

$$(3200 \times x) - \left(600 \times x \times \frac{x}{2}\right) = 0$$

$$3200x - 300x^2 = 0$$

$$x = 10.6 \text{ ft}$$

Examine the curve of the bending moment. You will see that the curve is symmetrical between the left reaction and the inflection point. Now, since the curve reaches its highest point at 5.3 ft from the left support, the inflection point occurs at 2×5.3, or 10.6 ft from R_1. This is the same result found by solving for x in the foregoing equation.

For the beam and load shown in Fig. 3-12 note that the value of the maximum vertical shear is -4000 lb. It occurs immediately to the left of the right-hand support.

Example 2.   Compute the maximum bending moment for the overhanging beam shown in Fig. 3-13.
Solution: (1) Computing the reactions,

$$12R_1 + (200 \times 2) = (800 \times 16) + (1000 \times 10) + (4000 \times 4)$$

$$R_1 = 3200 \text{ lb}$$

$$\text{total load} = 800 + 1000 + 4000 + 200 = 6000 \text{ lb}$$

Therefore
$$R_2 = 6000 - 3200 = 2800 \text{ lb}$$

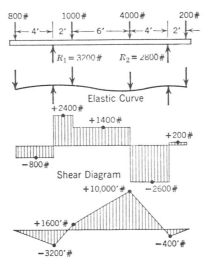

FIGURE 3-13

(2) Computing the values of the shear,

$$V_{(x=1)} = -800 \text{ lb}$$

$$V_{(x=4+)} = 3200 - 800 = +2400 \text{ lb}$$

$$V_{(x=6+)} = 3200 - (800 + 1000) = +1400 \text{ lb}$$

$$V_{(x=12+)} = 3200 - (800 + 1000 + 4000) = -2600 \text{ lb}$$

$$V_{(x=16+)} = (3200 + 2800) - (800 + 1000 + 4000) = +200 \text{ lb}$$

With these values we now plot the shear diagram. We note that the shear passes through zero at three points, R_1, R_2, and under the 4000-lb load. We expect the bending moment to reach maximum values at these points.

(3) Computing the values of the bending moment,

$$M_{(x=0)} = 0$$

$$M_{(x=4)} = -(800 \times 4) = -3200 \text{ ft-lb}$$

$$M_{(x=6)} = (3200 \times 2) - (800 \times 6) = +1600 \text{ ft-lb}$$

$$M_{(x=12)} = (3200 \times 8) - [(800 \times 12) + (1000 \times 6)]$$

$$= +10,000 \text{ ft-lb}$$

$$M_{(x=16)} = (3200 \times 12) - [(800 \times 16) + (1000 \times 10)$$

$$+ (4000 \times 4)] = -400 \text{ ft-lb}$$

$$M_{(x=18)} = 0$$

The maximum value of the bending moment (10,000 ft-lb) occurs under the 4000-lb load

The value of the maximum vertical shear is -2600 lb.

Note that the bending moment diagram changes from a plus value to a minus value at two points. There are two inflection points. If it were possible to scale the moment diagram of Fig. 3-13 with sufficient accuracy, we would find that one of these inflection points is located approximately 1.3 ft to the right of R_1 and the other approximately 0.15 ft to the left of R_2. The reader should check the accuracy of these locations by writing moment equations about each inflection point. If the locations are "exactly" correct, the value of the bending moment at each point will, by definition, equal zero.

Problems 3-7-A*-B-C*-D-E-F-G*-H-I-J. Draw the shear and bending moment diagrams for the beams shown in Fig. 3-14. Determine the maximum bending moment in each case.

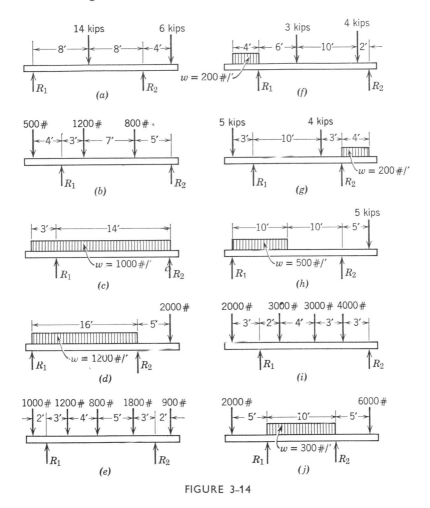

FIGURE 3-14

3-8 Cantilever Beams

When computing the shear and bending moment for cantilever beams, it is convenient to draw the fixed end at the right and to compute the shear and bending moment values to the left, as in the preceding examples.

Example 1.   The cantilever beam shown in Fig. 3-15 projects 12 ft from the face of the wall and has a concentrated load of 800 lb at the unsupported end. Draw the shear and moment diagrams. What are the values of the maximum shear and maximum bending moment?
Solution: (1) The value of the shear is -800 lb throughout the entire length of the beam.

(2) The bending moment is maximum at the wall; its value is -9600 ft-lb.

$$M_{(x=0)} = 0$$

$$M_{(x=1)} = -(800 \times 1) = -800 \text{ ft-lb}$$

$$M_{(x=2)} = -(800 \times 2) = -1600 \text{ ft-lb}$$

$$M_{(x=12)} = -(800 \times 12) = -9600 \text{ ft-lb}$$

Example 2.   Draw the shear and bending moment diagrams for the cantilever beam shown in Fig. 3-16, which carries a uniformly distributed load of 500 lb per lin ft over its full length.
Solution: (1) Computing the values of the shear,

$$V_{(x=1)} = -(500 \times 1) = -500 \text{ lb}$$

$$V_{(x=2)} = -(500 \times 2) = -1000 \text{ lb}$$

$$V_{(x=10)} = -(500 \times 10) = -5000 \text{ lb, the maximum value}$$

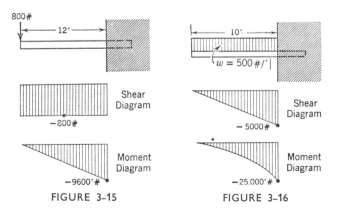

FIGURE 3-15   FIGURE 3-16

(2) Computing the moments,

$$M_{(x=0)} = 0$$

$$M_{(x=2)} = -(500 \times 2 \times 1) = -1000 \text{ ft-lb}$$

$$M_{(x=4)} = -(500 \times 4 \times 2) = -4000 \text{ ft-lb}$$

$$M_{(x=10)} = -(500 \times 10 \times 5) = -25,000 \text{ ft-lb, the maximum value}$$

Example 3. The cantilever beam indicated in Fig. 3-17 has a concentrated load of 2000 lb and a uniformly distributed load of 600 lb per lin ft at the positions shown. Draw the shear and bending moment diagrams. What are the magnitudes of the maximum shear and maximum bending moment?

Solution: (1) Computing the values of the shear,

$$V_{(x=1)} = -2000 \text{ lb}$$

$$V_{(x=8)} = -2000 \text{ lb}$$

$$V_{(x=10)} = -[2000 + (600 \times 2)] = -3200 \text{ lb}$$

$$V_{(x=14)} = -[2000 + (600 \times 6)] = -5600 \text{ lb}$$

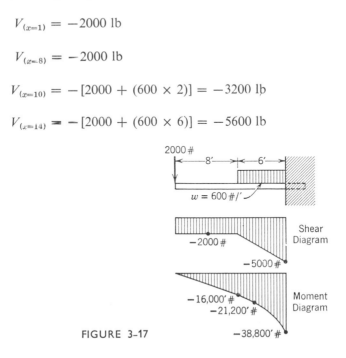

FIGURE 3-17

(2) Computing the moments,

$M_{(x=0)} = 0$

$M_{(x=4)} = -(2000 \times 4) = -8000$ ft-lb

$M_{(x=8)} = -(2000 \times 8) = -16,000$ ft-lb

$M_{(x=10)} = -[(2000 \times 10) + (600 \times 2 \times 1)] = -21,200$ ft-lb

$M_{(x=14)} = -[(2000 \times 14) + (600 \times 6 \times 3)] = -38,800$ ft-lb

The maximum shear is 5600 lb, and the maximum bending moment is 38,800 ft-lb. In cantilever beams the maximum values of both shear and moment occur at the support; the beams are concave downward and the bending moment is negative throughout the length.[1]

Problems 3-8-A-B*-C-D*.   Draw the shear and bending moment diagrams for the cantilever beams shown in Fig. 3-18. State the maximum shear and maximum bending moment values in each beam.

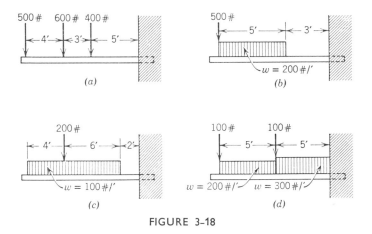

FIGURE 3-18

[1] The reader should satisfy himself that this statement holds when the fixed end of a cantilever beam is taken to the left. It is suggested that Example 3 be reworked with Fig. 3-17 reversed left for right. All numerical results will be the same but the shear diagram will be positive over its full length.

3-9   Bending Moment Formulas

The method of computing bending moments presented thus far in this chapter enables us to find the maximum value under a wide variety of loading conditions. However, certain conditions occur so frequently that it is convenient to use formulas that give the maximum values directly. Structural design handbooks contain many such formulas; two of the most commonly used formulas are derived in the following articles, and a few others are given in Art. 3-12.

3-10   Concentrated Load at Center of Span

A simple beam with a concentrated load at the center of the span occurs very frequently in practice. Call the load P and the span length between supports L, as indicated in the load diagram of Fig. 3-19. For this symmetrical loading each reaction is $P/2$ and it is readily apparent that the shear will pass through zero at distance $x = L/2$ from R_1. Therefore the maximum bending moment occurs at the center of the span, under the load. Now let us compute the value of the bending moment at this section:

$$M_{(x=L/2)} = \frac{P}{2} \times \frac{L}{2}$$

or

$$M = \frac{PL}{4}$$

Note that this value is given in Case 1, Table 3-1. This formula is well worth remembering. Observe how quickly bending moments are computed by its use.

Example.   A simple beam 20 ft in length has a concentrated load of 8000 lb at the center of the span. Compute the maximum bending moment.
Solution: The formula giving the value of the maximum bending moment for this condition is $M = PL/4$. Therefore

$$M = \frac{8000 \times 20}{4} = 40,000 \text{ ft-lb}$$

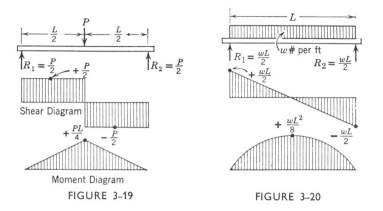

Shear Diagram

Moment Diagram

FIGURE 3-19

FIGURE 3-20

If we want the results in inch-pounds, as we frequently do in design, we simply multiply by 12; that is

$$M = 40,000 \text{ ft-lb} = (40,000 \times 12)$$

or

$$480,000 \text{ in-lb}$$

3-11 Simple Beam with Uniform Load

This is probably the most common beam loading. It occurs time and again. Let us call the span L and the load w lb per lin ft, as indicated in Fig. 3-20. The total load on the beam is wL; hence each reaction is $wL/2$. The maximum bending moment occurs at the center of the span at distance $L/2$ from R_1. Writing the value of M for this section, we have

$$M_{(x=L/2)} = \left(\frac{wL}{2} \times \frac{L}{2}\right) - \left(\frac{wL}{2} \times \frac{L}{4}\right)$$

$$M = \frac{wL^2}{4} - \frac{wL^2}{8}$$

$$M = \frac{wL^2}{8}$$

If, instead of being given the load per linear foot, we are given the total uniformly distributed load, we call the total load W. Because

$wL = W$, the value of the maximum bending moment can be written $M = wL^2/8$ or $WL/8$. See Case 2, Table 3-1. Remember this formula. You will use it many times. Its convenience is demonstrated in the practical example below.

Example.   A simple beam 14 ft long has a uniformly distributed load of 800 lb per lin ft. Compute the maximum bending moment. *Solution:*   The formula that gives the maximum bending moment for a simple beam with a uniformly distributed load is $M = wL^2/8$. Substituting these values,

$$M = \frac{800 \times 14 \times 14}{8}$$

$$M = 19{,}600 \text{ ft-lb}$$

or

$$M = 19{,}600 \times 12 = 235{,}200 \text{ in-lb}$$

Suppose in this problem we had been given the total load of 11,200 lb instead of 800 lb per ft. Then,

$$M = \frac{WL}{8}$$

$$M = \frac{11{,}200 \times 14}{8} = 19{,}600 \text{ ft-lb}$$

$$19{,}600 \text{ ft-lb} = 235{,}200 \text{ in-lb}$$

The result, of course, is the same.

3-12   Typical Loadings: Shear and Moment Formulas

Some of the most common beam loadings are shown in Table 3-1 In addition to the formulas for maximum shear V and maximum bending moment M, expressions for maximum deflection D are given also. (Discussion of deflection formulas will be deferred for the time being but will be considered under beam design in subsequent sections.)

In the table, if the loads P and W are in pounds or kips, the vertical shear V will also be in units of pounds or kips. When the loads are given in pounds or kips and the span in feet, the bending moment M will be in units of foot-pounds or kip-feet.

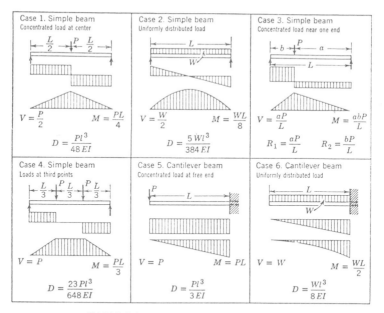

TABLE 3-1.   Beam Diagrams and Formulas

An extensive series of beam diagrams and formulas is contained in Part 2 of the *Manual of Steel Construction* published by the American Institute of Steel Construction.

Problem 3-12-A*.   A simple beam with a span of L ft has three concentrated loads of P lb each, located at the quarter points of the span. Compute the values of the maximum shear and maximum bending moment, using the terms P and L.

Problem 3-12-B*.   Four concentrated loads of P lb each are placed at the fifth points of the span of a simple beam. If the span length is L ft, compute the magnitudes of the maximum shear and maximum bending moment in terms of P and L.

Problem 3-12-C.   Draw the shear and moment diagrams for the beam and loading described in Problem 3-12-A, noting significant values in terms of P and L (similar to Fig. 3-19).

Problem 3-12-D.   For the beam and loading described in Problem 3-12-B, draw and label the shear and bending moment diagrams in terms of P and L.

4

Theory of Bending and Properties of Sections

||||||||||||||||||||||||||||||||||||||||||||||||||||||||||

4-1 Resisting Moment

We learned in the preceding chapter that bending moment is a measure of the tendency of the external forces on a beam to deform it by bending. We will now consider the action within the beam that resists bending and is called the *resisting moment*.

Figure 4-1*a* shows a simple beam, rectangular in cross section, supporting a single concentrated load P. Figure 4-1*b* is an enlarged sketch of the left-hand portion of the beam between the reaction and section X–X. From the preceding discussions we know that the reaction R_1 tends to cause a clockwise rotation about point A in the section under consideration; this we have defined as the bending moment at the section. In this type of beam the fibers in the upper part are in compression and those in the lower part are in tension. There is a horizontal plane separating the compressive and tensile stresses; it is called the *neutral surface*, and at this plane there are neither compressive nor tensile stresses with respect to bending. The

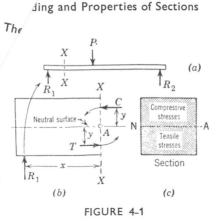

FIGURE 4-1

line in which the neutral surface intersects the beam cross section (Fig. 4-1c) is called the *neutral axis*, NA.

Call the sum of all the compressive stresses acting on the upper part of the cross section C, and the sum of all the tensile stresses acting on the lower part T. It is the sum of the moments of these stresses at the section that holds the beam in equilibrium; this is called the *resisting moment* and is equal to the bending moment in magnitude. The bending moment about point A is $R_1 \times x$, and the resisting moment about the same point is $(C \times y) + (T \times y)$. The bending moment tends to cause a clockwise rotation, and the resisting moment tends to cause a counterclockwise rotation. If the beam is in equilibrium, these moments are equal, or

$$R_1 \times x = (C \times y) + (T \times y)$$

that is, the bending moment equals the resisting moment. This is the theory of flexure (bending) in beams. For any type of beam we can compute the bending moment; and if we wish to design a beam to withstand this tendency to bend, we must select a member with a cross section of such shape, area, and material that it is capable of developing a resisting moment equal to the bending moment.

4-2   The Flexure Formula

The flexure formula $M = fS$ is an expression for resisting moment that involves the size and shape of the beam cross section (represented by S in the formula) and the material of which the beam is made (represented by f). It is used in the design of all homogeneous beams, i.e. beams made of one material only such as steel or wood. You will never need to derive this formula, but you will use it many times. The following brief derivation is presented to show the principles on which the formula is based.

Figure 4-2 represents a partial side elevation and the cross section of a homogeneous beam subjected to bending stresses. The cross section shown is unsymmetrical about the neutral axis but this discussion applies to a cross section of any shape. In Fig. 4-2a let c be the distance of the fiber farthest from the neutral axis, and let f be the unit stress on the fiber at distance c. If f, the extreme fiber stress, does not exceed the elastic limit of the material, the stresses in the other fibers are directly proportional to their distances from the neutral axis. That is to say, if one fiber is twice the distance from the neutral axis than another fiber, the fiber at the greater distance will have twice the stress. The stresses are indicated in the figure by the small lines with arrows, which represent the compressive and tensile stresses acting toward and away from the section, respectively. If c is in inches, the unit stress on a fiber at 1 in. distance is f/c. Now imagine an infinitely small area a at z distance from the neutral axis. The unit stress on this fiber is $(f/c) \times z$, and because this small area contains a square inches the total stress on fiber a is $(f/c) \times z \times a$. The *moment* of the stress on fiber a at z distance is

$$\frac{f}{c} \times z \times a \times z \quad \text{or} \quad \frac{f}{c} \times a \times z^2$$

We know, however, that there is an extremely large number of these minute areas, and if we use the symbol $\sum$ to represent the sum

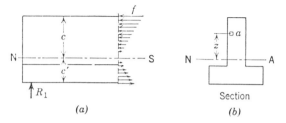

FIGURE 4-2

of this very large number we can write

$$\sum \frac{f}{c} \times a \times z^2$$

which means the sum of the moments of all the stresses in the cross section with respect to the neutral axis. This we know is the *resisting moment*, and it is equal to the bending moment.

Therefore

$$M = \frac{f}{c} \sum a \times z^2$$

The quantity $\sum a \times z^2$ may be read "the sum of the products of all the elementary areas times the square of their distances from the neutral axis." We call this the *moment of inertia* and represent it by the letter I. Therefore, substituting in the above, we have

$$M = \frac{f}{c} \times I \qquad \text{or} \qquad M = \frac{fI}{c}$$

This is known as the *flexure formula* or *beam formula*, and by its use we may design any beam that is composed of a single material. The expression may be simplified further by substituting S for I/c, called the *section modulus*, a term that is described more fully in Art. 4-6. Making this substitution, the formula becomes

$$M = fS$$

Use of the flexure formula in the design and investigation of beams is discussed in Art. 4-7.

4-3   Properties of Sections: Structural Shapes

Each of the terms, moment of inertia and section modulus, used in the preceding article represents a *property* of a particular beam cross section. Other properties are the area of the section, the radius of gyration, and the position of the centroid of the cross-sectional area. These properties and other useful design constants are tabulated for the commonly used structural steel shapes[1] in Tables 4-1 through 4-5, and the properties of structural lumber are given in Table 4-6.

---

[1] For W shapes the designations in the first column of the tables indicate the nominal depth and the weight per linear foot; for S shapes and channels, the actual depth and the nominal weight per foot are given.

Reference to the tables will show that structural steel shapes have two major axes, designated X–X and Y–Y. The position in which the member is placed determines the axis to be considered. For example, the wide flange sections and I-beams (Tables 4-1 and 4-2, respectively) are nearly always used with the web vertical; hence the X–X or horizontal axis determines the applicable properties. The rectangular cross sections characteristic of structural lumber also have a vertical as well as a horizontal axis, but Table 4-6 records properties about the X–X axis only since beams and joists are always used with the longer side vertical and planks with the shorter side vertical.

Several of the properties listed in the tables are discussed below and others are considered under the design of structural members in Sections II and III.

4-4   Centroids

The centroid of a plane surface is a point that corresponds to the center of gravity of a very thin homogeneous plate of the same area and shape. It can be shown that the neutral axis of a beam cross section passes through its centroid; consequently it is necessary to know its exact position. For symmetrical sections such as rectangles and I-shaped sections, it can be seen by inspection that the centroid lies at a point midway between the upper and lower surfaces of the section, at the intersection of the X–X and Y–Y axes shown at the head of Tables 4-1 and 4-2. The distance c from the extreme (most remote) fiber to the neutral axis is, therefore, half the depth of symmetrical sections. For unsymmetrical sections the position of the centroid must be computed. This is accomplished by using the principle of *statical moments.*

The statical moment of a plane area with respect to an axis is the area multiplied by the perpendicular distance from the centroid to the axis. *If an area is divided into a number of parts, the statical moment of the entire area is equal to the sum of the statical moments of the parts.* This is our key for locating the centroid. The following example shows how the distance c is computed.

Example.   Figure 4-3 is a beam cross section, unsymmetrical with respect to the horizontal axis. Find the value of c, the distance of the neutral axis from the most remote fiber.

TABLE 4-1.

Selected Wide Flange Sections*
W SHAPES
Properties for designing

Designation	Area A	Depth d	Flange Width b_f	Flange Thickness t_f	Web thickness t_w	Axis X-X I	Axis X-X S	Axis X-X r	Axis Y-Y I	Axis Y-Y S	Axis Y-Y r
	In.²	In.	In.	In.	In.	In.⁴	In.³	In.	In.⁴	In.³	In.
W 36 × 300	88.3	36.72	16.655	1.680	0.945	20300	1110	15.2	1300	156	3.83
× 135	39.8	35.55	11.945	0.794	0.598	7820	440	14.0	226	37.9	2.39
W 33 × 240	70.6	33.50	15.865	1.400	0.830	13600	813	13.9	933	118	3.64
× 118	34.8	32.86	11.484	0.738	0.554	5900	359	13.0	187	32.5	2.32
W 30 × 132	38.9	30.30	10.551	1.000	0.615	5760	380	12.2	196	37.2	2.25
× 99	29.1	29.64	10.458	0.670	0.522	4000	270	11.7	128	24.5	2.10
W 27 × 114	33.6	27.28	10.070	0.932	0.570	4090	300	11.0	159	31.6	2.18
× 84	24.8	26.69	9.963	0.636	0.463	2830	212	10.7	105	21.1	2.06
W 24 × 120	35.4	24.31	12.088	0.930	0.556	3650	300	10.2	274	45.4	2.78
× 110	32.5	24.16	12.042	0.855	0.510	3330	276	10.1	249	41.4	2.77
× 100	29.5	24.00	12.000	0.775	0.468	3000	250	10.1	223	37.2	2.75
× 76	22.4	23.91	8.985	0.682	0.440	2100	176	9.69	82.6	18.4	1.92
× 68	20.0	23.71	8.961	0.582	0.416	1820	153	9.53	70.0	15.6	1.87
W 21 × 73	21.5	21.24	8.295	0.740	0.455	1600	151	8.64	70.6	17.0	1.81
× 68	20.0	21.13	8.270	0.685	0.430	1480	140	8.60	64.7	15.7	1.80
× 62	18.3	20.99	8.240	0.615	0.400	1330	127	8.54	57.5	13.9	1.77
× 55	16.2	20.80	8.215	0.522	0.375	1140	110	8.40	48.3	11.8	1.73
× 49	14.4	20.82	6.520	0.532	0.368	971	93.3	8.21	24.7	7.57	1.31
× 44	13.0	20.66	6.500	0.451	0.348	843	81.6	8.07	20.7	6.38	1.27
W 18 × 85	25.0	18.32	8.838	0.911	0.526	1440	157	7.57	105	23.8	2.05
× 77	22.7	18.16	8.787	0.831	0.475	1290	142	7.54	94.1	21.4	3.04
× 55	16.2	18.12	7.532	0.630	0.390	891	98.4	7.42	45.0	11.9	1.67
× 50	14.7	18.00	7.500	0.570	0.358	802	89.1	7.38	40.2	10.7	1.65
× 45	13.2	17.86	7.477	0.499	0.335	706	79.0	7.30	34.8	9.32	1.62
W 16 × 50	14.7	16.25	7.073	0.628	0.380	657	80.8	6.68	37.1	10.5	1.59
× 45	13.3	16.12	7.039	0.563	0.346	584	72.5	6.64	32.8	9.32	1.57
× 40	11.8	16.00	7.000	0.503	0.307	517	64.6	6.62	28.8	8.23	1.56
× 36	10.6	15.85	6.992	0.428	0.299	447	56.5	6.50	24.4	6.99	1.52

* Compiled from data in the 7th Edition of the *Manual of Steel Construction*. Courtesy American Institute of Steel construction.

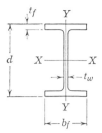

TABLE 4-1.

W SHAPES
Properties for designing

Nominal weight per ft.	r_T	$\dfrac{d}{A_f}$	Compact section criteria					Torsional constant J	Warping constant C_W	Plastic modulus	
			$\dfrac{b_f}{2t_f}$	F'_y	$\dfrac{d}{t_w}$	F''_y	F'''_y			Z_x	Z_y
Lb	In.	In.		Ksi		Ksi	Ksi	In.⁴	In.⁶	In.³	In.³
300	4.46	1.31	4.96	—	38.9	—	43.7	64.2	398000	1260	241
135	2.97	3.75	7.52	48.2	59.4	48.0	18.7	7.03	68300	510	59.9
240	4.23	1.51	5.67	—	40.4	—	40.5	36.6	240000	919	182
118	2.87	3.88	7.78	45.0	59.3	48.2	18.8	5.32	48200	415	51.3
132	2.72	2.87	5.28	—	49.3	—	27.2	9.72	42100	437	58.5
99	2.61	4.23	7.80	44.7	56.8	52.6	20.5	3.78	26900	313	38.7
114	2.62	2.91	5.40	—	47.9	—	28.8	7.36	27600	343	49.4
84	2.52	4.21	7.83	44.4	57.6	51.1	19.9	2.79	17800	244	33.0
120	3.22	2.16	6.50	64.5	43.7	—	34.5	8.27	37500	338	69.9
110	3.20	2.35	7.04	54.9	47.4	—	29.4	6.45	33800	309	63.6
100	3.18	2.58	7.74	45.5	51.3	64.5	25.1	4.87	30100	280	57.2
76	2.32	3.90	6.59	62.8	54.3	57.5	22.4	2.70	11100	201	28.7
68	2.28	4.55	7.70	46.0	57.0	52.3	20.3	1.86	9350	176	24.4
73	2.16	3.46	5.60	—	46.7	—	30.3	3.02	7410	172	26.6
68	2.15	3.73	6.04	—	49.1	—	27.4	2.45	6760	160	24.4
62	2.13	4.14	6.70	60.7	52.5	61.6	24.0	1.83	5960	144	21.7
55	2.10	4.85	7.87	44.0	55.5	55.2	21.5	1.24	4970	126	18.4
49	1.63	6.00	6.13	—	56.6	53.0	20.6	1.09	2540	108	10.2
44	1.59	7.05	7.21	52.5	59.4	48.2	18.7	0.768	2120	95.3	10.2
85	2.37	2.28	4.85	—	34.8	—	54.4	5.50	7960	178	36.8
77	2.36	2.49	5.29	—	38.2	—	45.2	4.16	7070	161	33.1
55	1.98	3.82	5.98	—	46.5	—	30.6	1.66	3440	112	18.6
50	1.96	4.21	6.58	63.0	50.3	—	26.1	1.25	3050	101	16.6
45	1.94	4.79	7.49	48.5	53.3	59.7	23.2	0.889	2620	89.7	14.5
50	1.87	3.66	5.63	—	42.8	—	36.1	1.51	2260	91.8	16.3
45	1.85	4.07	6.25	—	46.6	—	30.4	1.11	1980	82.1	14.4
40	1.84	4.54	6.96	56.3	52.1	62.5	24.3	0.790	1730	72.8	12.7
36	1.81	5.30	8.17	40.8	53.0	60.4	23.5	0.545	1450	64.0	10.8

TABLE 4-1.

Selected Wide Flange Sections (continued)
W SHAPES
Properties for Designing

Designation	Area A	Depth d	Flange Width b_f	Flange Thickness t_f	Web thickness t_w	Axis X-X I	Axis X-X S	Axis X-X r	Axis Y-Y I	Axis Y-Y S	Axis Y-Y r
	In.^{2}	In.	In.	In.	In.	In.^{4}	In.^{3}	In.	In.^{4}	In.^{3}	In.
W 14 × 87	25.6	14.00	14.500	0.688	0.420	967	138	6.15	350	48.2	3.70
× 74	21.8	14.19	10.072	0.783	0.450	797	112	6.05	133	26.5	2.48
× 68	20.0	14.06	10.040	0.718	0.418	724	103	6.02	121	24.1	2.46
× 53	15.6	13.94	8.062	0.658	0.370	542	77.8	5.90	57.5	14.3	1.92
× 48	14.1	13.81	8.031	0.593	0.339	485	70.2	5.86	51.3	12.8	1.91
× 43	12.6	13.68	8.000	0.528	0.308	429	62.7	5.82	45.1	11.3	1.89
× 38	11.2	14.12	6.776	0.513	0.313	386	54.7	5.88	26.6	7.86	1.54
× 34	10.0	14.00	6.750	0.453	0.287	340	48.6	5.83	23.3	6.89	1.52
× 30	8.83	13.86	6.733	0.383	0.270	290	41.9	5.74	19.5	5.80	1.49
W 12 × 65	19.1	12.12	12.000	0.606	0.390	533	88.0	5.28	175	29.1	3.02
× 58	17.1	12.19	10.014	0.641	0.359	476	78.1	5.28	107	21.4	2.51
× 53	15.6	12.06	10.000	0.576	0.345	426	70.7	5.23	96.1	19.2	2.48
× 50	14.7	12.19	8.077	0.641	0.371	395	64.7	5.18	56.4	14.0	1.96
× 45	13.2	12.06	8.042	0.576	0.336	351	58.2	5.15	50.0	12.4	1.94
× 40	11.8	11.94	8.000	0.516	0.294	310	51.9	5.13	44.1	11.0	1.94
× 36	10.6	12.24	6.565	0.540	0.305	281	46.0	5.15	25.5	7.77	1.55
× 31	9.13	12.09	6.525	0.465	0.265	239	39.5	5.12	21.6	6.61	1.54
× 27	7.95	11.96	6.497	0.400	0.237	204	34.2	5.07	18.3	5.63	1.52
W 10 × 89	26.2	10.88	10.275	0.998	0.615	542	99.7	4.55	181	35.2	2.63
× 60	17.7	10.25	10.075	0.683	0.415	344	67.1	4.41	116	23.1	2.57
× 49	14.4	10.00	10.000	0.558	0.340	273	54.6	4.35	93.0	18.6	2.54
× 45	13.2	10.12	8.022	0.618	0.350	249	49.1	4.33	53.2	13.3	2.00
× 39	11.5	9.94	7.990	0.528	0.318	210	42.2	4.27	44.9	11.2	1.98
× 33	9.71	9.75	7.964	0.433	0.292	171	35.0	4.20	36.5	9.16	1.94
× 25	7.36	10.08	5.762	0.430	0.252	133	26.5	4.26	13.7	4.76	1.37
× 21	6.20	9.90	5.750	0.340	0.240	107	21.5	4.15	10.8	3.75	1.32
W 8 × 67	19.7	9.00	8.287	0.933	0.575	272	60.4	3.71	88.6	21.4	2.12
× 40	11.8	8.25	8.077	0.558	0.365	146	35.5	3.53	49.0	12.1	2.04
× 31	9.12	8.00	8.000	0.433	0.288	110	27.4	3.47	37.0	9.24	2.01
× 28	8.23	8.06	6.540	0.463	0.285	97.8	24.3	3.45	21.6	6.61	1.62
× 24	7.06	7.93	6.500	0.398	0.245	82.5	20.8	3.42	18.2	5.61	1.61
× 20	5.89	8.14	5.268	0.378	0.248	69.4	17.0	3.43	9.22	3.50	1.25
× 17	5.01	8.00	5.250	0.308	0.230	56.6	14.1	3.36	7.44	2.83	1.22

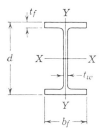

TABLE 4-1.

W SHAPES
Properties for Designing

Nominal weight per ft.	r_T	$\dfrac{d}{A_f}$	Compact section criteria					Torsional constant J	Warping constant C_W	Plastic modulus	
			$\dfrac{b_f}{2t_f}$	F'_y	$\dfrac{d}{t_w}$	F''_y	F'''_y			Z_x	Z_y
Lb.	In.			Ksi		Ksi	Ksi	In.⁴	In.⁶	In.³	In.³
87	4.02	1.40	10.5	24.5	33.3	—	59.4	3.68	15500	151	73.0
74	2.76	1.80	6.43	—	31.5	—	—	3.86	6000	126	40.5
68	2.74	1.95	6.99	55.7	33.6	—	58.4	3.01	5390	115	36.8
53	2.18	2.63	6.13	—	37.7	—	46.5	1.93	2540	87.1	21.9
48	2.16	2.90	6.77	59.4	40.7	—	39.8	1.44	2240	78.4	19.6
43	2.14	3.24	7.58	47.5	44.4	—	33.5	1.05	1950	69.7	17.3
38	1.80	4.06	6.60	62.5	45.1	—	32.5	0.796	1230	61.6	12.1
34	1.78	4.58	7.45	49.1	48.8	—	27.8	0.567	1070	54.6	10.6
30	1.75	5.37	8.79	35.3	51.3	64.4	25.1	0.376	886	47.2	8.95
65	3.31	1.67	9.90	27.8	31.1	—	—	2.19	5790	97.0	44.1
58	2.75	1.90	7.81	44.7	34.0	—	57.3	2.10	3580	86.5	32.6
53	2.74	2.09	8.68	36.2	35.0	—	54.1	1.59	3170	78.1	29.2
50	2.19	2.35	6.30	—	32.9	—	61.2	1.79	1880	72.5	21.4
45	2.18	2.60	6.98	55.9	35.9	—	51.3	1.32	1650	64.8	19.0
40	2.16	2.89	7.75	45.3	40.6	—	40.0	0.956	1440	57.5	16.8
36	1.77	3.45	6.08	—	40.1	—	41.0	0.830	873	51.6	11.9
31	1.75	3.98	7.02	55.4	45.6	—	31.7	0.536	728	44.1	10.1
27	1.74	4.60	8.12	41.3	50.5	—	25.9	0.351	611	38.0	8.62
89	2.88	1.06	5.15	—	17.7	—	—	7.74	4410	114	53.6
60	2.80	1.49	7.38	50.1	24.7	—	—	2.49	2670	75.0	35.1
49	2.77	1.79	8.96	33.9	29.4	—	—	1.38	2070	60.3	28.2
45	2.21	2.04	6.49	64.7	28.9	—	—	1.50	1200	54.9	20.2
39	2.19	2.36	7.57	47.6	31.3	—	—	0.971	995	46.9	17.1
33	2.16	2.83	9.20	32.2	33.4	—	59.2	0.580	792	38.8	14.0
25	1.56	4.07	6.70	60.7	40.0	—	41.3	0.373	320	29.6	7.30
21	1.53	5.06	8.46	38.1	41.3	—	38.8	0.210	246	24.1	5.77
67	2.33	1.16	4.44	—	15.7	—	—	5.05	1440	70.2	32.7
40	2.24	1.83	7.24	52.0	22.6	—	—	1.12	725	39.8	18.5
31	2.21	2.31	9.24	31.9	27.8	—	—	0.534	529	30.4	14.0
28	1.80	2.66	7.06	54.6	28.3	—	—	0.533	312	27.1	10.1
24	1.78	3.07	8.17	40.9	32.4	—	63.0	0.343	259	23.1	8.54
20	1.42	4.09	6.97	56.1	32.8	—	61.3	0.245	139	19.1	5.37
17	1.40	4.95	8.52	37.5	34.8	—	54.6	0.147	110	15.9	4.36

TABLE 4-2.

American Standard I-Beams*
S SHAPES
Properties for Designing

Designation	Area A	Depth d	Flange Width b_f	Flange Thickness t_f	Web thickness t_w	Axis X-X I	Axis X-X S	Axis X-X r	Axis Y-Y I	Axis Y-Y S	Axis Y-Y r
	In.²	In.	In.	In.	In.	In.⁴	In.³	In.	In.⁴	In.³	In.
S 24 × 120	35.3	24.00	8.048	1.102	0.798	3030	252	9.26	84.2	20.9	1.54
× 105.9	31.1	24.00	7.875	1.102	0.625	2830	236	9.53	78.2	19.8	1.58
S 24 × 100	29.4	24.00	7.247	0.871	0.747	2390	199	9.01	47.8	13.2	1.27
× 90	26.5	24.00	7.124	0.871	0.624	2250	187	9.22	44.9	12.6	1.30
× 79.9	23.5	24.00	7.001	0.871	0.501	2110	175	9.47	42.3	12.1	1.34
S 20 × 95	27.9	20.00	7.200	0.916	0.800	1610	161	7.60	49.7	13.8	1.33
× 85	25.0	20.00	7.053	0.916	0.653	1520	152	7.79	46.2	13.1	1.36
S 20 × 75	22.1	20.00	6.391	0.789	0.641	1280	128	7.60	29.6	9.28	1.16
× 65.4	19.2	20.00	6.250	0.789	0.500	1180	118	7.84	27.4	8.77	1.19
S 18 × 70	20.6	18.00	6.251	0.691	0.711	926	103	6.71	24.1	7.72	1.08
× 54.7	16.1	18.00	6.001	0.691	0.461	804	89.4	7.07	20.8	6.94	1.14
S 15 × 50	14.7	15.00	5.640	0.622	0.550	486	64.8	5.75	15.7	5.57	1.03
× 42.9	12.6	15.00	5.501	0.622	0.411	447	59.6	5.95	14.4	5.23	1.07
S 12 × 50	14.7	12.00	5.477	0.659	0.687	305	50.8	4.55	15.7	5.74	1.03
× 40.8	12.0	12.00	5.252	0.659	0.472	272	45.4	4.77	13.6	5.16	1.06
S 12 × 35	10.3	12.00	5.078	0.544	0.428	229	38.2	4.72	9.87	3.89	0.980
× 31.8	9.35	12.00	5.000	0.544	0.350	218	36.4	4.83	9.36	3.74	1.00
S 10 × 35	10.3	10.00	4.944	0.491	0.594	147	29.4	3.78	8.36	3.38	0.901
× 25.4	7.46	10.00	4.661	0.491	0.311	124	24.7	4.07	6.79	2.91	0.954
S 8 × 23	6.77	8.00	4.171	0.425	0.441	64.9	16.2	3.10	4.31	2.07	0.798
× 18.4	5.41	8.00	4.001	0.425	0.271	57.6	14.4	3.26	3.73	1.86	0.831
S 7 × 20	5.88	7.00	3.860	0.392	0.450	42.4	12.1	2.69	3.17	1.64	0.734
× 15.3	4.50	7.00	3.662	0.392	0.252	36.7	10.5	2.86	2.64	1.44	0.766
S 6 × 17.25	5.07	6.00	3.565	0.359	0.465	26.3	8.77	2.28	2.31	1.30	0.675
× 12.5	3.67	6.00	3.332	0.359	0.232	22.1	7.37	2.45	1.82	1.09	0.705
S 5 × 14.75	4.34	5.00	3.284	0.326	0.494	15.2	6.09	1.87	1.67	1.01	0.620
× 10	2.94	5.00	3.004	0.326	0.214	12.3	4.92	2.05	1.22	0.809	0.643
S 4 × 9.5	2.79	4.00	2.796	0.293	0.326	6.79	3.39	1.56	0.903	0.646	0.569
× 7.7	2.26	4.00	2.663	0.293	0.193	6.08	3.04	1.64	0.764	0.574	0.581
S 3 × 7.5	2.21	3.00	2.509	0.260	0.349	2.93	1.95	1.15	0.586	0.468	0.516
× 5.7	1.67	3.00	2.330	0.260	0.170	2.52	1.68	1.23	0.455	0.390	0.522

* Taken from the 7th Edition of the *Manual of Steel Construction*. Courtesy American Institute of Steel Construction.

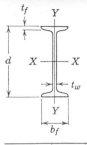

TABLE 4-2.

S SHAPES
Properties for Designing

Nominal weight per ft.	r_T	$\dfrac{d}{A_f}$	$\dfrac{b_f}{2t_f}$	F_y'	$\dfrac{d}{t_w}$	F_y''	F_y'''	Torsional constant J	Warping constant C_W	Z_x	Z_y
Lb.	In.			Ksi		Ksi	Ksi	In.⁴	In.⁶	In.³	In.³
120	1.93	2.71	3.65	—	30.1	—	—	13.0	11000	299	36.4
105.9	1.93	2.77	3.57	—	38.4	—	44.8	10.4	10200	274	33.5
100	1.65	3.80	4.16	—	32.1	—	64.0	7.63	6390	240	24.0
90	1.65	3.87	4.09	—	38.5	—	44.6	6.05	6010	222	22.3
79.9	1.66	3.94	4.02	—	47.9	—	28.8	4.90	5660	205	20.7
95	1.70	3.03	3.93	—	25.0	—	—	8.46	4520	194	24.7
85	1.69	3.09	3.85	—	30.6	—	—	6.63	4200	179	22.8
75	1.46	3.96	4.05	—	31.2	—	—	4.60	2730	153	16.6
65.4	1.45	4.05	3.96	—	40.0	—	41.3	3.50	2530	138	15.2
70	1.41	4.17	4.52	—	25.3	—	—	4.15	1810	125	14.4
54.7	1.41	4.34	4.34	—	39.0	—	43.3	2.37	1560	105	12.1
50	1.31	4.28	4.53	—	27.3	—	—	2.12	811	77.1	9.97
42.9	1.31	4.38	4.42	—	36.5	—	49.6	1.54	743	69.3	9.02
50	1.31	3.32	4.15	—	17.5	—	—	2.82	404	61.2	10.3
40.8	1.25	3.46	3.98	—	26.0	—	—	1.76	436	53.1	8.85
35	1.20	4.34	4.67	—	28.0	—	—	1.08	324	44.8	6.79
31.8	1.20	4.41	4.60	—	34.3	—	56.2	0.901	307	42.0	6.40
35	1.15	4.12	5.03	—	16.8	—	—	1.29	189	35.4	6.22
25.4	1.13	4.37	4.74	—	32.2	—	63.9	0.604	153	28.4	4.96
23	0.987	4.51	4.90	—	18.1	—	—	0.551	61.8	19.3	3.68
18.4	0.973	4.70	4.70	—	29.5	—	—	0.336	53.5	16.5	3.16
20	0.914	4.63	4.92	—	15.6	—	—	0.451	34.6	14.5	2.96
15.3	0.894	4.88	4.67	—	27.8	—	—	0.241	28.8	12.1	2.44
17.25	0.845	4.69	4.97	—	12.9	—	—	0.374	18.4	10.6	2.36
12.5	0.817	5.02	4.64	—	25.9	—	—	0.168	14.5	8.47	1.85
14.75	0.761	4.66	5.03	—	10.1	—	—	0.323	9.09	7.42	1.88
10	0.741	5.10	4.60	—	23.4	—	—	0.114	6.64	5.67	1.37
9.5	0.684	4.88	4.77	—	12.3	—	—	0.120	3.10	4.04	1.13
7.7	0.662	5.13	4.54	—	20.7	—	—	0.073	2.62	3.51	0.964
7.5	0.621	4.60	4.83	—	8.60	—	—	0.091	1.10	2.36	0.826
5.7	0.585	4.95	4.48	—	17.6	—	—	0.044	0.854	1.95	0.653

TABLE 4-3.

American Standard Channels*
Properties for Designing

Designation	Area A	Depth d	Flange Width b_f	Flange Average thickness t_f	Web thickness t_w	$\dfrac{d}{A_f}$	Axis X-X I	S	r
	In.²	In.	In.	In.	In.		In.⁴	In.³	In.
C 15 × 50	14.7	15.00	3.716	0.650	0.716	6.21	404	53.8	5.24
× 40	11.8	15.00	3.520	0.650	0.520	6.56	349	46.5	5.44
× 33.9	9.96	15.00	3.400	0.650	0.400	6.79	315	42.0	5.62
C 12 × 30	8.82	12.00	3.170	0.501	0.510	7.55	162	27.0	4.29
× 25	7.35	12.00	3.047	0.501	0.387	7.85	144	24.1	4.43
× 20.7	6.09	12.00	2.942	0.501	0.282	8.13	129	21.5	4.61
C 10 × 30	8.82	10.00	3.033	0.436	0.673	7.55	103	20.7	3.42
× 25	7.35	10.00	2.886	0.436	0.526	7.94	91.2	18.2	3.52
× 20	5.88	10.00	2.739	0.436	0.379	8.36	78.9	15.8	3.66
× 15.3	4.49	10.00	2.600	0.436	0.240	8.81	67.4	13.5	3.87
C 9 × 20	5.88	9.00	2.648	0.413	0.448	8.22	60.9	13.5	3.22
× 15	4.41	9.00	2.485	0.413	0.285	8.76	51.0	11.3	3.40
× 13.4	3.94	9.00	2.433	0.413	0.233	8.95	47.9	10.6	3.48
C 8 × 18.75	5.51	8.00	2.527	0.390	0.487	8.12	44.0	11.0	2.82
× 13.75	4.04	8.00	2.343	0.390	0.303	8.75	36.1	9.03	2.99
× 11.5	3.38	8.00	2.260	0.390	0.220	9.08	32.6	8.14	3.11
C 7 × 14.75	4.33	7.00	2.299	0.366	0.419	8.31	27.2	7.78	2.51
× 12.25	3.60	7.00	2.194	0.366	0.314	8.71	24.2	6.93	2.60
× 9.8	2.87	7.00	2.090	0.366	0.210	9.14	21.3	6.08	2.72
C 6 × 13	3.83	6.00	2.157	0.343	0.437	8.10	17.4	5.80	2.13
× 10.5	3.09	6.00	2.034	0.343	0.314	8.59	15.2	5.06	2.22
× 8.2	2.40	6.00	1.920	0.343	0.200	9.10	13.1	4.38	2.34
C 5 × 9	2.64	5.00	1.885	0.320	0.325	8.29	8.90	3.56	1.83
× 6.7	1.97	5.00	1.750	0.320	0.190	8.93	7.49	3.00	1.95
C 4 × 7.25	2.13	4.00	1.721	0.296	0.321	7.84	4.59	2.29	1.47
× 5.4	1.59	4.00	1.584	0.296	0.184	8.52	3.85	1.93	1.56
C 3 × 6	1.76	3.00	1.596	0.273	0.356	6.87	2.07	1.38	1.08
× 5	1.47	3.00	1.498	0.273	0.258	7.32	1.85	1.24	1.12
× 4.1	1.21	3.00	1.410	0.273	0.170	7.78	1.66	1.10	1.17

* Taken from the 7th Edition of the *Manual of Steel Construction*. Courtesy American Institute of Steel Construction.

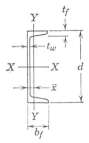

TABLE 4-3.

Properties for Designing

Nominal weight per ft.	Axis Y-Y			$\bar{x}$	Shear center location E_0	Torsional constant J	Warping constant C_w
	I	S	r				
	In.^{4}	In.^{3}	In.	In.	In.	In.^{4}	In.^{6}
50	11.0	3.78	0.867	0.799	0.941	2.66	492
40	9.23	3.36	0.886	0.778	1.03	1.46	410
33.9	8.13	3.11	0.904	0.787	1.10	1.01	358
30	5.14	2.06	0.763	0.674	0.873	0.865	151
25	4.47	1.88	0.780	0.674	0.940	0.541	131
20.7	3.88	1.73	0.799	0.698	1.01	0.371	112
30	3.94	1.65	0.669	0.649	0.705	1.22	79.5
25	3.36	1.48	0.676	0.617	0.757	0.690	68.4
20	2.81	1.32	0.691	0.606	0.826	0.370	57.0
15.3	2.28	1.16	0.713	0.634	0.916	0.211	45.5
20	2.42	1.17	0.642	0.583	0.739	0.429	39.5
15	1.93	1.01	0.661	0.586	0.824	0.209	31.0
13.4	1.76	0.962	0.668	0.601	0.859	0.169	28.2
18.75	1.98	1.01	0.599	0.565	0.674	0.436	25.1
13.75	1.53	0.853	0.615	0.553	0.756	0.187	19.3
11.5	1.32	0.781	0.625	0.571	0.807	0.131	16.5
14.75	1.38	0.779	0.564	0.532	0.651	0.268	13.1
12.25	1.17	0.702	0.571	0.525	0.695	0.161	11.2
9.8	0.968	0.625	0.581	0.541	0.752	0.100	9.16
13	1.05	0.642	0.525	0.514	0.599	0.241	7.21
10.5	0.865	0.564	0.529	0.500	0.643	0.131	5.94
8.2	0.692	0.492	0.537	0.512	0.699	0.075	4.73
9	0.632	0.449	0.489	0.478	0.590	0.109	2.93
6.7	0.478	0.378	0.493	0.484	0.647	0.055	2.22
7.25	0.432	0.343	0.450	0.459	0.546	0.082	1.24
5.4	0.319	0.283	0.449	0.458	0.594	0.040	0.923
6	0.305	0.268	0.416	0.455	0.500	0.073	0.463
5	0.247	0.233	0.410	0.438	0.521	0.043	0.380
4.1	0.197	0.202	0.404	0.437	0.546	0.027	0.307

TABLE 4-4.

Selected Angles, Equal Legs—Properties for Designing*

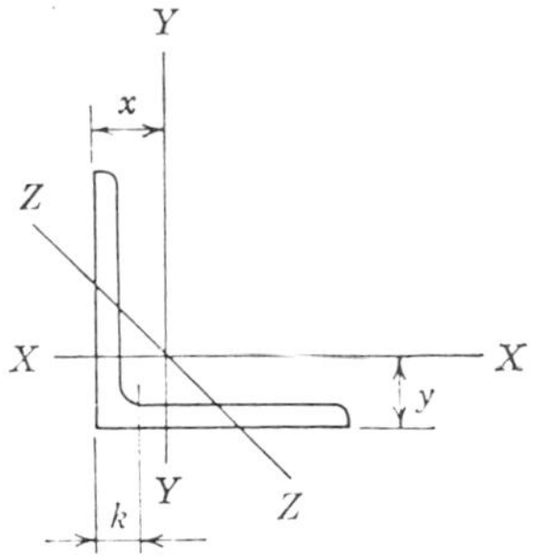

Size and thickness	k	Weight per foot	Area	Axis X-X and axis Y-Y				Axis Z-Z
				I	S	r	x or y	r
In.	In.	Lb.	In.^{2}	In.^{4}	In.^{3}	In.	In.	In.
L 8 × 8 × 1⅛	1¾	56.9	16.7	98.0	17.5	2.42	2.41	1.56
1	1⅝	51.0	15.0	89.0	15.8	2.44	2.37	1.56
¾	1⅜	38.9	11.4	69.7	12.2	2.47	2.28	1.58
L 6 × 6 × 1	1½	37.4	11.0	35.5	8.57	1.80	1.86	1.17
⅞	1⅜	33.1	9.73	31.9	7.63	1.81	1.82	1.17
¾	1¼	28.7	8.44	28.2	6.66	1.83	1.78	1.17
⅝	1⅛	24.2	7.11	24.2	5.66	1.84	1.73	1.18
L 5 × 5 × ⅞	1⅜	27.2	7.98	17.8	5.17	1.47	1.57	0.973
¾	1¼	23.6	6.94	15.7	4.53	1.51	1.52	0.975
⅝	1⅛	20.0	5.86	13.6	3.86	1.52	1.48	0.978
½	1	16.2	4.75	11.3	3.16	1.54	1.43	0.983
L 4 × 4 × ¾	1⅛	18.5	5.44	7.67	2.81	1.19	1.27	0.778
⅝	1	15.7	4.61	6.66	2.40	1.20	1.23	0.779
½	⅞	12.8	3.75	5.56	1.97	1.22	1.18	0.782
⅜	¾	9.8	2.86	4.36	1.52	1.23	1.14	0.788
5/16	11/16	8.2	2.40	3.71	1.29	1.24	1.12	0.791
L 3½ × 3½ × ½	⅞	11.1	3.25	3.64	1.49	1.06	1.06	0.683
⅜	¾	8.5	2.48	2.87	1.15	1.07	1.01	0.687
5/16	11/16	7.2	2.09	2.45	0.976	1.08	0.990	0.690
L 3 × 3 × ½	13/16	9.4	2.75	2.22	1.07	0.898	0.932	0.584
⅜	11/16	7.2	2.11	1.76	0.833	0.913	0.888	0.587
5/16	⅝	6.1	1.78	1.51	0.707	0.922	0.869	0.589
L 2½ × 2½ × ½	13/16	7.7	2.25	1.23	0.724	0.739	0.806	0.487
⅜	11/16	5.9	1.73	0.984	0.566	0.753	0.762	0.487
5/16	⅝	5.0	1.46	0.849	0.482	0.761	0.740	0.489
¼	9/16	4.1	1.19	0.703	0.394	0.769	0.717	0.491
L 2 × 2 × ⅜	⅞	4.7	1.36	0.479	0.351	0.594	0.636	0.389
5/16	13/16	3.92	1.15	0.416	0.300	0.601	0.614	0.390
¼	¾	3.19	0.938	0.348	0.247	0.609	0.592	0.391

* Compiled from data in the 7th Edition of the *Manual of Steel Construction.* Courtesy American Institute of Steel Construction.

TABLE 4-5.

Selected Angles, Unequal Legs—Properties for Designing*

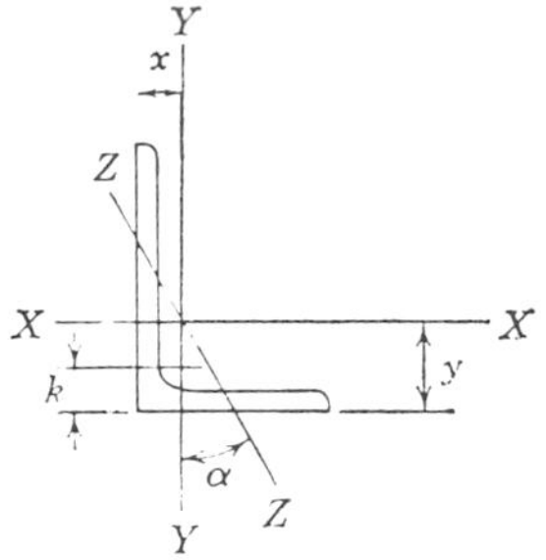

Size and thickness	k	Weight per foot	Area	Axis X-X				Axis Y-Y				Axis Z-Z	
				I	S	r	y	I	S	r	x	r	Tan α
In.	In.	Lb.	In.^{2}	In.^{4}	In.^{3}	In.	In.	In.^{4}	In.^{3}	In.	In.	In.	
L 6×4×7/8	1 3/8	27.2	7.98	27.7	7.15	1.86	2.12	9.75	3.39	1.11	1.12	0.857	0.421
3/4	1 1/4	23.6	6.94	24.5	6.25	1.88	2.08	8.68	2.97	1.12	1.08	0.860	0.428
1/2	1	16.2	4.75	17.4	4.33	1.91	1.99	6.27	2.08	1.15	0.987	0.870	0.440
3/8	7/8	12.3	3.61	13.5	3.32	1.93	1.94	4.90	1.60	1.17	0.941	0.877	0.446
L 5×3½×3/4	1 1/4	19.8	5.81	13.9	4.28	1.55	1.75	5.55	2.22	0.977	0.996	0.748	0.464
5/8	1 1/8	16.8	4.92	12.0	3.65	1.56	1.70	4.83	1.90	0.991	0.951	0.751	0.472
1/2	1	13.6	4.00	9.99	2.99	1.58	1.66	4.05	1.56	1.01	0.906	0.755	0.479
L 5×3×1/2	1	12.8	3.75	9.45	2.91	1.59	1.75	2.58	1.15	0.829	0.750	0.648	0.357
3/8	7/8	9.8	2.86	7.37	2.24	1.61	1.70	2.04	0.888	0.845	0.704	0.654	0.364
5/16	13/16	8.2	2.40	6.26	1.89	1.61	1.68	1.75	0.753	0.853	0.681	0.658	0.368
L 4×3×5/8	1 1/16	13.6	3.98	6.03	2.30	1.23	1.37	2.87	1.35	0.849	0.871	0.637	0.534
1/2	15/16	11.1	3.25	5.05	1.89	1.25	1.33	2.42	1.12	0.864	0.827	0.639	0.543
5/16	3/4	7.2	2.09	3.38	1.23	1.27	1.26	1.65	0.734	0.887	0.759	0.647	0.554
L 3½×3×1/2	15/16	10.2	3.00	3.45	1.45	1.07	1.13	2.33	1.10	0.881	0.875	0.621	0.714
3/8	13/16	7.9	2.30	2.72	1.13	1.09	1.08	1.85	0.851	0.897	0.830	0.625	0.721
5/16	3/4	6.6	1.93	2.33	0.954	1.10	1.06	1.58	0.722	0.905	0.808	0.627	0.724
L 3½×2½×1/2	15/16	9.4	2.75	3.24	1.41	1.09	1.20	1.36	0.760	0.704	0.705	0.534	0.486
3/8	13/16	7.2	2.11	2.56	1.09	1.10	1.16	1.09	0.592	0.719	0.660	0.537	0.496
5/16	3/4	6.1	1.78	2.19	0.927	1.11	1.14	0.939	0.504	0.727	0.637	0.540	0.501
L 3×2½×1/2	7/8	8.5	2.50	2.08	1.04	0.913	1.00	1.30	0.744	0.722	0.750	0.520	0.667
5/16	11/16	5.6	1.62	1.42	0.688	0.937	0.933	0.898	0.494	0.744	0.683	0.525	0.680
1/4	5/8	4.5	1.31	1.17	0.561	0.945	0.911	0.743	0.404	0.753	0.661	0.528	0.684
L 3×2×1/2	13/16	7.7	2.25	1.92	1.00	0.924	1.08	0.672	0.474	0.546	0.583	0.428	0.414
5/16	5/8	5.0	1.46	1.32	0.664	0.948	1.02	0.470	0.317	0.567	0.516	0.432	0.435
L 2½×2×3/8	11/16	5.3	1.55	0.912	0.547	0.768	0.831	0.514	0.363	0.577	0.581	0.420	0.614
5/16	5/8	4.5	1.31	0.788	0.466	0.776	0.809	0.446	0.310	0.584	0.559	0.422	0.620
1/4	9/16	3.62	1.06	0.654	0.381	0.784	0.787	0.372	0.254	0.592	0.537	0.424	0.626

* Compiled from data in the 7th Edition of the *Manual of Steel Construction*. Courtesy, American Institute of Steel Construction.

TABLE 4-6. Properties of Structural Lumber* Standard Dressed (S4S) Sizes

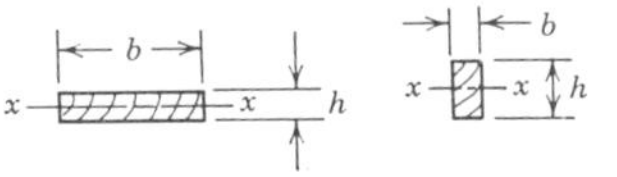

Nominal size (in.) b h	Dressed size (in.) b h	Area of section (sq in.) A	Moment of inertia (in.⁴) I	Section modulus (in.³) S	Weight per linear foot†
2 × 4	1½ × 3½	5.250	5.359	3.063	1.458
2 × 6	1½ × 5½	8.250	20.797	7.563	2.292
2 × 8	1½ × 7¼	10.875	47.635	13.141	3.021
2 × 10	1½ × 9¼	13.875	98.932	21.391	3.854
2 × 12	1½ × 11¼	16.875	177.979	31.641	4.688
2 × 14	1½ × 13¼	19.875	290.775	43.891	5.521
3 × 2	2½ × 1½	3.750	0.703	0.938	1.042
3 × 4	2½ × 3½	8.750	8.932	5.104	2.431
3 × 6	2½ × 5½	13.750	34.661	12.604	3.819
3 × 8	2½ × 7¼	18.125	79.391	21.901	5.035
3 × 10	2½ × 9¼	23.125	164.886	35.651	6.424
3 × 12	2½ × 11¼	28.125	296.631	52.734	7.813
3 × 14	2½ × 13¼	33.125	484.625	73.151	9.201
3 × 16	2½ × 15¼	38.125	738.870	96.901	10.590
4 × 2	3½ × 1½	5.250	0.984	1.313	1.458
4 × 3	3½ × 2½	8.750	4.557	3.646	2.431
4 × 4	3½ × 3½	12.250	12.505	7.146	3.403
4 × 6	3½ × 5½	19.250	48.526	17.646	5.347
4 × 8	3½ × 7¼	25.375	111.148	30.661	7.049
4 × 10	3½ × 9¼	32.375	230.840	49.911	8.933
4 × 12	3½ × 11¼	39.375	415.283	73.828	10.938
4 × 14	3½ × 13¼	46.375	678.475	102.411	12.877
4 × 16	3½ × 15¼	53.375	1,034.418	135.66	14.828
6 × 2	5½ × 1½	8.250	1.547	2.063	2.292
6 × 3	5½ × 2½	13.750	7.161	5.729	3.819
6 × 4	5½ × 3½	19.250	19.651	11.229	5.347
6 × 6	5½ × 5½	30.250	76.255	27.729	8.403
6 × 8	5½ × 7½	41.250	193.359	51.563	11.458
6 × 10	5½ × 9½	52.250	392.963	82.729	14.514
6 × 12	5½ × 11½	63.250	697.068	121.229	17.569
6 × 14	5½ × 13½	74.250	1,127.672	167.063	20.625
6 × 16	5½ × 15½	85.250	1,706.776	220.229	23.681

* Compiled from data in the 1973 Edition of the *National Design Specification for Stress-Grade Lumber and Its Fastenings.* Courtesy National Forest Products Association.
† Based on an assumed average weight of 40 lb per cu ft.

Nominal size (in.) b h	Dressed size (in.) b h	Area of section (sq in.) A	Moment of inertia (in.^{4}) I	Section modulus (in.^{3}) S	Weight per linear foot†
8 × 2	$7\frac{1}{4} \times 1\frac{1}{2}$	10.875	2.039	2.719	3.021
8 × 3	$7\frac{1}{4} \times 2\frac{1}{2}$	18.125	9.440	7.552	5.035
8 × 4	$7\frac{1}{4} \times 3\frac{1}{2}$	25.375	25.904	14.802	7.049
8 × 6	$7\frac{1}{2} \times 5\frac{1}{2}$	41.250	103.984	37.813	11.458
8 × 8	$7\frac{1}{2} \times 7\frac{1}{2}$	56.250	263.672	70.313	15.625
8 × 10	$7\frac{1}{2} \times 9\frac{1}{2}$	71.250	535.859	112.813	19.792
8 × 12	$7\frac{1}{2} \times 11\frac{1}{2}$	86.250	950.547	165.313	23.958
8 × 14	$7\frac{1}{2} \times 13\frac{1}{2}$	101.250	1,537.734	227.813	28.125
8 × 16	$7\frac{1}{2} \times 15\frac{1}{2}$	116.250	2,327.422	300.313	32.292
10 × 2	$9\frac{1}{4} \times 1\frac{1}{2}$	13.875	2.602	3.469	3.854
10 × 3	$9\frac{1}{4} \times 2\frac{1}{2}$	23.125	12.044	9.635	6.424
10 × 4	$9\frac{1}{4} \times 3\frac{1}{2}$	32.375	33.049	18.885	8.993
10 × 6	$9\frac{1}{2} \times 5\frac{1}{2}$	52.250	131.714	47.896	14.514
10 × 8	$9\frac{1}{2} \times 7\frac{1}{2}$	71.250	333.984	89.063	19.792
10 × 10	$9\frac{1}{2} \times 9\frac{1}{2}$	90.250	678.755	142.896	25.069
10 × 12	$9\frac{1}{2} \times 11\frac{1}{2}$	109.250	1,204.026	209.396	30.347
10 × 14	$9\frac{1}{2} \times 13\frac{1}{2}$	128.250	1,947.797	288.563	35.625
10 × 16	$9\frac{1}{2} \times 15\frac{1}{2}$	147.250	2,948.068	380.396	40.903
10 × 18	$9\frac{1}{2} \times 17\frac{1}{2}$	166.250	4,242.836	484.896	46.181
12 × 2	$11\frac{1}{4} \times 1\frac{1}{2}$	16.875	3.164	4.219	4.688
12 × 3	$11\frac{1}{4} \times 2\frac{1}{2}$	28.125	14.648	11.719	7.813
12 × 4	$11\frac{1}{4} \times 3\frac{1}{2}$	39.375	40.195	22.969	10.938
12 × 6	$11\frac{1}{2} \times 5\frac{1}{2}$	63.250	159.443	57.979	17.569
12 × 8	$11\frac{1}{2} \times 7\frac{1}{2}$	86.250	404.297	107.813	23.958
12 × 10	$11\frac{1}{2} \times 9\frac{1}{2}$	109.250	821.651	172.979	30.347
12 × 12	$11\frac{1}{2} \times 11\frac{1}{2}$	132.250	1,457.505	253.479	36.736
12 × 14	$11\frac{1}{2} \times 13\frac{1}{2}$	155.250	2,357.859	349.313	43.125
12 × 16	$11\frac{1}{2} \times 15\frac{1}{2}$	178.250	3,568.713	460.479	49.514
14 × 16	$13\frac{1}{2} \times 15\frac{1}{2}$	209.250	4,189.359	540.563	58.125
14 × 18	$13\frac{1}{2} \times 17\frac{1}{2}$	236.250	6,029.297	689.063	65.625
14 × 20	$13\frac{1}{2} \times 19\frac{1}{2}$	263.250	8,341.734	855.563	73.125
14 × 22	$13\frac{1}{2} \times 21\frac{1}{2}$	290.250	11,180.672	1,040.063	80.625

† Based on an assumed average weight of 40 lb per cu ft.

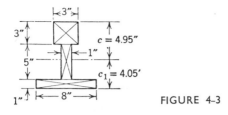

FIGURE 4-3

Solution: (1) It is not always possible to tell by observation whether the centroid is nearer the top or bottom surface of the area. In this instance let us write an equation of moments about an axis through the *uppermost* surface. First divide the section into any number of convenient areas as shown by the diagonals, in this case three rectangles. The area of the upper flange is 9 sq in., and its centroid is $1\frac{1}{2}$ in. from the assumed axis; the web has an area of 5 sq in., and its centroid is $5\frac{1}{2}$ in. from the same axis. Likewise, the bottom flange contains 8 sq in., and $8\frac{1}{2}$ in. is the distance of its centroid to the same axis. The entire area is 22 sq in., and its centroid is c in. from the axis. Then, because *the sum of the moments of the parts is equal to the moment of the whole area*, we may write

$$(9 \times 1\frac{1}{2}) + (5 \times 5\frac{1}{2}) + (8 \times 8\frac{1}{2}) = 22 \times c$$

$$109 = 22 \times c$$

$$c = 4.95 \text{ in.}$$

(2) The depth of the section is 9 in.; hence the distance of the centroid to the lowest surface is

$$9 - 4.95 = 4.05 \text{ in.}$$

Call this distance c_1. This last value may be checked by writing an equation of moments about an axis taken through the *bottom* surface; thus

$$(8 \times 0.5) + (5 \times 3.5) + (9 \times 7.5) = 22 \times c_1$$

$$c_1 = 4.05 \text{ in.}$$

Remember that the centroid is a *point* through which the neutral axis passes. With respect to the vertical axis of the section shown in Fig. 4-3, the value of c is 4 in. The position of the centroid for structural steel angle sections may be found directly by reference to

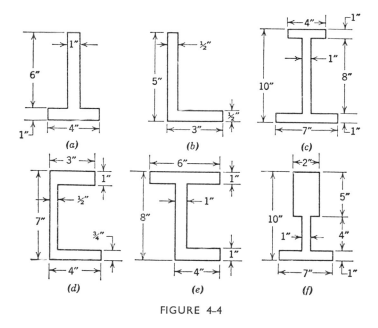

FIGURE 4-4

Tables 4-4 and 4-5. The locating dimensions from the backs of the angle legs are given as y and x in the sketches at the top of the tables. As Table 4-4 shows, the y and x distances are the same for equal leg angles.

Problems 4-4-A*-B-C-D*-E-F.   Compute the value of c with respect to the horizontal axes, for the sections shown in Fig. 4-4.

4-5   Moment of Inertia

In developing the flexure formula, we found the quantity $\sum a \times z^2$. This is the *moment of inertia*, which may be defined as *the sum of the products obtained by multiplying all the infinitely small areas by the square of their distances to the neutral axis*. It is represented by I. The elementary areas, though extremely small, are in units of square inches, and square inches multiplied by a distance squared gives inches to the fourth power. For instance, 24 in.⁴ is read "24 inches to the fourth power."

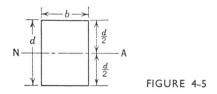

FIGURE 4-5

In Fig. 4-5 we have a rectangular cross section of breadth b and depth d. The neutral axis passes through the centroid of the cross section; it is represented by the line NA and is a distance $d/2$ from the upper and lower edges. It can be shown that *the moment of inertia of a rectangular cross section about an axis passing through its centroid, parallel to the base, is $bd^3/12$.* By the use of this formula, $I = bd^3/12$, the value of the moment of inertia for rectangular cross sections is quickly found. The values of I for standard timber sizes are given in Table 4-6; they are computed for actual or dressed sizes (Art. 10-2).

Example 1.   Compute the value of the moment of inertia of a nominal 6 × 12 in. wood beam about an axis through its centroid parallel to the base of the section.
Solution: Referring to Table 4-6, we find that a nominal 6 × 12 in. section has standard dressed dimensions of $5\frac{1}{2}$ × $11\frac{1}{2}$ in. Then

$$I = \frac{bd^3}{12} = \frac{5.5 \times 11.5 \times 11.5 \times 11.5}{12} = 697.07 \text{ in.}^4$$

This is in agreement with the value of I listed in the table.

Another proposition relating to rectangular cross sections states that the moment of inertia of a rectangle *about an axis through its base* is $bd^3/3$. This expression is often convenient for computing I of unsymmetrical sections such as angles.

Example 2.   Compute the moment of inertia of a 6 × 4 × $\frac{1}{2}$ in. angle (Fig. 4-6) about its neutral axis, assuming the neutral axis is 2 in. above the base of the short leg.
Solution: (1) Make a sketch showing all pertinent dimensions, as has been done in Fig. 4-6.

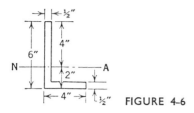

FIGURE 4-6

(2) Take the portion of the long leg above the neutral axis and write the value of I about the base of the rectangle.

$$I = \frac{bd^3}{3} = \frac{0.5 \times 4 \times 4 \times 4}{3} = 10.66 \text{ in.}^4$$

In the same manner, the value of I for the rectangle below NA, having a breadth of 4 in. and depth of 2 in., is

$$I = \frac{bd^3}{3} = \frac{4 \times 2 \times 2 \times 2}{3} = 10.66 \text{ in.}^4$$

From this last quantity we must substract I for the rectangle of width 3.5 in. and depth 1.5 in. in which no material occurs and which was included in the preceding equation. Therefore

$$I = \frac{bd^3}{3} = \frac{3.5 \times 1.5 \times 1.5 \times 1.5}{3} = 3.93 \text{ in.}^4$$

Then $10.66 - 3.93 = 6.73 \text{ in.}^4$, the moment of inertia of the area of the angle section below NA. This quantity added to the I for the section above NA $= 10.66 + 6.73 = 17.39 \text{ in.}^4$, the value of I for the entire section taken about the neutral axis. Check this value by referring to Table 4-5.

The preceding example illustrates one method of finding the moment of inertia of an unsymmetrical section, but it is not necessary to compute I for the commonly used structural steel angles, since these may be found directly from Tables 4-4 and 4-5, and from the more extensive tables in the AISC Manual.

Problems 4-5-A-B*-C-D-E-F*. Compute the moment of inertia about the neutral axes of the beam sections shown in Fig. 4-7.

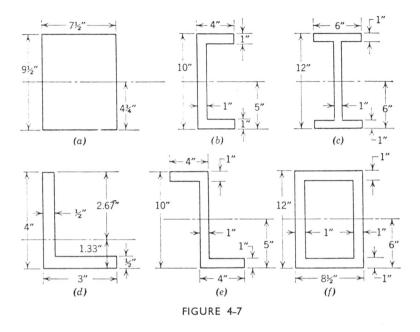

FIGURE 4-7

4-6 Section Modulus

As noted in Art. 4-2, the term I/c in the flexure formula is called the *section modulus*. It is defined as the moment of inertia divided by the distance of the most remote fiber from the neutral axis, and is denoted by the symbol S. Since I and c always have the same values for any given cross section, values of S may be computed and tabulated for structural shapes. With I expressed in inches to the fourth power and c a linear dimension in inches, S is in units of inches to the third power, written *inches*³. Section moduli are among the properties tabulated for structural steel shapes in Tables 4-1 through 4-5, and for structural lumber cross sections in Table 4-6.

Referring to Fig. 4-5, a rectangular cross section of breadth b and depth d, we know that the moment of inertia about the X–X axis is $bd^3/12$ and that $c = d/2$. Therefore

$$\frac{I}{c} \quad \text{or} \quad S = \frac{bd^3}{12} \div \frac{d}{2} = \frac{bd^3}{12} \times \frac{2}{d}$$

or $S = bd^2/6$. It is often convenient to use this formula directly. It applies, of course, only to rectangular cross sections.

Example 1. Verify the tabulated value of the section modulus of a 6 × 12 in. wood beam.

Solution: Referring to Table 4-6, we find that the dressed dimensions of this section are $5\frac{1}{2} \times 11\frac{1}{2}$ in. Then

$$S = \frac{bd^2}{6} = \frac{5.5 \times 11.5 \times 11.5}{6} = 121.23 \text{ in.}^3$$

Compare this with the tabulated value of S in Table 4-6.

Example 2. Verify the tabulated value of the section modulus with respect to the X–X axis, for the wide flange shape designated W 18 × 45.

Solution: Referring to Table 4-1, we find that the moment of inertia for this section is 706 in.^{4} and its actual depth is 17.86 in. Since it is a symmetrical section, $c = d \div 2$ or $17.86 \div 2 = 8.93$ in. Therefore,

$$S = \frac{I}{c} = \frac{706}{8.93} = 79.0 \text{ in.}^3$$

which checks with the value given in the table.

Problem 4-6-A. Compute the section modulus of a 6 × 8 in. timber beam and compare your result with the value given in Table 4-6.

Problem 4-6-B. Verify the value of S_{X-X} given in Table 4-2 for the beam shape designated S 12 × 31.8.

Problem 4-6-C. Verify the value of S_{X-X} given in Table 4-5 for the structural angle designated L 5 × $3\frac{1}{2}$ × $\frac{1}{2}$.

Problem 4-6-D. Verify the value of S_{Y-Y} given in Table 4-4 for the structural angle designated L 4 × 4 × $\frac{1}{2}$.

4-7 Application of the Flexure Formula

Now that we have discussed the properties of structual sections in some detail, let us return to the flexure formula and consider its application. The expression $M = fS$ may be stated in three different forms depending upon the information desired. These are given below using a nomenclature which makes a distinction with respect to f as the *allowable* bending stress (F_b) and f as the *computed* bending stress (f_b).

$$(1) \ M = F_b S \qquad (2) \ f_b = \frac{M}{S} \qquad (3) \ S = \frac{M}{F_b}$$

Form (1) gives the maximum potential resisting moment when the section modulus of the beam and the maximum allowable bending stress are known. Form (2) gives the computed bending stress when the maximum bending moment due to the loading is known, together with the section modulus of the beam. These are the two forms used when investigating the adequacy of given beams.

Form (3) is the one used in design. It gives the *required* section modulus when the maximum bending moment and the allowable bending stress are known. When the required section modulus has been determined, a beam having an S equal to or greater than the computed value is selected from tables giving properties of the various structural shapes.

When using the beam formula, care must be exercised with respect to the units in which the terms are expressed. Bending stress values F_b and f_b may be written in pounds per square inch (psi) or kips per square inch (ksi); S is stated in inches^3 and I in inches^4. Therefore, M must be written in inch-pounds or kip-inches. As customarily computed from the loads and reactions, M is given in foot-pounds or kip-feet, and must be converted to inch-pounds or kip-inches; this is accomplished by multiplying its value by 12 before it is used in the formula.

Example. Select a wide flange steel beam to support a uniformly distributed load of 48 kips, including its own weight, on a span of 16 ft. The allowable extreme fiber stress (bending stress) in the steel is 24 ksi.

Solution: (1) Computing the bending moment (Case 2, Table 3-1),

$$M = \frac{WL}{8} = \frac{48 \times 16}{8} = 96 \text{ kip-ft}$$

(2) Substituting in the flexure formula,

$$S = \frac{M}{F_b} = \frac{96 \times 12}{24} = 48 \text{ in.}^3$$

(3) Referring to Table 4-1, the section designated W 14 × 34 has an S of 48.6 in.[3]; therefore this beam may be used. Any steel beam having a section modulus equal to or greater than 48 in.[3] will safely support this 48-kip load. A W 12 × 40 ($S = 51.9$) is also acceptable and is not quite so deep. In general, the lightest weight beam is the most economical.

4-8 Transferring Moments of Inertia

When rolled steel shapes are combined to form built-up structural sections similar to those illustrated in Fig. 4-10, it is necessary to determine the moment of inertia of the built-up section about its neutral axis. This requires transferring the moments of inertia of some of the individual parts from one axis to another, and is accomplished by means of the transfer-of-axis equation which may be stated as follows:

The moment of inertia of a cross section about any axis parallel to an axis through its own centroid is equal to the moment of inertia of the cross section about its own gravity axis, plus its area times the square of the distance between the two axes. Expressed mathematically,

$$I = I_0 + Az^2$$

In this formula,

$I =$ moment of inertia of the cross section about the required axis,

$I_0 =$ moment of inertia of the cross section about its own gravity axis parallel to the required axis,

$A =$ area of the cross section,

$z =$ distance between the two parallel axes.

These relationships are indicated in Fig. 4-8 where X–X is the gravity axis of the angle (passing through its centroid) and Y–Y is the axis about which the moment of inertia is to be found.

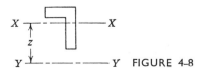

FIGURE 4-8

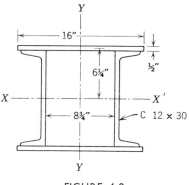

FIGURE 4-9

To illustrate the use of the equation, we may prove the proposition stated in Art. 4-5 that the value of I for a rectangle about an axis through its base is $bd^3/3$. Since I for a rectangle about its gravity axis is known to be $bd^3/12$, and z in this instance is $d/2$, we may write

$$I = I_0 + Az^2$$

$$I = \frac{bd^3}{12} + \left[bd \times \left(\frac{d}{2} \right)^2 \right]$$

$$I = \frac{bd^3}{12} + \frac{bd^3}{4} = \frac{bd^3}{3}$$

The application of the transfer formula to the steel built-up section shown in Fig. 4-9 is illustrated in the following example.

Example.   Compute the moment of inertia of a built-up section composed of two 12-in. 30-lb channels (C 12 × 30) and two 16 × $\frac{1}{2}$ in. plates, about the X–X axis. See Fig. 4-9.
Solution: (1) Referring to Table 4-3, we find that I for a C 12 × 30 is 162 in.[4] about the X–X axis. Because there are two channels, the moment of inertia for both is 2 × 162 = 324 in.[4]

(2) For one 16 × $\frac{1}{2}$ in. plate, the moment of inertia about an axis through *its* centroid is

$$I_0 = \frac{bd^3}{12} = \frac{16 \times 0.5 \times 0.5 \times 0.5}{12} = 0.166 \text{ in.}^4$$

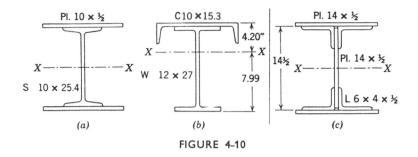

FIGURE 4-10

The distance between the centroid of the plate and the X–X axis is 6.25 in., and the area of one plate is 8 sq in. Therefore the I of one plate about X–X is

$$I = I_0 + Az^2 = 0.166 + (8 \times 6.25^2) = 312.67 \text{ in.}^4$$

The value of I for the two plates is $2 \times 312.67 = 625.34$ in.^{4}

(3) Adding the moments of inertia of the channels and plates, we have $324 + 625.34 = 949.34$ in.^{4}, the value of I for the entire cross section with respect to the X–X axis.

Problems 4-8-A*-B*-C. Compute I with respect to the X–X axes for the built-up sections shown in Fig. 4-10. Make use of any appropriate data given in Tables 4-1 through 4-5.

4-9 Radius of Gyration

This property of a cross section is related to the design of compression members rather than beams, and will be discussed in more detail under the design of columns in subsequent sections of the book. Radius of gyration will be considered briefly here, however, because it is a property listed in Tables 4-1 through 4-5.

Just as the section modulus is a measure of the resistance of a beam section to bending, the radius of gyration (which is also related to the size and shape of the cross section) is an index of the stiffness of a structural section when used as a column or other compression member. The radius of gyration is found from the formula

$$r = \sqrt{\frac{I}{A}}$$

and is expressed in inches since the moment of inertia is in inches[4] and the cross-sectional area is in square inches.

If a section is symmetrical about both major axes, the moment of inertia, and consequently the radius of gyration, is the same for each axis. But most column sections, particularly steel columns, are not symmetrical about the two major axes, and *in the design of columns the least moment of inertia, and therefore the least radius of gyration, is the one used in computations*. By *least* we mean the smallest in magnitude. Note in Tables 4-4 and 4-5 that the least radius of gyration of angle sections occurs about the Z–Z axes.

Example.   Verify the tabulated values of radii of gyration for the wide flange section designated W 12 × 65 in Table 4-1.
Solution: (1) The table shows the area of this section to be 19.1 sq in., and I with respect to the X–X axis equal to 533 in.[4] Then

$$r_{X-X} = \sqrt{\frac{I}{A}} = \sqrt{\frac{533}{19.1}} = \sqrt{27.9} = 5.28 \text{ in.}$$

(2) The value of I with respect to the Y–Y axis is given as 175 in.[4] Therefore

$$r_{Y-Y} = \sqrt{\frac{I}{A}} = \sqrt{\frac{175}{19.1}} = \sqrt{9.17} = 3.02 \text{ in.}$$

(3) Compare these values with those listed for r in the table.

Problem 4-9-A.   With respect to Table 4-2, verify the value of the least radius of gyration tabulated for the section designated S 12 × 40.8.

Problem 4-9-B.   Referring to Table 4-5, note the value of the radius of gyration about the X–X axis for the angle designated L 4 × 3 × ⅝; verify this value.

Problem 4-9-C*.   Compute the radius of gyration with respect to the X–X axis for the built-up section shown in Fig. 4-9.

II

STEEL CONSTRUCTION

||||||||||||||||||||||||||||||||||||||||||||||||||||||||||||||

5

Steel Beams

||||||||||||||||||||||||||||||||||||||||||||||||||||||||

5-1 Structural Steel

The grade of structural steel commonly used in building construction is designated ASTM A36, indicating that it meets the requirements of the American Society for Testing and Materials Specification A36. This specification requires that the steel have an ultimate strength of 58 ksi to 80 ksi and a minimum yield point of 36 ksi. It may be used for bolted, riveted, and welded fabrication.

In addition to A36 as the all-purpose grade carbon steel, other special steels are permitted by the American Institute of Steel Construction's 1969 *Specification for the Design, Fabrication & Erection of Structural Steel for Buildings*. Several of these steels are listed below.

Structural Steel, ASTM A36
Structural Steel with 42,000 psi Minimum Yield Point, ASTM 529
High-Strength Structural Steel, ASTM A440
High-Strength Low-Alloy Structural Manganese Vanadium Steel, ASTM A441
High-Strength Low-Alloy Columbium-Vanadium Steels of Structural Quality, ASTM A572

91

High-Strength Low-Alloy Structural Steel, ASTM A242
*High-Strength Low-Alloy Structural Steel with 50,000 psi Minimum
Yield Point to 4 in. Thick,* ASTM A588

Table 5-1 summarizes certain information about these structural steels. Note in particular that F_y, the minimum yield-point stress, is given for each type of steel. This stress is important because allowable unit stresses are specified as percentages of the yield stress for a particular grade of steel. The table also shows that all of the steels listed are suitable for welding except A440 which is not recommended for welding by the AISC.

It will be noted that the yield points for the high-strength steels are higher than those for the structural carbon steels; consequently, their allowable unit stresses are higher as well. Also, the thickness of the material is a factor in establishing the yield stress. In addition to their greater strengths, these steels have varying degrees of increased atmospheric corrosion resistance. ASTM A588 steel is now being used without painting in exposed frame construction, and weathers to a deep rust color.

Because of their individual characteristics, each of the steels mentioned is employed when required by specific conditions. The steel most commonly used, however, is A36.[1] Unless otherwise noted, all examples and problems throughout this book are based on the use of A36 steel.

5-2 Structural Shapes

The rolled structural steel shapes that are most commonly used in building construction are the wide flange (W shapes), American Standard I-beams (S shapes), channels, angles, and plates. Tables 4-1 through 4-5, some of which are necessarily condensed, give the dimensions and weights of some of these shapes, together with other properties which were discussed in Chapter 4. Complete tables of structural shapes are contained in the AISC *Manual of Steel Construction.*

---

[1] Prior to promulgation of the 1963 AISC Specification, ASTM A7 was the basic structural steel for buildings. Its yield point was 33 ksi. This steel has been withdrawn from production and is no longer available.

TABLE 5-1. Structural Steels for Buildings—1969 AISC Specification

| Steel type | ASTM number | Strength | | Fabrication* |
		Thickness (in.)	Yield F_y (ksi)	
Carbon	A36		36	B R W
	A529		42	B R W
High-Strength	A440	$1\frac{1}{2}$ to 4	42	B R
		$\frac{3}{4}$ to $1\frac{1}{2}$	46	
		to $\frac{3}{4}$	50	
High-Strength Low-Alloy	A441	4 to 8	40	B R W
		$1\frac{1}{2}$ to 4	42	
		$\frac{3}{4}$ to $1\frac{1}{2}$	46	
		to $\frac{3}{4}$	50	
	A572		42	B R W
Corrosion-Resistant High-Strength Low-Alloy	A242	$1\frac{1}{2}$ to 4	42	B R W
		$\frac{3}{4}$ to $1\frac{1}{2}$	46	
		to $\frac{3}{4}$	50	
	A588	5 to 8	42	B R W
		4 to 5	46	
		to 4	50	

* B, R, and W indicate that the steel is suitable for fabrication by means of bolting, riveting, or welding, respectively.

93

Study of the tables will reveal that wide flange shapes have greater flange widths and thinner webs than Standard I-beams, and are characterized by parallel flange surfaces as contrasted with the tapered inside flange surfaces of the S shapes. In both shapes, most of the material in the cross section is contained in the flanges, and for bending this condition is ideal, because the greatest bending stresses occur in the flange areas. The tables also show that the actual depths of wide flange sections vary from the nominal depths, while the nominal and actual depths of Standard I-beams are the same.

5-3   Designations of Rolled Steel Shapes

In 1970 the Committee of Structural Steel Producers of the American Iron and Steel Institute promulgated a new standard system for designating structural shapes. The new designations are used in this book and in the Seventh Edition of the AISC Manual; they supersede the designations formerly used in practice and in the Sixth and previous editions of the Manual. In the new system, wide flange shapes are designated by the letter W followed by the *nominal* depth

Type of shape	New designation	Old designation
Wide flange shapes	W 12 × 27	12 W 27
American Standard beams	S 12 × 35	12 I 35
Miscellaneous shapes	M 8 × 18.5	8 M 18.5
American Standard channels	C 10 × 20	10 [20
Miscellaneous channels	MC 12 × 45	12 × 4 [45.0
Angles—equal legs	L 4 × 4 × $\frac{1}{2}$	∠ 4 × 4 × $\frac{1}{2}$
Angles—unequal legs	L 5 × 3$\frac{1}{2}$ × $\frac{1}{2}$	∠ 5 × 3$\frac{1}{2}$ × $\frac{1}{2}$
Structural trees—cut from wide flange shapes	WT 6 × 53	ST 6 W 53
Structural trees—cut from American Standard beams	ST 9 × 35	ST 9 I 35
Structural trees—cut from miscellaneous shapes	MT 4 × 9.25	ST 4 M 9.25
Plate	PL $\frac{1}{2}$ × 12	PL 12 × $\frac{1}{2}$
Structural tubing: square	TS 4 × 4 × 0.375	Tube 4 × 4 × 0.375
Pipe	Pipe 4 Std.	Pipe 4 Std.

in inches and the weight in pounds per linear foot. Thus the designation W 16 × 50 indicates a wide flange shape of nominal 16-in. depth weighing 50 lb per lin ft. (In the old system its designation was 16 W⁻ 50.) An American Standard I-beam (S shape) 10 in. deep and weighing 35 lb per lin ft is designated S 10 × 35. (In the former system it was 10 I 35.)

The accompanying table gives designations for certain structural shapes, both in the current and superseded systems. The latter are for reference only since they are now obsolete. A similar but more complete table is contained in Part 1 of the AISC Manual.

5-4 Nomenclature

The symbols given in the general nomenclature of the AISC Specification differ somewhat from those commonly used in structural mechanics. In general, the former are employed throughout this section in design discussions, but the latter are sometimes used in theoretical discussions. An abridgment of the general nomenclature presented in the AISC Manual is given below for ready reference. In this listing the units for stress values are indicated as kips per square inch (ksi) or pounds per square inch (psi), although the 1969 AISC Specification gives all values in kips per square inch.

A Cross-sectional area (sq in.)
        Area of beam or column base plate (sq in.)

A_f Area of compression flange (sq in.)

A_w Area of girder web (sq in.)

B_x, B_y Bending factor with respect to the X–X axis and Y–Y axis, respectively, for determining the equivalent axial load in columns subjected to combined loading conditions; equal to A/S_x and A/S_y, respectively

C_c Column slenderness ratio dividing elastic and inelastic buckling; equal to

$$\sqrt{\frac{2\pi^2 E}{F_y}}$$

E Modulus of elasticity of steel (29,000 ksi)

F_a Axial stress permitted in the absence of bending moment (ksi or psi)

F_{as} Axial compressive stress, permitted in the absence of bending moment, for bracing and other secondary members (ksi or psi)

F_b Bending stress permitted in the absence of axial force (ksi or psi)

F_p Allowable bearing stress (ksi or psi)
          Allowable bearing pressure on support (ksi or psi)

F_t Allowable tensile stress (ksi or psi)

F_v Allowable shear stress (ksi or psi)

F_{vw} Allowable shear stress in welds (ksi or psi)

F_y Specified minimum yield stress of the type of steel being used (ksi or psi). As used in AISC Specification, "yield stress" denotes either the specified minimum yield point (for those steels that have a yield point) or specified minimum yield strength (for those steels that do not have a yield point).

F_y' The theoretical maximum yield stress (ksi) based on the width–thickness ratio of one half the unstiffened compression flange, beyond which a particular shape is not "compact" (AISC Specification Sect. 1.5.1.4.1, subparagraph b)

F_y'' The theoretical maximum yield stess (ksi) based on the depth–thickness ratio of the web, beyond which a shape is not "compact" (AISC Specification Sect. 1.5.1.4.1, subparagraph d); it is only applicable for cases of pure bending; i.e., $f_a = 0$.

F_y''' The theoretical maximum yield stress (ksi) based on the depth–thickness ratio of the web below which a particular shape may be considered "compact" for any condition of combined bending and axial stresses (AISC Specification Sect. 1.5.4.1, subparagraph d)

I Moment of inertia of a section (in.^{4})

K Effective length factor

L Span length (ft)

L_b Unbraced length of compression flange (ft)

L_c Maximum unbraced length of the compression flange at which the allowable bending stress may be taken at $0.66F_y$ (ft)

L_u	Maximum unbraced length of the compression flange at which the allowable bending stress may be taken at $0.60F_y$ (ft)
M	Moment (kip-ft or kip-in.)
M_D	Moment produced by dead load (kip-ft or kip-in.)
M_L	Moment produced by live load (kip-ft or kip-in.)
M_p	Plastic moment (kip-ft)
M_R	Beam resisting moment (kip-ft or kip-in.)
N	Length of bearing of applied load (in.)
N_e	Length at end bearing to develop maximum web shear (in.)
P	Applied load (kips)
P'	Equivalent axial load due to bending in members subject to axial compression and bending (kips)
R	Reaction (kips)
	Maximum end reaction for $3\frac{1}{2}$ in. of bearing (kips)
R_i	Increase in reaction (R) in kips for each additional inch of bearing
S	Elastic section modulus (in.³)
V	Statical (vertical) shear on beams (kips)
Z	Plastic section modulus (in.³)
b_f	Flange width of rolled beam or plate girder (in.)
c	Distance from neutral axis to extreme fiber of beams (in.)
d	Depth of beam or girder (in.)
f_a	Computed axial stress (ksi or psi)
f_b	Computed bending stress (ksi or psi)
f_p	Actual bearing pressure on support (ksi or psi)
f_t	Computed tensile stress (ksi or psi)
f_v	Computed shear stress (ksi or psi)
l	Actual unbraced length (in.)
l_b	Actual unbraced length in plane of bending (in.)
r	Governing radius of gyration (in.)
r_b	Radius of gyration about axis of concurrent bending (in.)
r_T	Radius of gyration of a section comprising the compression flange plus one third of the compression web area, taken about an axis in the plane of the web (in.)
r_x	Radius of gyration with respect to the X–X axis (in.)
r_y	Radius of gyration with respect to the Y–Y axis (in.)

t Girder, beam, or column web thickness (in.)

t_f Flange thickness (in.)

t_w Web thickness (in.)

$\bar{x}$ Distance from the outside of the web to the minor (Y–Y) axis of a channel section (in.)

y Distance from neutral axis to the outermost fibers of cross section (in.)

      Distance from the back of the flange to the major (X–X) axis of a tee section (in.)

Δ Beam deflection (in.)

kip   1000 pounds

5-5   Allowable Stresses for Structural Steel

The allowable stresses for use in the design of structural steel for buildings are given in the current AISC Specification as percentages of F_y, the yield-point stress. As an example F_t, the allowable tension on net sections, except at pinholes, is given as $F_t = 0.60 \, F_y$. Thus, for A36 steel, $F_t = 0.60 \times 36{,}000$, or 21,600 psi; in tables of allowable stresses, to simplify computations, this stress is rounded off and we see $F_t = 22{,}000$ psi. See Table 5-2. This table contains a selected list of stresses; it is compiled from the AISC Specification. These stresses appear in the examples and problems given in this book. Because it is used so extensively, the A36 steel is used in all the examples and problems unless otherwise noted. If it is desirable to use one of the other steels, the allowable unit stresses are the specified percentages of F_y, the yield-point stress. These stresses are found in Table 5-2. It will be noted that the first column of the table contains several qualifying terms such as net section , gross section, and compact beams, that have not been introduced in this book so far. These will be identified in subsequent articles dealing with the design of members to which they apply.

5-6   Materials and Stresses for Connectors

In addition to the steels used for structural shapes, the AISC Specification designates the type of material permitted in bolted,

TABLE 5-2. Allowable Unit Stresses for Structural Steel

Note: Stresses are given in kips per sq in. (ksi) and are rounded off to conform with the values given in Appendix A of the 1969 AISC Specification

	AISC Specification 1969	A36	A242, A440, A441		
			Section thickness		
			Over 1½ in. and up to 4 in., incl.	Over ¾ in. and up to 1½ in., incl.	Up to ¾ in., incl.
		$F_y = 36.0$	$F_y = 42.0$	$F_y = 46.0$	$F_y = 50.0$
Tension					
Tension on net section, except at pin holes	$F_t = 0.60F_y$	22.0	25.2	27.5	30.0
Tension on net section at pin holes	$F_t = 0.45F_y$	16.2	19.0	20.5	22.5
Shear					
Shear on gross section	$F_v = 0.40F_y$	14.5	17.0	18.5	20.5
Compression					
See Chapter 10					
Bending					
Tension and compression on extreme fibers of compact members braced laterally, symmetrical about and loaded in the plane of their minor axis	$F_b = 0.66F_y$	24.0	28.0	30.5	33.0
Tension and compression on extreme fibers of unsymmetrical rolled shapes, except channels, continuously braced in the region under compression stress	$F_b = 0.60F_y$	22.0	25.2	27.5	30.0
Tension on extreme fibers of other rolled shapes, built-up members and plate girders	$F_b = 0.60F_y$	22.0	25.2	27.5	30.0
Compression on extreme fibers of channels		See	Art. 5-16		
Tension and compression on extreme fibers of rectangular bearing plates	$F_b = 0.75F_y$	27.0	31.5	34.5	37.5
Bearing					
Bearing on contact area of milled surfaces	$F_p = 0.90F_y$	33.0	38.0	41.5	45.0

riveted, and welded connections. Material specifications and allowable stresses for bolts and rivets are discussed in Chapter 7 under design of bolted and riveted connections; filler material for welding and allowable stresses in welds are considered in Chapter 8.

5-7 Factors in Beam Design

Complete design of a beam includes consideration of bending strength, shear resistance, deflection, lateral support of the compression flange, and web crippling. The AISC Specification includes special factors that influence design for bending; the availability of higher strength steels with higher allowable stresses has led to the classification of structural shapes as *compact* or *noncompact* sections. Consequently, the shape employed, the laterally unsupported length of span, and the grade of steel must be known in order to determine the beam size. These factors are considered individually in the next several articles.

5-8 Compact and Noncompact Sections

The principal criterion used in classifying a particular W, S, or M shape as a compact section is that the *width–thickness ratio* of the projecting compression flange (half-flange) does not exceed $52.2/\sqrt{F_y}$. The width–thickness ratio is denoted by $b_f/2t_f$ and its limiting value is computed as follows for A36 steel:

$$\frac{b_f}{2t_f} = \frac{52.2}{\sqrt{F_y}} = \frac{52.2}{\sqrt{36}} = 8.70$$

To see how this determination is made, refer to Fig. 5-1. Two beam sections are shown with their flange and web dimensions as given in Table 4-1. The width–thickness ratio of the compression half-flange of the W 10 × 45 is

$$\frac{b_f}{2t_f} = \frac{8.022}{2 \times 0.618} = 6.49$$

Because this value is less than the limiting value of 8.70, the W 10 × 45 is a *compact section* for A36 steel.

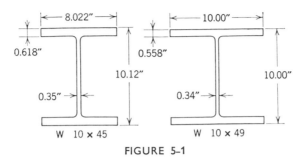

FIGURE 5-1

Now making the same test for the W 10 × 49,

$$\frac{b_f}{2t_f} = \frac{10}{2 \times 0.558} = 8.96 > 8.70$$

Since the value of 8.96 exceeds the limiting value of 8.70, the W 10 × 49 is a *noncompact section*. This classification of beams is a controlling factor in determining the allowable bending stress. For A36 steel, $F_b = 24$ ksi for compact sections (see Table 5-2). For noncompact shapes, F_b may be taken as 22 ksi.[2]

It is not necessary in practice to make the arithmetical test demonstrated above to determine whether a section is compact or noncompact. Reference to Tables 4-1 and 4-2 will show that values for $b_f/2t_f$ are given under Compact Section Criteria for each of the structural shapes listed. When working with A36 steel, a section will be identified as compact if the tabular value of this ratio *does not exceed* 8.70. This will be the case for most W and all S sections in A36 steel.

5-9   Lateral Support of Beams

Consider a W or S section used as a simple beam. The upper flange is in compression and there is a tendency for it to fail by sidewise

[2] However, the current AISC Specification provides a formula for graded reduction of the allowable bending stress from 24 ksi to 22 ksi. The AISC formula (1.5-5) is given in Section 1.5.1.4.2 of the 1969 Specification and may be used when $b_f/2t_f$ exceeds $52.2/\sqrt{F_y}$ but is less than $95.0/\sqrt{F_y}$. Tables to facilitate use of the formula are given in Appendix A of the Specification.

buckling if lateral deflection is not prevented. The tendency to buckle increases as the compressive bending stress in the flange increases, and also as the unbraced length of span increases. Consequently the full value of the allowable extreme fiber stress ($F_b =$ 24 ksi for A36 steel) can be used only when the compression flange is "adequately braced." The upper flange may be laterally supported by the floor construction for its entire length; for example, when it is encased in a concrete slab as indicated in Fig. 5-4. Very often, however, lateral support occurs only at certain points in the length of the beam, and the designer must give consideration to this distance *between points of lateral support.*

The designation L_c is given to the maximum unbraced length of compression flange for which loads may be computed (for a compact section) with a bending stress of 24 ksi. The unbraced length beyond which the allowable stress would be less than 22 ksi is L_u. Values of L_c and L_u are not given in Tables 4-1 and 4-2, "Properties for designing", because they depend upon the grade of steel employed, represented by F_y, as well as upon the dimensions of the section. However, these values are given in connection with safe load tables (discussed in Art. 5-15) and are recorded in the third and fourth columns of Table 5-4 for several of the W and S shapes that are listed in Tables 4-1 and 4-2. The design of beams in which the unsupported length exceeds L_u is discussed in Art. 5-18.

5-10 Flexure or Bending

As demonstrated in the example of Art. 4-7, the design of a beam for bending consists of applying form (3) of the flexure formula, $S = M/F_b$. Before starting computations, it is necessary to decide on the grade of steel to be used, and to determine the laterally unsupported length of the compression flange. Because the allowable bending stress given in the example is 24 ksi, we may assume that A36 steel was specified and that adequate lateral support for the beam was provided. With these items established, the design procedure followed involved the steps listed below:

(1) Determine the maximum bending moment.
(2) Compute the required section modulus.

(3) Refer to the tables of properties of steel sections and select a beam with a section modulus equal to or greater than that which is required. In general, the lightest weight section is the most economical.[3]

The beam selected in the Art. 4-7 example (W 14 × 34) was designed for bending only. A complete design would require investigation for shear and deflection. In the following example, the procedure outlined above is applied to a simple beam with a more complicated loading.

Example. A girder has a span of 18 ft with floor beams framing into it from both sides at 6-ft intervals, similar to the arrangement shown in Fig. 5-6. The reaction of each floor beam is 4000 lb, so the girder receives two concentrated loads of 8000 lb each at the third points of the span. The beams are connected to the girder as shown in Fig. 7-4c, thereby furnishing lateral support to the girder at 6-ft intervals. In addition, the girder has a uniformly distributed load of 400 lb per lin ft, including its weight, extending over the full span. Design the girder for bending, assuming that A36 steel is used.

Solution: (1) Because of the symmetry of loading, we know that the maximum bending moment will occur at the center of the span and that each reaction will equal half the total load or $\frac{1}{2}$[8000 + 8000 + (400 × 18)] = 11,600 lb. Then

$$M_{(x=9)} = (11{,}600 \times 9) - [(8000 \times 3) + (400 \times 9 \times 4.5)]$$

$$= 64{,}200 \text{ ft-lb} = 770{,}400 \text{ in-lb}$$

(2) Assuming a compact shape will be used, and therefore $F_b = 24{,}000$ psi (Table 5-2), the required section modulus is

$$S = \frac{M}{f} = \frac{770{,}400}{24{,}000} = 32.1 \text{ in.}^3$$

---

[3] Table 5-4 will be found very useful in this operation because the beams are listed in the order of increasing section moduli. This arrangement, and the listing of corresponding values of L_c and L_u, is presented in the AISC Manual in a much more extensive table called Allowable Stress Design Selection Table for Shapes Used as Beams. The reader should verify the beam sections selected in step (3) of the Art 4-7 example by checking them out in Table 5-4.

(3) Referring to the second column of Table 5-4, we find that a section modulus of 34.2 in.³ is provided by a W 12 × 27. In addition, we see that L_c for this beam is 6.9 ft, which is greater than the 6-ft intervals between points of lateral support. Consequently the W 12 × 27 is accepted.

Occasionally, adjacent construction limits the depth of a beam and the selection must be made with depth, rather than minimum weight per foot, as the controlling factor. The reader should satisfy himself that a W 10 × 39 would also be satisfactory.

Note: In the following problems assume that A36 steel will be used and that each beam is supported laterally for its full length. This, then, results in an allowable bending stress of $F_b = 24,000$ psi or 24 ksi.

Problem 5-10-A. Design for flexure a simple beam 14 ft in length and having a total uniformly distributed load of 19,800 lb.

Problem 5-10-B*. Design for flexure a beam having a span of 16 ft with a concentrated load of 12,500 lb at the center of the span.

Problem 5-10-C. A beam 15 ft in length has three concentrated loads of 4 kips, 5 kips, and 6 kips at 4 ft, 10 ft, and 12 ft, respectively, from the left-hand support. Design the beam for flexure.

Problem 5-10-D*. A beam 30 ft long has concentrated loads of 9 kips each at the third points and also a total uniformly distributed load of 30 kips. Design the beam for flexure.

Problem 5-10-E. Design for flexure a beam 12 ft in length, having a uniformly distributed load of 2000 lb per lin ft and a concentrated load of 8400 lb a distance of 5 ft from a support.

Problem 5-10-F. A beam 19 ft in length has concentrated loads of 6000 lb and 9000 lb at 5 ft and 13 ft, respectively, from the left-hand support. In addition, there is a uniformly distributed load of 1200 lb per lin ft beginning 5 ft from the left support and continuing to the right support. Design the beam for flexure.

Problem 5-10-G*. A steel beam 16 ft 0 in. long has two uniformly distributed loads, one of 200 lb per lin ft extending 10 ft 0 in. from the left support and the other of 100 lb per lin ft extending from a point 10 ft 0 in. from the left support to the right support. In addition, there is a concentrated load of 8 kips a distance of 10 ft 0 in. from the left support. Design the beam for flexure.

Problem 5-10-H. Design for flexure a simple beam 12 ft 0 in. in length, having two concentrated loads of 12,000 lb each, one 4 ft 0 in. from the right end and the other 4 ft 0 in. from the left end.

Problem 5-10-I. A cantilever beam 8 ft 0 in. long has a uniformly distributed load of 1600 lb per lin ft. Design the beam for flexure.

Problem 5-10-J*. A cantilever beam 6 ft 0 in. in length has a concentrated load of 12.3 kips at its unsupported end. Design the beam for flexure.

5-11 Shear

After a beam is designed for bending, it should be investigated for shear. Most beams large enough to resist flexure are also large enough to resist shear; consequently this step is often omitted. However, short beams or beams with relatively large loads near the supports should always be investigated for shear. Occasionally the beam must be increased in size to resist shearing stresses.

As discussed in Art. 3-1 and illustrated in Fig. 3-1b and d, a beam has a tendency to fail in shear by the fibers sliding past each other both vertically and horizontally. It can be shown that at any section of a beam the intensities of the vertical and horizontal shearing stresses are equal. The shearing stresses are not distributed evenly over the cross section but are greatest at the neutral axis and are zero at the extreme fibers. Because of this fact, *it is assumed that the web of steel beams is the only portion of the cross section resisting shear.* If the average unit shearing stress does not exceed F_v, the allowable unit shearing stress, 14,500 psi for A36 steel, the beam is safe with respect to shear.

To find the actual unit shearing stress the following formula may be used:

$$f_v = \frac{V}{A_w} = \frac{V}{dt_w}$$

in which f_v = actual unit shearing stress in psi or ksi,

V = maximum vertical shear in pounds or kips,

A_w = gross area of the web in square inches,

d = overall depth of the beam in inches,

t_w = thickness of the beam web in inches.

Remember that the maximum vertical shear is equal to the greater reaction for simple beams and that for symmetrically loaded beams each reaction is equal to one half the total load on the beam. To find the actual depth and web thickness of beams refer to Tables 4-1 and 4-2.

Example 1. An S 15 × 42.9 with a length of 14 ft has a total uniformly distributed load of 50 kips. Investigate the shear.

Solution: (1) Each reaction is 50 ÷ 2 = 25 kips; therefore the maximum vertical shear V = 25 kips.

(2) Referring to Table 4-2, we find that the actual depth of this beam is 15 in. and the web thickness = 0.411 in. Then

$$f_v = \frac{V}{dt_w} = \frac{25}{15 \times 0.411} = 4.07 \text{ ksi}$$

(3) Comparing this value with the allowable unit shearing stress F_v = 14.5 ksi (Table 5-2), we see that the beam is more than adequate to resist the shear.

Example 2. With respect to shear, what is the maximum concentrated load that may be placed at the center of the span of a W 18 × 55, used as a simple beam?

Solution: (1) Referring to Table 4-1, we find that d = 18.12 in. and t_w = 0.390 in., making A_w = 18.12 × 0.390 = 7.06 sq in.

(2) In this example, the actual unit shearing stress will be equal to the allowable value F_v = 14.5 ksi. Therefore, transposing the shear stress formula and making the substitutions,

$$V = f_v \times A_w = F_v \times A_w = 14.5 \times 7.06 = 102 \text{ kips}$$

which is the maximum permissible end shear for this beam.

(3) Because V is the magnitude of each of the two reactions under the given loading, P, the magnitude of the concentrated load is $2 \times V$. Thus, $P = 2 \times 102 = 204$ kips. It should be noted that the weight of the beam has been neglected in this example.

Note: In the following problems, neglect the beam weight. Keep in mind that investigating the shear in a beam means comparing the computed value of the unit shearing stress with the allowable value. All beams in these problems are rolled from A36 steel.

Problem 5-11-A*. An S 10 × 35 supports a total uniformly distributed load of 40 kips on a span of 10 ft. Investigate the shear.

Problem 5-11-B.   A W 10 × 39 supports the same load as the beam in the preceding problem. Investigate the shear. To what do you attribute the higher f_v value in this beam which weighs more per linear foot than the S 10 × 35?

Problem 5-11-C.   An S 15 × 42.9 has concentrated loads of 60 kips each located at the third points of its span. Investigate the shear.

Problem 5-11-D.   A W 14 × 34 is used as a simple beam on a span of 15 ft. It supports a uniform load of 20,000 lb distributed over the span length, and a concentrated load of 10,000 lb at the center of the span. Investigate the shear.

Problem 5-11-E.   With respect to shear, what is the maximum uniformly distributed load that may be placed on an S 20 × 65.4 used as a simple beam?

Problem 5-11-F*.   A W 12 × 27 is used as a simple beam on a span of 20 ft. It carries three concentrated loads of 3 kips, 4 kips, and 5 kips located at distances of 2 ft, 8 ft, and 11 ft, respectively, from the left support. Investigate the beam for shear.

5-12   Deflection

In addition to resisting bending and shear, beams must not deflect excessively. Floor and ceiling cracks may result if the beams are not stiff enough, and beams should be investigated to see that the deflection does not exceed $\frac{1}{360}$ of the span, the generally accepted limit with plastered ceilings. The current AISC Specification requirement is that steel beams and girders supporting plastered ceilings be of such dimensions that the maximum *live load* deflection will not exceed $\frac{1}{360}$ of the span. It frequently happens that a beam may be of adequate dimensions to resist bending and shear but, on investigation, may be found to deflect more than the maximum permitted by building codes.

For typical beams and loads the actual deflection may be computed from the formulas given in Table 3–1, but in using these formulas note carefully that *l*, in the term l^3, is in inches, not feet. For a simple beam with a uniformly distributed load the deflection is found by the formula

$$D = \frac{5}{384} \times \frac{Wl^3}{EI} \qquad \text{(Case 2, Table 3-1)}$$

in which D = maximum deflection in inches,[4]

> W = total uniformly distributed load in pounds or kips,
> l = length of the span *in inches*,
> E = modulus of elasticity of the beam in psi or ksi (for structural steel E = 29,000,000 psi),
> I = moment of inertia of the cross section of the beam in inches to the fourth power.

For a beam on which the loading is not typical we may find W, the uniformly distributed load that would produce the same bending moment, and then apply the foregoing formula to find the approximate deflection. When the maximum deflection occurs at the center of the span, it is sometimes convenient to compute the deflections due to individual loads on the beam; their sum will be the total deflection.

Example.   A simple beam of A36 steel has a span of 20 ft, and will carry a total uniformly distributed load of 40,000 lb including its own weight. Deflection is limited to $\frac{1}{360}$ of the span, and full lateral support is provided. Design the beam.

Solution: (1) To determine the required section modulus, we first compute the bending moment using Case 2, Table 3-1:

$$M = \frac{Wl}{8} = \frac{40,000 \times 20 \times 12}{8} = 1,200,000 \text{ in-lb}$$

The required section modulus is then

$$S = \frac{M}{f} = \frac{1,200,000}{24,000} = 50 \text{ in.}^3$$

(2) Referring to Table 5-4, select a W 14 × 38 (S = 54.7) as a trial section to be investigated for shear and deflection.

(3) Since the total load on the beam is 40,000 lb, each reaction—and consequently the maximum vertical shear—is equal to 20,000 lb. Table 4-1 shows that the depth of the W 14 × 38 is 14.12 in. and the web thickness is 0.313 in. Then

$$f_v = \frac{V}{dt_w} = \frac{20,000}{14.12 \times 0.313} = 4420 \text{ psi}$$

---

[4] The Greek letter *delta* (Δ) is also used as a symbol for beam deflection.

which is acceptable because it is less than the allowable value of 14.5 ksi (Table 5-2).

(4) By data, the allowable deflection is $(20 \times 12)/360 = 0.67$ in. The deflection formula that applies to this beam and loading is given in Case 2, Table 3-1:

$$D = \frac{5}{384} \times \frac{Wl^3}{EI}$$

The moment of inertia of the W 14 × 38 is 386 in.[4] (Table 4-1) and the span in inches is $20 \times 12 = 240$ in. Therefore,

$$D = \frac{5}{384} \times \frac{40,000 \times 240^3}{29,000,000 \times 386} = 0.64 \text{ in.}$$

Because the computed deflection of 0.64 in. does not exceed the permissible value of 0.67 in., the beam is acceptable for deflection, as well as for shear and bending. Therefore, use the W 14 × 38.

5-13   Deflection Coefficients

A simple method of computing the deflection of steel beams, uniformly loaded, is to use Table 5-3 and the formula

$$D = \frac{coefficient \ in \ table}{depth \ of \ beam}$$

To find the maximum deflection, first select from the table the coefficient corresponding to the span length and f_b, the actual extreme fiber stress; then divide this coefficient by the depth of the beam in inches. Note that this table is to be used directly for uniformly distributed loads, but that deflections for other types of loading may be found by multiplying the coefficients by the factors given in the footnote to the table.

Example 1.   A simple beam of A36 steel has a span of 16 ft and a uniformly distributed load of 39,000 lb, including its own weight. The section used for this loading is a W 12 × 31, and the extreme fiber stress is 24,000 psi. Does the actual deflection exceed the allowable $\frac{1}{360}$ of the span?

TABLE 5-3. Deflection Coefficients for Uniformly Distributed Loads

Span, Feet	Bending Stress Pounds per Square Inch		Span, Feet	Bending Stress Pounds per Square Inch		Span, Feet	Bending Stress Pounds per Square Inch	
	22,000	24,000		22,000	24,000		22,000	24,000
10	2.274	2.481	18	7.365	8.035	26	15.369	16.766
11	2.750	3.000	19	8.206	8.952	27	16.572	18.079
12	3.273	3.571	20	9.094	9.921	28	17.824	19.448
13	3.840	4.190	21	10.025	10.936	29	19.120	20.858
14	4.455	4.860	22	11.002	12.002	30	20.462	22.322
15	5.115	5.580	23	12.027	13.120	31	21.848	23.834
16	5.821	6.350	24	13.094	14.284	32	23.280	25.396
17	6.571	7.169	25	14.209	15.506	33	24.758	27.009

Coefficient for concentrated load at center of span = 0.8 of above values.
Coefficient for triangular loading, apex at center of span = 0.96 of above values.
Coefficient for equal concentrated loads at ⅓ points = 1.02 of above values.
Coefficient for irregular loading = 0.92 of above values (approximate).

Solution: (1) The coefficient in Table 5-3 corresponding to a bending stress of 24,000 psi and a span of 16 ft is 6.35. The depth of the W 12 × 31 is 12.09 in. (Table 4-1). Then, using the formula stated above,

$$D = \frac{6.35}{12.09} = 0.526 \text{ in.}$$

which is the actual deflection. The allowable deflection is (16 × 12)/360 = 0.533 in. Since this value is greater than the computed deflection, the beam is satisfactory.

(2) To check the actual deflection of 0.526 in., we may use the formula given in Case 2, Table 3-1:

$$D = \frac{5}{384} \times \frac{Wl^3}{EI}$$

and

$$D = \frac{5}{384} \times \frac{39,000 \times (16 \times 12)^3}{29,000,000 \times 239} = 0.526 \text{ in.}$$

In this example, it was stated that the uniform load on the W 12 × 31 resulted in an extreme fiber stess of 24,000 psi; this made it possible to use Table 5-3 directly. In the following example, we will investigate deflection for the same section with a slightly larger load but with the span substantially reduced.

Example 2. A simple beam of A36 steel supports a uniformly distributed load of 42,000 lb, including its own weight. The span length is 12 ft, and the section is a W 12 × 31. Compute the actual deflection. *Solution:* (1) First, compute the actual extreme fiber stress. This is readily accomplished by use of the flexure formula once the maximum bending moment is known. Then

$$M = \frac{WL}{8} = \frac{42,000 \times 12}{8} = 63,000 \text{ ft-lb}$$

Referring to Table 4-1, we find that S for a W 12 × 31 is 39.5 in.[3] Then, using form (2) of the flexure formula (Art. 4-7),

$$f_b = \frac{M}{S} = \frac{63,000 \times 12}{39.5} = 19,200 \text{ psi}$$

which is the *actual* extreme fiber (bending) stress.

(2) In Table 5-3 we find only bending stresses 22,000 psi and 24,000 psi. For a stress of 24,000 psi and a span of 12 ft, the coefficient in Table 5-3 is 3.571. However, we know that the deflection is in direct proportion to the magnitude of the extreme fiber stress. For this reason, the coefficient to use in this problem is a fractional part of 3.571. Actually it is $(19,200/24,000) \times 3.75$. Since the depth of the W 12 × 31 is 12.09 (Table 4-1), the deflection is

$$D = \frac{19,200}{24,000} \times \frac{3.571}{12.09} = 0.24 \text{ in.}$$

This value does not exceed the allowable deflection which is $(12 \times 12)/360 = 0.4$ in.

5-14 A Convenient Deflection Formula

If a designer is computing the size of simple beams with uniformly distributed loads and is using an allowable bending stress of 24,000 psi, the actual deflection may be determined by the convenient formula[5]

$$D = \frac{0.02483 \times L^2}{d}$$

in which D = maximum deflection of the beam in inches,
L = span length of the beam *in feet*,
d = depth of the beam in inches.

Example.   A simple beam has a span of 16 ft with a uniformly distributed load of 34,000 lb, including its own weight. The beam used for this loading is a W 12 × 27, and the extreme fiber stress developed by the load is 24,000 psi. Compute the deflection.
Solution: (1) Referring to Table 4-1, the depth of a W 12 × 27 is found to be 11.96 in. Then

$$D = \frac{0.02483 \times L^2}{d} = \frac{0.02483 \times 16 \times 16}{11.69} = 0.53 \text{ in.}$$

(2) Checking this value by coefficient from Table 5-3,

$$D = \frac{6.35}{11.96} = 0.53 \text{ in.}$$

Deflection is usually the controlling factor in the design of beams carrying light loads on relatively long spans. The procedure for handling this situation is discussed in Art. 5-19.

Note: In the following problems, assume that the steel is A36 and that each beam is laterally supported throughout its full length. To simplify the computations, neglect the uniformly distributed load due to the weight of the beam.

Problem 5-14-A*.   Design a steel beam to carry a uniformly distributed load of 36 kips on a span of 18 ft. Deflection is limited to ⅟₃₆₀ of the span.
Problem 5-14-B.   A steel beam 15 ft long has two concentrated loads of 14 kips each at the third ponts of the span. Design the beam and compute its deflection. (See footnote to Table 5-3.)

---

[5] For the derivation of this formula see *Simplified Design of Structural Steel*, by Harry Parker, Fourth Edition (New York: Wiley, 1974), Art 7-3.

Problem 5-14-C*. Design a beam 12 ft long to carry a concentrated load of 22,000 lb at the center of the span. Compute the deflection.

Problem 5-14-D. Compute the deflection of a W 16 × 36, having a span of 18 ft with a concentrated load of 14,000 lb at the center of the span.

Problem 5-14-E*. Design a steel beam 20 ft long to support three concentrated loads of 5 kips, 6 kips, and 7 kips located at 6 ft, 9 ft, and 14 ft, respectively, from the left reaction. Deflection is limited to 1/360 of the span.

Problem 5-14-F. Compute the deflection of an S 15 × 42.9 simple beam having two concentrated loads of 12 kips each located at the third points of a 21-ft span.

Problem 5-14-G. Design a steel beam 20 ft long to support a total uniformly distributed load of 29,000 lb. Investigate for shear and deflection.

5-15 Safe Load Table for W and S Shapes

Because simple beams with uniformly distributed loads occur so frequently in practice, tables of maximum loads for specific spans are of great convenience to the designer. Table 5-4 is such a table compiled from data in the extensive series of safe load tables presented in the AISC Manual. The loads are given in kips and are based on an allowable bending stress of 24 ksi for compact sections that are laterally braced at intervals not greater than L_c.

Referring to Table 5-4, we see that an S 12 × 31.8 will safely support a uniform load of 36 kips on a span of 16 ft. Obviously this beam will support any load of smaller magnitude on this span, but the resulting extreme fiber stress will be less than 24 ksi. For beams supported laterally at intervals greater than L_c but not greater than L_u, the allowable bending stress is 22 ksi. If this lower stress is to be used, the loads shown in the table must be multiplied by 22/24, a reduction of $8\frac{1}{2}\%$.

In order to use Table 5-4 directly when $F_b = 22$ ksi, we multiply the uniform *load to be carried* by 1.09. (This is equivalent to reducing all the loads in the table by 22/24.) For example, if we wish to select a beam that will carry 36 kips on a 16-ft span with $F_b = 22$ ksi, the design load becomes 36 × 1.09 = 39.2 kips. Scanning the values listed for a 16-ft span, we find that 39.5 kips is given for a W 12 × 31 and 41 kips for a W 14 × 30. Therefore, either of these sections could be used.

TABLE 5-4. Allowable Uniform Loads in kips for Selected W and

Shape	S in.³	L_c ft	L_u ft	Span in feet						
				8	9	10	11	12	13	14
W 8 × 17	14.1	5.5	9.4	28.2	25.1	22.6	20.5	18.8	17.4	16.1
S 8 × 18.4	14.4	4.2	9.9	28.8	25.6	23.0	20.9	19.2	17.7	16.5
S 8 × 23	16.2	4.4	10.3	32.4	28.8	25.9	23.6	21.6	19.9	18.5
W 8 × 20	17.0	5.6	11.3	34.0	30.2	27.2	24.7	22.7	20.9	19.4
W 8 × 24	20.8	6.9	15.1	41.6	37.0	33.3	30.3	27.7	25.6	23.8
W 10 × 21	21.5	6.1	9.1	43.0	38.2	34.4	31.3	28.7	26.5	24.6
W 8 × 28	24.3	6.9	17.4	48.6	43.2	38.9	35.3	32.4	29.9	27.8
S 10 × 25.4	24.7	4.9	10.6	49.4	43.9	39.5	35.9	32.9	30.4	28.2
W 10 × 25	26.5	6.1	11.4	53.0	47.1	42.4	38.5	35.3	32.6	30.3
S 10 × 35	29.4	5.2	11.2	58.8	52.3	47.0	42.8	39.2	36.2	33.6
W 10 × 29	30.8	6.1	13.2	61.6	54.8	49.3	44.8	41.4	37.9	35.2
W 12 × 27	34.2	6.9	10.1	68.4	60.8	54.7	49.7	45.6	42.1	39.1
S 12 × 31.8	36.4	5.3	10.5	73	65	58	53	49	45	42
S 12 × 35	38.2	5.4	10.7	76	68	61	56	51	47	44
W 12 × 31	39.5	6.9	11.6	79.0	70.2	63.2	57.5	52.7	48.6	45.1
W 14 × 30	41.9	7.1	8.6	83	74	66	60	55	51	47
W 10 × 39	42.2	8.4	19.6	84	75	68	61	56	52	48
S 12 × 40.8	45.4	5.5	13.4	91	81	73	66	61	56	52
W 12 × 36	46.0	6.9	13.4	92.0	81.8	73.6	66.9	61.3	56.6	52.6
W 14 × 34	48.6	7.1	10.1	97	86	78	71	65	60	56
S 12 × 50	50.8	5.8	13.9	102	90	81	74	68	63	58
W 12 × 40	51.9	8.4	16.0	102	92	83	75	69	64	59
W 14 × 38	54.7	7.2	11.4	109	97	88	80	73	67	63
W 16 × 36	56.5	7.4	8.7	113	100	90	82	75	70	65
W 12 × 45	58.2	8.5	17.8	116	103	93	85	78	72	67
S 15 × 42.9	59.6	5.8	10.6	119	106	95	87	79	73	68
W 14 × 43	62.7	8.4	14.3	122	111	100	91	84	77	72

Attention is called to the solid heavy lines in Table 5-4; loads to the right of these lines produce deflections in excess of $\frac{1}{360}$ of the span.

If the load and span length are known, the proper size section is selected directly by referring to the table. Keep in mind that if depth is not a factor, the lightest weight beam is generally the most economical.

S shapes Used as Beams Laterally Supported, Based on $F_y = 36$ ksi*

15	16	17	18	19	20	21	22	23	24	25	26
15.0	14.1	13.3									
15.4	14.4	13.6									
17.3	16.2	15.2									
18.1	17.0	16.0									
22.2	20.8	19.6									
22.9	21.5	20.2	19.1	18.1	17.2	16.4	15.6				
25.9	24.3	22.9									
26.3	24.7	23.2	22.0	20.8	19.8	18.8					
28.3	26.5	24.9	23.6	22.3	21.2	20.2	19.3				
31.4	29.4	27.7	26.1	24.8	23.5	22.4					
32.9	30.8	29.0	27.4	25.9	24.6	23.5	22.4				
36.5	34.2	32.2	30.4	28.8	27.4	26.1	24.9	23.8	22.8	21.9	
39	36	34	32	31	29	28	26	25	24	23	
41	38	36	34	32	31	29	28	27	25	24	
42.1	39.5	37.2	35.1	33.3	31.6	30.1	28.7	27.5	26.3	25.3	
44	41	39	37	35	33	32	30	29	28	27	25
45	42	40	38	36	34	32	31				
48	45	43	40	38	36	35	33	32	30	29	
49.1	46.0	43.3	40.9	37.8	36.8	35.0	33.5	32.0	30.7	29.4	
52	49	46	43	41	39	37	35	34	32	31	30
54	51	48	45	43	41	39	37	35	34	33	
55	52	49	46	44	42	40	38	36	35	33	
58	55	51	49	46	44	42	40	38	36	35	34
60	57	53	50	48	45	43	41	39	38	36	35
62	58	55	52	49	47	44	42	40	39	37	
64	60	56	53	50	48	45	43	41	40	38	37
67	63	59	56	53	50	48	46	44	42	40	39

Loads to right of heavy vertical lines produce deflections exceeding $\frac{1}{360}$ of the span.
* Compiled from data in the 7th Edition of the *Manual of Steel Construction.* Courtesy American Institute of Steel Construction.

Example. A beam of A36 steel supports a uniformly distributed load of 40 kips on a span of 18 ft, with deflection limited to $\frac{1}{360}$ of the span. Assuming full lateral support, select from Table 5-4 the lightest weight suitable section.
Solution: Scanning the values listed for an 18-ft span, we see that an S 12 × 40.8 and a W 12 × 36 will each carry 40 kips but will

TABLE 5-4. Allowable Uniform Loads in kips for Selected W and

Shape	S	L_c	L_u	Span in feet						
	in.^{3}	ft	ft	15	16	17	18	19	20	21
W 16 × 40	64.6	7.4	10.2	69	65	61	57	54	52	49
W 12 × 50	64.7	8.5	19.7	69	65	61	58	54	52	49
W 14 × 48	70.2	8.5	16.0	75	70	66	62	59	56	53
W 16 × 45	72.5	7.4	11.4	77	73	68	64	61	58	55
W 14 × 53	77.8	8.5	17.6	83	78	73	69	66	62	59
W 18 × 50	89.1	7.9	11.0	95	89	84	79	75	71	68
S 18 × 54.7	89.4	6.3	10.7	95	89	84	79	75	72	68
W 16 × 58	94.4	8.9	15.9	101	94	89	84	79	76	72
W 18 × 55	98.4	8.0	12.1	105	98	93	87	83	79	75
S 18 × 70	103	6.6	11.1	110	103	97	92	87	82	78
W 16 × 64	104	9.0	17.6	111	104	98	92	88	83	79
W 21 × 55	110	8.7	9.5	117	110	104	98	93	88	84
W 18 × 64	118	9.2	15.5	126	118	111	105	99	94	90
W 21 × 62	127	8.7	11.2	135	127	120	113	107	102	96
W 18 × 70	129	9.2	16.9	138	129	121	115	109	103	98
W 21 × 68	140	8.7	12.4	149	140	132	124	118	112	107
W 24 × 68	153	9.5	10.2	163	153	144	136	129	122	117
W 21 × 82	169	9.5	15.8	180	169	159	150	142	135	129
W 24 × 76	176	9.5	11.9	188	176	166	156	148	141	134
W 18 × 96	185	12.4	24.9	197	185	174	164	156	148	141
W 27 × 84	212	10.5	11.2	226	212	200	188	179	170	162
W 27 × 94	243	10.5	12.8	259	243	229	216	205	194	185
W 24 × 100	250	12.7	17.9	267	250	235	222	211	200	190
W 21 × 112	250	13.7	24.8	267	250	235	222	211	200	190
W 30 × 99	270	10.9	11.6	288	270	254	240	227	216	206
W 24 × 110	276	12.7	19.7	294	276	260	245	232	221	210
W 21 × 127	284	13.8	28.1	303	284	267	252	239	227	216
W 30 × 108	300	11.1	12.4	320	300	282	267	253	240	229
W 24 × 130	332	14.8	24.1	354	332	312	295	280	266	253

have excessive deflection. The lightest weight suitable section is the W 14 × 34 which will carry 43 kips and not exceed the permissible deflection.

Note: In the following problems, determine by use of Table 5-4 the lightest weight beams for the loads and spans indicated. The

S Shapes Used as Beams Laterally Supported, Based on $F_y = 36$ ksi* (*continued*)

22	23	24	25	26	27	28	29	30	31	32	33
▌47	45	43	41	40	38	37	36	34	33	32	31
47	45	43	41								
51	49	47	45	43	42	40	39	37			
▌53	50	48	46	45	43	41	40	39	37	36	35
57	54	52	50	48	46	44	43	41			
65	62	59	▌57	55	53	51	49	48	46	45	43
65	62	60	57	55	53	51	49	48	46	45	43
▌69	66	63	60	58	56	54	52	50	49	47	46
72	68	66	▌63	61	58	56	54	52	51	49	48
75	72	69	▌66	63	61	59	57	55	53	52	50
▌76	72	69	67	64	62	59	57	55	54	52	50
80	76	73	70	68	65	63	▌61	59	57	55	53
86	82	79	▌76	73	70	67	65	63	61	59	57
92	88	85	81	78	75	73	▌70	68	65	64	61
94	90	86	▌83	79	76	74	71	69	67	65	63
102	97	93	90	86	83	80	▌77	75	72	70	68
111	106	102	98	94	91	87	84	82	79	77	▌74
123	118	113	108	104	100	97	▌93	90	87	85	82
128	122	117	113	108	104	101	97	94	91	88	▌85
135	129	123	▌118	114	110	106	102	99	95	93	90
154	147	141	136	130	126	121	117	113	109	106	103
177	169	162	156	150	144	139	134	130	125	122	118
182	174	167	160	154	148	143	138	133	129	125	▌121
182	174	167	160	154	148	143	▌138	133	129	125	121
196	188	180	173	166	160	154	149	144	135	127	120
201	192	184	177	170	164	158	152	147	142	138	▌134
207	198	189	182	175	168	162	▌157	151	147	142	138
218	209	200	192	185	178	171	166	160	155	150	145
241	231	221	212	204	197	190	183	177	171	166	▌161

beams are laterally supported for their entire lengths and the deflection is not to exceed $\frac{1}{360}$ of the span.

Problem 5-15-A-B-C*.   A uniformly distributed load of 25 kips on spans of (a) 8 ft, (b) 10 ft, (c) 13 ft.

Problems 5-15-D-E*-F.   A uniformly distributed load of 35 kips on spans of (d) 12 ft, (e) 14 ft, (f) 16 ft.

5-16   Safe Load Table for Channels

Table 5-5 is a safe load table for American Standard channels used as beams. The allowable bending stress for channels of A36 steel is 22 ksi when the compression flange is braced laterally at intervals not greater than L_u. Table 5-5, as well as the more extensive channel safe load tables in the AISC Manual, are computed on this basis.

5-17   Equivalent Tabular Loads

The safe loads shown in Tables 5-4 and 5-5 are for uniformly distributed loads on simple beams. By the use of coefficients we can convert other types of loading to equivalent uniform loads and thereby greatly extend the usefulness of the tables.

The maximum bending moment for a simple beam with a uniformly distributed load over the entire span is $M = WL/8$. For a simple beam with equal concentrated loads at the third points of the span the maximum moment is $M = PL/3$. (See Cases 2 and 4, Table 3-1.) Equating these two bending moments,

$$\frac{WL}{8} = \frac{PL}{3} \quad \text{and} \quad W = 2.67 \times P$$

which shows that if the value of one of the concentrated loads in Case 4 is multiplied by the coefficient 2.67, we will have an equivalent uniform load that produces the same bending moment as the concentrated loads. Because of their use with safe load tables, equivalent uniform loads are usually called *equivalent tabular loads*, abbreviated *ETL*.

Coefficients for a few other types of loading are shown in Fig. 5-2, and several others are given in Part 2 of the AISC Manual. It is important to remember that an *ETL* does not include the weight of the beam, for which an estimated amount should be added. Beams found by this method should be investigated for shear and deflection; it is assumed that they are adequately supported laterally. Also, when recording the beam reactions and values of the vertical shear (V), these must be determined from the *actual* loading conditions without regard to the *ETL*.

Example 1. A simple beam has a span of 19 ft with concentrated loads of 11.5 kips each located at the quarter points of the span. Using Table 5-4, and allowing 1 kip for the beam weight, select the most economical section.

Solution: Referring to Fig. 5-2c, we find that $ETL = 4 \times P$ for this loading, or $ETL = 4 \times 11.5 = 46$ kips. Adding 1 kip for the estimated beam weight, the total $ETL = 46 + 1 = 47$ kips. Scanning the values listed for a 19-ft span, we find that a W 16 × 36 will safely carry 48 kips. Therefore use W 16 × 36.

Example 2. A cantilever beam has a length of 11 ft with a concentrated load of 7 kips at the unsupported end. By the use of Table 5-4, select a beam to support this load. Neglect the beam weight.

Solution: Referring to Fig. 5-2e, we find that the ETL for this condition is $8 \times P$, or $ETL = 8 \times 7 = 56$ kips. Table 5-4 shows that a W 14 × 30 is the lightest weight section that will support this load.

Note: In each of the following problems, determine the equivalent tabular load and select the lightest weight beam section by the use of Table 5-4. Check each of your selections by use of the flexure formula. (Neglect beam weight.)

Problem 5-17-A*. A simple beam has a span of 15 ft with two concentrated loads of 8 kips each located at the third points of the span.

Problem 5-17-B. A cantilever beam 12 ft long has a uniformly distributed load of 8 kips extending over its entire length.

Problem 5-17-C. A simple beam 24 ft long has three concentrated loads of 9 kips each located at the quarter points of the span.

Problem 5-17-D. A simple beam with a span of 18 ft has a concentrated load of 15 kips at the center of the span.

Problem 5-17-E*. A simple beam with a span of 21 ft has a single concentrated load of 22 kips placed 7 ft from one end.

5-18 Laterally Unsupported Beams

As discusses in Arts. 5-9 and 5-15, the allowable bending stress is 24 ksi when the compression flanges of compact beams are braced laterally at intervals not greater than L_c. For beams supported laterally at intervals greater than L_c but not greater than L_u, the allowable

TABLE 5-5. Allowable Uniform Loads in kips for Selected Channels

Shape	S	L_u	Span in feet							
	in.³	ft	10	11	12	13	14	15	16	17
C 6 × 8.2	4.38	5.1	6.4	5.8	5.4	4.9	4.6	4.3		
C 6 × 10.5	5.06	5.4	7.4	6.7	6.2	5.7	5.3	4.9		
C 6 × 13	5.80	5.7	8.5	7.7	7.1	6.5	6.1	5.7		
C 7 × 9.8	6.08	5.1	8.9	8.1	7.4	6.9	6.4	5.9	5.6	5.2
C 7 × 12.25	6.93	5.3	10.2	9.2	8.5	7.8	7.3	6.8	6.4	6.0
C 7 × 14.75	7.78	5.6	11.4	10.4	9.5	8.8	8.2	7.6	7.1	6.7
C 8 × 11.5	8.14	5.1	11.9	10.9	9.9	9.2	8.5	8.0	7.5	7.0
C 8 × 13.75	9.03	5.3	13.2	12.0	11.0	10.2	9.5	8.8	8.3	7.8
C 9 × 13.4	10.6	5.2	15.5	14.1	13.0	12.0	11.1	10.4	9.7	9.1
C 8 × 18.75	11.0	5.7	16.1	14.7	13.4	12.4	11.5	10.8	10.1	9.5
C 9 × 15	11.3	5.3	16.6	15.1	13.8	12.7	11.8	11.0	10.4	9.7
C 9 × 20	13.5	5.6	19.8	18.0	16.5	15.2	14.1	13.2	12.4	11.6
C 10 × 15.3	13.5	5.3	19.8	18.0	16.5	15.2	14.1	13.2	12.4	11.6
C 10 × 20	15.8	5.5	23.2	21.1	19.3	17.8	16.6	15.4	14.5	13.6
C 10 × 25	18.2	5.8	26.7	24.3	22.2	20.5	19.1	17.8	16.7	15.7
C 10 × 30	20.7	6.1	30.4	27.6	25.3	23.4	21.7	20.2	19.0	17.9
C 12 × 20.7	21.5	5.7	31.5	28.7	26.3	24.3	22.5	21.0	19.7	18.5
C 12 × 25	24.1	5.9	35.3	32.1	29.5	27.2	25.2	23.6	22.1	20.8
C 12 × 30	27.0	6.1	39.6	36.0	33.0	30.5	28.3	26.4	24.8	23.3
C 15 × 33.9	42.0	6.8	61.6	56.0	51.3	47.4	44.0	41.1	38.5	36.2
C 15 × 40	46.5	7.1	68.2	62.0	56.8	52.5	48.7	45.5	42.6	40.1
C 15 × 50	53.8	7.5	78.9	71.7	65.8	60.7	56.4	52.6	49.3	46.4

bending stress is 22 ksi (except as higher stresses may be permitted under the conditions noted in the footnote to Art. 5-8). For shapes that do not qualify as compact sections, the allowable bending stress for A36 steel is 22 ksi for all laterally unbraced lengths up to L_u.

When the unbraced length of the compression flange exceeds L_u, the allowable bending stress must be reduced in accordance with certain formulas and provisions set forth in the AISC Specification. The design of beams by use of the specified provisions is not a simple matter, and the AISC Manual gives supplementary charts to aid the designer. These beam charts give "Allowable Moments in Beams with Unbraced Lengths Greater than L_u," and provide a workable approach to this otherwise rather cumbersome problem. One of

(C) Used as Beams Laterally Supported, Based on $F_y = 36$ ksi*

18	19	20	21	22	23	24	25	26	27	28	29
5.0											
5.6											
6.3											
6.6	6.3	6.0									
7.4	7.0	6.6									
8.6	8.2	7.8	7.4	7.1	6.8						
9.0	8.5	8.1									
9.2	8.7	8.3	7.9	7.5	7.2						
11.0	10.4	9.9	9.4	9.0	8.6						
11.0	10.4	9.9	9.4	9.0	8.6	8.3	7.9				
12.9	12.2	11.6	11.0	10.5	10.1	9.7	9.3				
14.8	14.0	13.3	12.7	12.1	11.6	11.1	10.7				
16.9	16.0	15.2	14.5	13.8	13.2	12.7	12.1				
17.5	16.6	15.8	15.0	14.3	13.7	13.1	12.6	12.1	11.7	11.3	10.9
19.6	18.6	17.7	16.8	16.1	15.4	14.7	14.1	13.6	13.1	12.6	12.2
22.0	20.8	19.8	18.9	18.0	17.2	16.5	15.8	15.2	14.7	14.1	13.7
34.2	32.4	30.8	29.3	28.0	26.8	25.7	24.6	23.7	22.8	22.0	21.2
37.9	35.9	34.1	32.5	31.0	29.7	28.4	27.3	26.2	25.3	24.4	23.5
43.8	41.5	39.5	37.6	35.9	34.3	32.9	31.6	30.3	29.2	28.2	27.2

Loads to right of heavy vertical lines produce deflections exceeding $\frac{1}{360}$ of the span.
* Compiled from data in the 7th Edition of the *Manual of Steel Construction*. Courtesy American Institute of Steel Construction.

these charts from the 7th Edition of the AISC Manual is reproduced in Fig. 5-3 to demonstrate their nature and method of use.

After determining the maximum bending moment in kip-ft, and noting the longest unbraced length of compression flange, these two coordinates are located on the sides of the chart and traced to their intersection. Any beam whose curve lies above and to the right of this intersection satisfies the bending stress requirement. The nearest curve that is a solid line represents the most economical section in terms of beam weight. If there is any question about deflection being critical, it should be investigated; if found excessive, another selection should be made from the chart.

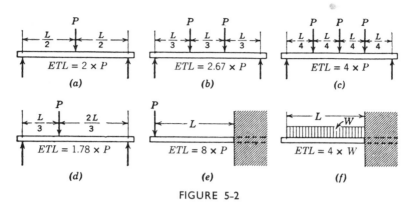

FIGURE 5-2

Example. A simple beam carries a total uniform load of 19.6 kips, including its own weight, over a span of 20 ft. It has no lateral support except at the ends of the span. Assuming A36 steel, select a beam from the chart of Fig. 5-3 that will meet bending strength requirements and will not deflect more than 0.75 in. under the full load.

Solution: (1) The maximum bending moment is

$$M = \frac{WL}{8} = \frac{19.6 \times 20}{8} = 49 \text{ kip-ft}$$

(2) Entering the chart on the bottom scale with an unbraced length of 20 ft, proceed vertically to intersection with the horizontal line representing a moment of 49 kip-ft on the scale to the left. The nearest solid-line curve above and to the right of the intersection represents a W 8 × 31.

(3) Observing that this is quite a shallow beam for the 20-ft span length, a deflection check is called for. Since we know the allowable deflection (0.75 in.), it will be convenient to solve the deflection formula for the moment of inertia required to maintain this limit. Referring to Case 2, Table 3-1, and making the transformation,

$$I = \frac{5}{384} \times \frac{Wl^3}{E\Delta} = \frac{5 \times 19.6 \times (20 \times 12)^3}{384 \times 29{,}000 \times 0.75} = 162 \text{ in.}^4$$

For the W 8 × 31 selected under Step (2), $I = 110$ in.^{4} (Table 4-1). Since this is less than the 162 in.^{4} required. we return to the

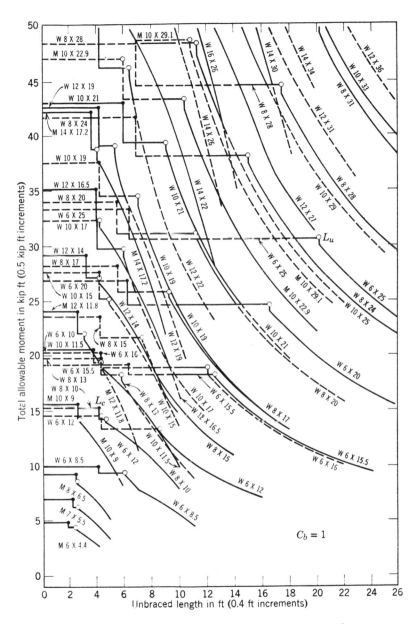

FIGURE 5-3. Allowable moments in beams ($F_y = 36$ ksi). Taken from a more complete set of charts in the 7th Edition of the *Manual of Steel Construction.* Courtesy American Institute of Steel Construction.

chart of Fig. 5-3 and select the W 10 × 33. Table 4-1 shows $I = 171$ in.[4] for this section, which is adequate. Use the W 10 × 33.

5-19 Beams with Light Loads

A problem that occurs very often is the design of a beam whose span is long and whose load is relatively light. The example in Art. 5-18 applies to this situation where the beam is laterally unsupported, but the condition also arises with some of the lighter weight floor construction systems that furnish full lateral support. The following example illustrates this more usual case.

Example. A simple beam adequately braced has a length of 22 ft and supports a total uniform load of 8 kips, including its own weight. Determine the size of the lightest weight steel beam that will be sufficiently strong and whose deflection will not exceed $\frac{1}{360}$ of the span length under the total load; i.e. $(22 \times 12)/360 = 0.74$ in.
Solution: (1) Inspection of Table 5-4 shows that the loads tabulated for a span of 22 ft far exceed 8 kips. Bear in mind that the loads in the table produce a bending stress of 24 ksi. The beam we select will probably have a bending stress considerably smaller.
  (2) The maximum bending moment is

$$M = \frac{WL}{8} = \frac{8 \times 22}{8} = 22 \text{ kip-ft or } 264 \text{ kip-in.}$$

and the required section modulus is

$$S = \frac{M}{F_b} = \frac{264}{24} = 11.0 \text{ in.}^3$$

  (3) The lightest weight section listed in Table 5-4 which has a section modulus equal to or greater than 11.0 in.[3] is a W 8 × 17, but this beam will have to be tested for deflection (allowable $\Delta = 0.74$ in.). As we have seen before, this may be accomplished most readily by solving the deflection formula (Case 2, Table 3-1) for the I required to maintain the specified limit. Then, if the W 8 × 17 proves inadequate, another beam can be selected directly from the

table on the basis of required moment of inertia. Making the transformation and substituting,

$$I = \frac{5}{384} \times \frac{Wl^3}{E\Delta} = \frac{5 \times 8 \times (22 \times 12)^3}{384 \times 29,000 \times 0.74} = 89.3 \text{ in.}^4$$

which is the *required moment of inertia*.

(4) From Table 4-1 we find that I for the W 8 × 17 is only 56.6 in.⁴ so this section is not adequate. Scanning Table 4-1 further, the lightest weight section with a moment of inertia of 89.3 or larger is a W 10 × 21 ($I = 107$ in.⁴). Since the section modulus of this beam is 21.5 in.³ and only 11.0 in.³ is required, the W 10 × 21 meets the requirements for both strength and deflection.

If the thickness of the floor construction were restricted and a beam 10 in. deep could not be accepted, Table 4-1 shows that a W 8 × 28 has an I of 97.8 in.⁴ which is also satisfactory. Thus we see that for long spans and light loads the size of the beam is determined by the moment of inertia rather than by the section modulus; deflection is the controlling factor. The foregoing procedure should be thoroughly understood since the design of beams with relatively light loads occurs very often in practice.

Problem 5-19-A*.   A simple beam of A36 steel has a span of 16 ft and supports a total uniform load of 10 kips. The beam is laterally supported throughout its length and its deflection must not exceed ¹⁄₃₆₀ of the span. Design the beam

Problem 5-19-B.   A simple beam of A36 steel has a span of 22 ft and carries a total uniform load of 14 kips. The beam is fully supported laterally and its deflection is limited to ¹⁄₃₆₀ of the span. Determine the lightest weight section that is acceptable.

5-20   Beam Safe Load Table Based on Deflection

To assist in determining the proper size of lightly loaded beams, Table 5-6, based on *maximum allowable deflections*, is presented. As noted earlier, Table 5-4, which gives allowable uniform loads for W and S shapes, is based on an allowable bending stress of 24 ksi. However, if a beam is to be designed with the provision that the deflection must not exceed ¹⁄₃₆₀ of the span, Table 5-4 is of no assistance *when the loads are relatively light.* Consequently, Table

TABLE 5-6. Beam Safe Load Table Based on Allowable Deflection*
Allowable Loads in kips

Shape	I in.⁴	S in.³	10	11	12	13	14	15	16	17	18	19	20	21	22	23	24	25	26
									Span in feet									
S 7 × 15.3	36.7	10.5	15.6	12·8	10.8	9.21	7.94	6.92	6.08	5.38	4.80	4.31	3.90	3.52	3.21	2.94	2.70	2.48	2.30
W 8 × 17	56.6	14.1	22.6	20.0	16.8	14.3	12.4	10.8	9.45	8.38	7.47	6.70	6.05	5.53	5.00	4.58	4.20	3.88	3.59
S 8 × 18.4	56.7	14.4	23.0	20.2	16.9	14.4	12.5	10.9	9.57	8.46	7.53	6.77	6.12	5.54	5.05	4.63	4.25	3.91	3.61
W 8 × 20	69.4	17.0	27.2	24.6	20.6	17.6	15.2	13.2	11.6	10.3	9.21	8.25	7.43	6.74	6.15	5.62	5.17	4.77	4.39
W 10 × 21	107	21.5	34.4	31.3	28.7	26.5	23.2	20.3	17.8	15.8	14.1	12.6	11.4	10.3	9.43	8.65	7.91	7.32	6.75
S 10 × 25.4	124	24.7	39.5	35.9	32.9	30.4	26.8	23.4	20.5	18.2	16.2	14.5	13.1	11.9	10.8	9.91	9.13	8.41	7.76
W 10 × 25	133	26.5	42.4	38.5	35.3	32.6	29.2	25.4	22.4	19.9	17.7	15.9	14.3	13.0	11.8	10.8	9.95	9.18	8.48
W 10 × 29	158	30.8	49.3	44.8	41.4	37.9	34.4	30.0	26.4	23.4	20.8	18.7	16.9	15.3	14.0	12.8	11.7	10.8	10.1
W 12 × 27	204	34.2	54.7	49·7	45.6	42.1	39.1	36.5	34.2	30.4	27.1	24.2	21.9	19.9	18.1	16.6	15.2	14.0	13.0
S 12 × 31.8	218	36.4	58	53	49	45	42	39	36	32.0	28.6	25.6	23.1	20.9	19.1	17.5	16.0	14.8	13.7
W 12 × 31	239	39.5	63.2	57.5	52.7	48.6	45.1	42.1	39.5	35.4	31.6	28.4	25.6	23.2	21.2	19.4	17.8	16.4	15.1
W 12 × 36	281	46.0	73.6	66.9	61.3	56.6	52.6	49.1	46.0	41.7	37.2	33.4	30.1	27.4	24.4	22.8	20.9	19.2	17.8
W 14 × 30	290	41.9	66	60	55	51	47	44	41	39	37	34.4	31.1	28.2	25.7	23.5	21.6	19.8	18.3
W 14 × 34	340	48.6	78	71	65	60	56	52	49	46	43	40.4	36.4	33.0	30.1	27.6	25.2	23.3	21.5
W 14 × 38	386	54.7	88	80	73	67	63	58	55	51	49	45.8	41.3	37.5	34.2	31.3	28.2	26.4	24.4
S 15 × 42.9	447	59.6	95	87	79	73	68	64	60	56	53	50	48	43.1	39.2	35.9	33.0	30.4	28.1
W 16 × 36	447	56.5	90	82	75	70	65	60	57	53	50	48	45	43	39.6	36.2	33.2	30.6	28.4
W 16 × 40	517	64.6	103	94	86	80	74	69	65	61	57	54	52	49	45.8	41.9	38.4	35.5	32.8

* Loads to the *right* of the heavy vertical lines will produce deflections equal to ½₆₀ of the span. Loads to the *left* of these lines are based on an extreme fiber stress of 24 ksi and produce deflections less than ½₆₀ of the span. See Art. 5-20.

5-6 has been prepared for use with this type of problem. The magnitudes of the loads *to the right* of the heavy vertical lines in this table are computed by the formula

$$W_D = \frac{43,000 \times I}{L^2}$$

in which W_D = the uniformly distributed load that will result in a
deflection of $\frac{1}{360}$ of the span in pounds,
I = moment of inertia of the beam in inches^4,
L = span length in *feet*.

These loads are in units of kips; they will result in deflections of $\frac{1}{360}$ of the span, and their actual extreme fiber stress is less than 24 ksi. Note that the beam sections are arranged in the increasing order of their moments of inertia.

The loads in the table *to the left of the heavy vertical lines* will result in deflections of less than $\frac{1}{360}$ of the span; the actual extreme fiber stresses for these loads is 24 ksi. They are the same loads that are given in Table 5-4 for compact beams laterally supported throughout their lengths.

The use of Table 5-6 avoids the necessity of computing the minimum required moment of inertia for relatively light loads, as explained in Art. 5-19. To realize fully the value of Table 5-6, the reader should examine Art. 5-19 again.

Example.   A simple beam of A36 steel has a span of 20 ft and a uniformly distributed load of 30 kips. Full lateral support is provided and deflection is limited to $\frac{1}{360}$ of the span. Determine the size of the beam by use of Table 5-6.
Solution: (1) Scanning the loads listed for a 20-ft span, we see that a W 12 × 36 will support 30.1 kips. Its deflection will be $\frac{1}{360}$ of the span. This section is acceptable but we observe that a W 14 × 30 is also adequate; the load for this beam that will produce a deflection of $\frac{1}{360}$ of the span is 31.1 kips. It is preferable because it is lighter in weight than the W 12 × 36.

(2) Referring to Table 5-4, we note that the W 14 × 30 will carry 33 kips on a 20-ft span but that its deflection will be greater than the allowable value. There is no way by which we can tell accurately

from Table 5-4 whether the deflection of this beam under a 30-kip load would be within the prescribed limit.

Problems 5-20-A-B-C-D*-E. By the use of Table 5-6, determine the lightest weight beams for the loads and spans listed below. These are simple beams of A36 steel with uniformly distributed loads, and are laterally supported throughout their lengths. The deflection of each beam is limited to ⅟₃₆₀ of its span.

(a) Load = 19 kips,   span = 17 ft
(b) Load = 20 kips,   span = 18 ft
(c) Load = 30 kips,   span = 22 ft
(d) Load = 20 kips,   span = 21 ft
(e) Load = 20 kips,   span = 16 ft

Problems 5-20-F*. An S 10 × 25.4 has a span of 18 ft 6 in. It is fully supported laterally. Comute the uniformly distributed load for this beam that will produce a deflection of ⅟₃₆₀ of its span.

5-21 Provision for Beam Weight

In the problems and examples previously discussed, the given uniformly distributed loads have included the weight of the beam. This was done to simplify the explanations. In practice, however, the weight of the beam must be considered as a separate load. One method is to make an estimate of the beam weight, and fireproofing if required, and add this to the weight of the floor construction (dead load) and the load due to occupancy (live load). The other method is to neglect the weight of the beam in the computations and to select a beam with a section modulus slightly larger than that required by the combined live and dead loads. The latter method is used in the example given below.

In many types of construction the structural steel must be covered with a protecting material to serve as *fireproofing*. Among the various materials used for this purpose is concrete, and Fig. 5-4 shows the minimum dimensions often required for this covering. In this figure d and b represent the depth of beam and width of flange in inches. The fireproofing (not including the slab) is indicated by the hatched area; it is $[d \times (b + 3)]$ sq in. The number of cubic feet of fireproofing *per linear foot of beam* is $[d \times (b + 3)]/144$. Taking the

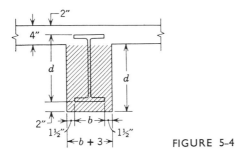

FIGURE 5-4

weight of unreinforced concrete as 144 lb per cu ft, the weight of fireproofing per linear foot of beam becomes

$$W_{FP} = \frac{d \times (b + 3)}{144} \times 144 = [d \times (b + 3)] \text{ lb per lin ft}$$

For obvious reasons this is an approximation, but this expression is sufficiently accurate to make a preliminary allowance for the weight of the concrete fireproofing in the design of a beam. Table 5-7, based on this formula, gives the weight of concrete fireproofing for a number of beam sections. When the beam section has finally been determined, the true weight of the beam and its fireproofing can be computed to see that an adequate weight allowance has been provided for.

TABLE 5-7. Approximate Weight of Concrete Fireproofing for Beams, in Pounds per Linear Foot

Wide flange beams		Standard I-beams	
Section	Weight	Section	Weight
W 8 × 17	66	S 6 × 12.5	38
W 10 × 21	86	S 7 × 15.3	47
W 12 × 27	114	S 8 × 18.4	56
W 14 × 30	136	S 10 × 25.4	77
W 16 × 36	159	S 12 × 31.8	96
W 18 × 50	189	S 15 × 42.9	128
W 21 × 62	236	S 18 × 54.7	162
W 24 × 76	287	S 20 × 65.4	185
W 27 × 94	350	S 24 × 79.9	240

Example.   The span length of a simple beam is 16 ft and the uniformly distributed load, *not including the beam weight*, is 38 kips. If the beam is laterally supported throughout its length, determine the size of the beam with respect to bending stresses. A36 steel is to be used, and the fireproofing must conform to the arrangement shown in Fig. 5-4.

Solution: (1) For the present we will neglect the weight of the beam and its fireproofing. The bending moment due to the applied load is

$$M = \frac{WL}{8} = \frac{38 \times 16}{8} = 76 \text{ kip-ft}$$

and this requires a section modulus of

$$S = \frac{M}{F_b} = \frac{76 \times 12}{24} = 38 \text{ in.}^3$$

(2) Referring to Table 5-4, select a W 14 × 30 with an S of 41.9 in.³, slightly larger than the minimum 38 in.³ required. Note that this excess section modulus is $41.9 - 38 = 3.9$ in.³.

(3) The fireproofing for a W 14 × 30 weights approximately 136 lb per lin ft (Table 5-7). Then, because the beam weighs 30 lb, the uniformly distributed load due to the weight of the beam and its fireproofing is $30 + 136 = 166$ lb per lin ft. The bending moment is

$$M = \frac{wL^2}{8} = \frac{166 \times 16 \times 16}{8} = 5310 \text{ ft-lb}$$

and the section modulus required is

$$S = \frac{M}{F_b} = \frac{5310 \times 12}{24,000} = 2.66 \text{ in.}^3$$

Because this value of 2.66 is less than the excess of 3.9 provided, the W 14 × 30 is accepted.

(4) It should be noted that the reactions the accepted beam transmits to its supports are each equal to half the total load or

$$R_1 = R_2 = \frac{38 + (0.166 \times 16)}{2} = \frac{38 + 2.66}{2} = 20.3 \text{ kips}$$

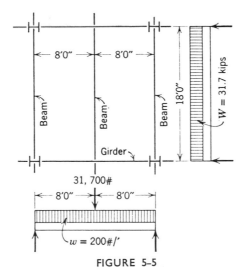

31, 700#

FIGURE 5-5

5-22   Floor Framing

Figures 5-5 and 5-6 illustrate two common layouts of framing when one of the shorter span floor systems is used, such as the solid reinforced concrete slab (Fig. 5-4) or certain of the ribbed steel deck systems. In both of these layouts, the floor load is carried directly by the beams as uniformly distributed loads; the loads on the girders are those that are transferred from the beams, and are received as concentrated loads.

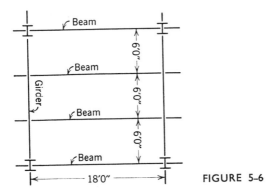

FIGURE 5-6

The total load on the framing is composed of the *dead load* and *live load*. Dead load consists of the weight of the construction: walls, partitions, columns, floor systems, etc. Table 5-8 gives the weights of various building materials for use in determining the dead load. The live load to be used in computations depends upon the use or occupancy of the building. Building codes of cities differ in their requirements for minimum live loads. Table 5-9 is a compilation of live load requirements assembled from a number of building codes.

In office buildings and certain other building types the partition layout may not be fixed, but erected or moved from one position to another to satisfy the space requirements of a specific occupant. In order to provide for this flexibility, it is customary to require that an allowance of 15 to 20 pounds per square foot (psf) be added to the design load to cover movable partitions. Since this requirement appears under dead loads in some codes and under live loads in others, care must be exercised to determine whether such a provision is mandatory. An example of the design of a typical bay (area between four columns) of floor framing follows.

Example.   Design the steel framing of A36 steel for the floor bay shown in Fig. 5-5. The floor construction consists of a 5-in. reinforced concrete slab, 2 in. of cinder concrete fill, ⅞-in. wood underflooring, and ⅞-in. finished flooring. An 18 psf allowance for movable partitions is to be made and the live load is 100 psf. The beams are laterally supported throughout their lengths by the concrete floor slab which is arranged in a manner similar to that shown in Fig. 5-4.

Solution for Beam Design: (1) The design load is computed with the aid of Table 5-8.

> *Dead load*
>
> | | | |
> |---|---|---|
> | 5-in. concrete slab | = | 60 psf |
> | 2-in. cinder concrete fill | = | 16 |
> | underflooring | = | 3 |
> | finished flooring | = | 3 |
> | movable partitions | = | 18 |
> | *Live load* | | = 100 |
> | *Total design load* | | = 200 psf of floor area |

TABLE 5-8. Weights of Building Materials

Floors	Pounds per square foot
Board flooring, per inch of thickness	3
Granolithic flooring, per inch of thickness	12
Floor tile, per inch of thickness	10
Asphalt mastic, per inch of thickness	12
Wood block, per inch of thickness	4
Cinder-concrete fill, per inch of thickness	8
Stone-concrete slab, per inch of thickness	12
Slag-concrete slab, per inch of thickness	10
Ceiling, suspended, metal lath and plaster	10
Ceiling, pressed steel	2

Roofs	Pounds per square foot
Three-ply roofing felt and gravel	5½
Five-ply roofing felt and gravel	6½
Roofing tile, cement	15 to 20
Roofing tile, clay, shingle type	12 to 14
Roofing tile, Spanish	8 to 10
Slate, ¼ in. thick	9½
Slate, ⅜ in. thick	12 to 14½
2 in. Book tile	12
Sheathing, wood, 1 in. thick	3
Skylight, ⅜ in. glass in galvanized iron frame	7½

Walls and partitions	Pounds per square foot
8 in. Brick wall	80
12 in. Brick wall	120
17 in. Brick wall	160
4 in. Brick, 8 in. tile backing	75
9 in. Brick, 4 in. tile backing	100
8 in. Wall tile	35
12 in. Wall tile	45
3 in. Clay-tile partition	18
4 in. Clay-tile partition	19
6 in. Clay-tile partition	25
4 in. Glass-block	18
3 in. Gypsum-block partition	11
4 in. Gypsum-block partition	13
2 in. Solid plaster partition	20
4 in. Stud partition, plastered both sides	20
Steel sash, glazed	10

Masonry	Pounds per cubic foot
Ashlar masonry, granite	165
Ashlar masonry, limestone	160
Ashlar masonry, sandstone	140
Brick masonry, common	120
Brick masonry, pressed	140
Concrete, plain stone	145
Concrete, reinforced stone	150
Concrete, cinder	110
Rubble masonry, limestone	150
Rubble masonry, sandstone	130

TABLE 5-9. Minimum Live Loads

Occupancy or use	Live load, pounds per square foot
Apartments	
Private suites	40
Corridors	100
Rooms for assembly	100
Buildings for public assembly	
Corridors	100
Rooms with fixed seats	60
Rooms with movable seats	100
Dwellings	40
Factories	125
Garages	100
Hotels	
Private rooms	40
Public rooms	100
Office buildings	
Offices	80
Public spaces	100
Restaurants	100
Schools	
Assembly rooms	100
Classrooms with fixed seats	40
Classrooms with movable seats	80
Corridors	100
Stairways and firetowers	100
Stores	
First floors	125
Upper floors	75
Theatres	
Corridors, aisles, and lobbies	100
Fixed seats areas	60
Stage	150

(2) Because the beams are 8 ft apart, the floor load brought to each linear foot of beam is $200 \times 8 = 1600$ lb. We estimate the weight of the beam and its fireproofing to be 160 lb per lin ft, making a total of $1600 + 160 = 1760$ lb per lin ft. Then, the total load on one beam is $W = 1760 \times 18 = 31,700$ lb or 31.7 kips. The load diagram for the beam may now be constructed as shown at the right of Fig. 5-5.

(3) To design the beam for bending, we note that the maximum bending moment for this loading is $M = WL/8$ (Case 2, Table 3-1) or

$$M = \frac{WL}{8} = \frac{31.7 \times 18}{8} = 71.3 \text{ kip-ft}$$

Since the beam is supported laterally, $F_b = 24$ ksi and

$$S = \frac{M}{F_b} = \frac{71.3 \times 12}{24} = 35.6 \text{ in.}^3$$

(4) Table 5-4 is of great convenience at this point. We note that an S 12×31.8 ($S = 36.4$) will support a uniform load of 32 kips on an 18-ft span, but with an excessive deflection. A W 14×30, on the other hand, has a section modulus of 41.9 in.³ (35.6 in.³ is required), and it will support a uniform load of 37 kips for the same span without excessive deflection. Hence, we accept temporarily a W 14×30.

(5) Checking the shear, we find from Table 4-1 that the depth of a W 14×30 is 13.86 in. and the web thickness is 0.27 in. Then, because we know that $V = 31.7 \div 2 = 15.85$ kips,

$$f_v = \frac{V}{d \times t_w} = \frac{15.85}{13.86 \times 0.27} = 4.2 \text{ ksi}$$

Because this value does not exceed the allowable shearing stress $F_v = 14.5$ ksi (Table 5-2), the W 14×30 is acceptable for shear.

(6) Checking deflection, we have already observed [Step (4)] that the W 14×30 is satisfactory from its position in Table 5-4; hence, it is not necessary to solve the deflection equations. Since this section is acceptable for bending, shear, and deflection, use the W 14×30.

Solution for Girder Design: (1) Figure 5-5 shows that each girder receives a concentrated load at the center of its span where the floor beams frame into it. The total load on one of the beams was found to be 31.7 kips, the reaction at each end of the beam being 15.85 kips. However, the concentrated load at the center of the girder span is the sum of the reactions of the beams framing into it on *both* sides. Since adjacent floor bays are assumed to be identical with the one under consideration, the concentrated load brought to the girder is 2 × 15.85 = 31.7 kips. The only uniformly distributed load on the girder will be its own weight plus the weight of the fireproofing, which we will estimate as 200 lb per lin ft. The load diagram for the girder may now be drawn as shown at the bottom of Fig. 5-5.

(2) To design the girder for bending, we note that the total load is 31.7 kips plus the estimated weight (200 × 16 = 3200 lb) or 31.7 + 3.2 = 34.9 kips. Because the girder is symmetrically loaded, $R_1 = R_2 = 34.9 \div 2 = 17.45$ kips. The maximum bending moment occurs at the center of the span and is equal to

$$M = (17.45 \times 8) - (200 \times 8 \times 4) = 133 \text{ kip-ft}$$

(3) The girder is laterally supported by the beams at the center of the span; therefore 8 ft is the laterally unsupported length of span. The size of the girder has not been determined, but we will assume that 8 ft will probably exceed L_c for the beam we select. This means that we will employ a bending stress of $F_b = 22$ ksi instead of 24 ksi. (Refer to Art. 5-9.) When the required size has been determined, we can see whether our assumption is correct. Then

$$S = \frac{M}{F_b} = \frac{133 \times 12}{22} = 72.5 \text{ in.}^3$$

which is the minimum required section modulus. Refer to Table 5-4 and select a W 16 × 45 as a trial section ($S = 72.5$).

(4) To investigate the shear, we note from Table 4-1 that this section has a depth of 16.12 in. and a web thickness of 0.346 in. The maximum vertical shear = $R_1 = R_2 = 17.45$ kips.

$$f_v = \frac{V}{d \times t_w} = \frac{17.45}{16.12 \times 0.346} = 3.12 \text{ ksi}$$

Since this value of the unit shearing stress is less than $F_v = 14.5$ ksi (Table 5-2), the W 16 × 45 is acceptable for shear.

(5) Investigating for deflection, we note that this loading is not one of the typical cases shown in Table 3-1; hence, there is no single formula that may be used to determine the deflection. However, we observe from the girder load diagram in Fig. 5-5 that the maximum deflections produced by both the uniform load and the concentrated load will occur at the center of the span. Therefore we can compute each deflection separately and their sum will be the total deflection. For the concentrated load (Case 1, Table 3-1),

$$D = \frac{Pl^3}{48EI} = \frac{31.7 \times (16 \times 12)^3}{48 \times 29{,}000 \times 584} = 0.276 \text{ in.}$$

and for the uniform load (Case 2, Table 3-1),

$$D = \frac{5Wl^3}{384EI} = \frac{5 \times 3.2 \times (16 \times 12)^3}{384 \times 29{,}000 \times 584} = 0.174 \text{ in.}$$

Therefore, the total deflection at the center of the span is 0.276 + 0.174 = 0.45 in. Since this value is less than $\frac{1}{360}$ of the span (0.53 in.), the deflection is not excessive.

(6) Returning to Table 5-4, we see that $L_c = 7.4$ ft and $L_u = 11.4$ ft for a W 16 × 45. Consequently, the 22 ksi allowable bending stress used for determining the girder size for bending [Step (3)] was correct. Note also that the 200 lb per lin ft estimated for the weight of the girder and its fireproofing is slightly more than its true weight. Since this section is acceptable for bending, shear, and deflection, use the W 16 × 45.

Problem 5-22-A. A floor panel in a steel-frame building has columns spaced 18 ft in both directions. The beams are spaced 6 ft on centers and they frame into the girders at the third points of the span as shown in Fig. 5-6. Adjacent bays are identical to the one shown. The floor construction consists of a 4½-in. reinforced concrete slab on which are placed 2-in. wood blocks. The concrete slab provides lateral support for the beams, and the fireproofing is substantially the same as that indicated in Fig. Fig. 5-4. There is a suspended ceiling of metal lath and plaster and an 18-psf allowance is to be made for movable partitions. Design both the beams and the girders of A36 steel, the live load being 60 psf and the deflection limited to $\frac{1}{360}$ of the span.

5-23 Bearing Plates for Beams

When a steel beam is supported by a masonry wall or pier, it is usually necessary to provide a steel bearing plate to distribute the beam reaction over an ample bearing area. The plate also helps to seat the beam at its proper elevation. It is assumed that the plate provides a uniform distribution of the beam reaction over the area of contact with the supporting masonry.

Referring to Fig. 5-7a, we see that the area of the bearing plate is $B \times N$. It is found by dividing the beam reaction by F_p, the allowable bearing pressure on the masonry. Thus

$$A = \frac{R}{F_p}$$

in which $A = B \times N$, the area of the plate in square inches,

R = reaction of beam in pounds or kips,

F_p = allowable bearing pressure on the masonry in psi or ksi (see Table 5-10).

The thickness of the wall generally determines N, the bearing length or dimension of the plate parallel to the length of the beam; B is the dimension of the plate parallel to the length of the wall. The plate dimensions B and N are usually in even inches, and a great variety of thicknesses is available. Usually the beam flange is neither welded

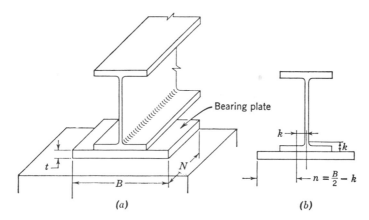

Bearing plate

k

$\frac{1}{2}k$

$n = \frac{B}{2} - k$

t

B

N

(a)

(b)

FIGURE 5-7

TABLE 5-10.   Allowable Bearing Pressure on Masonry Walls in psi

Stone concrete, depending on quality	600 or 800
Common brick, lime mortar	100
Common brick, lime-cement mortar	200
Common brick, cement mortar	250
Hard brick, cement mortar	300
Rubble, cement mortar	150
Rubble, lime-cement mortar	100
Hollow T. C. blocks, cement mortar	80
Hollow cinder blocks, cement mortar	80

nor riveted to the bearing plate, but the plate is shipped separately and grouted in place before erection of the beam.

The thickness of the plate is determined by considering the projection n (Fig. 5-7b) as an inverted cantilever, with the uniform bearing pressure on the bottom of the plate tending to curl it upward about the beam flange. The required thickness may be computed readily by the formula given below which does not involve direct computation of bending moment and section modulus:[6]

$$t = \sqrt{\frac{3f_p n^2}{F_b}}$$

in which t = thickness of plate in inches,

f_p = *actual* bearing pressure of the plate on the masonry psi or ksi,

F_b = allowable bending stress in the plate (the AISC Specification gives the value of F_b as $0.75F_y$; for A36 steel $F_y = 36$ ksi, therefore $F_b = 0.75 \times 36 = 27$ ksi,

$n = \dfrac{B}{2} - k$ in inches (see Fig. 5-7b),

k = distance from bottom of beam to web toe of fillet in inches (values of k for various beam sizes may be found in the AISC Manual "Dimensions for Detailing" tables).

[6] For the derivation of this formula see *Simplified Design of Structural Steel*, by Harry Parker, Fourth Edition (New York: Wiley, 1974), Art 8-14.

Example. An S 15 × 42.9 (k = 1.375 in.) has an end reaction of 31 kips. It is to be supported on a masonry wall laid up with common brick in cement mortar. If the width of the plate parallel to the length of the beam is 8 in., design a bearing plate made of A36 steel.

Solution: (1) Referring to Table 5-10, the allowable bearing pressure on this type of masonry is found to be 250 psi (0.25 ksi). The required area of the plate is

$$A = \frac{R}{F_p} = \frac{31}{0.25} = 124 \text{ sq in.}$$

(2) By data, $N = 8$ in. so $B = 124 \div 8 = 15.5$ in. Accept an 8 × 16 in. plate. By data also, $k = 1.375$ in. Then

$$n = \frac{B}{2} - k = \frac{16}{2} - 1.375 = 6.63 \text{ in.}$$

(3) The actual bearing pressure on the masonry wall is equal to the reaction divided by the area of the plate supplied, or f_p = 31 ÷ (8 × 16) = 0.242 ksi. Then

$$t = \sqrt{\frac{3f_p n^2}{F_b}} = \sqrt{\frac{3 \times 0.242 \times 6.63^2}{27}}$$

and

$$t = \sqrt{1.18} = 1.09 \text{ in.}$$

Working to the next greatest $\frac{1}{16}$-in. increment for plate thickness, we accept a plate 8 × 10 × 1⅛ in.

Problem 5-23-A*. Design a bearing plate for a W 14 × 38 (k = 1⅛ in.) with a reaction of 45 kips. The beam is supported on a wall of common brick laid in cement mortar and the bearing length (dimension N) is limited to 12 in.

Problem 5-23-B. A wall of concrete with an allowable bearing pressure of 500 psi supports an S 18 × 54.7 (k = 1¼ in.) with a reaction of 52 kips. Design the bearing plate if the bearing length is limited to 8 in. (N)

Problem 5-23-C. A masonry wall laid up with common brick in lime-cement mortar supports one end of an S 12 × 31.8 (k = 1³⁄₁₆ in.) on which there is a total uniformly distributed load of 57 kips. Compute the size of the bearing plate if the bearing length is limited to 12 in.

5-24 Crippling of Beam Webs

An excessive end reaction on a beam, or an excessive concentrated load at some point along the interior of the span, may cause crippling or buckling of the web. The AISC Specification requires that end reactions or concentrated loads for beams without stiffeners shall not exceed the following (Fig. 5-8):

$$\text{maximum end reaction} = 0.75F_y t(N + k)$$

$$\text{maximum concentrated load} = 0.75F_y t(N + 2k)$$

in which t = thickness of beam web in inches,

N = length of bearing or length of concentrated load (not less than k for end reactions) in inches,

k = distance from outer face of flange to web toe of fillet in inches (see Fig. 5-7b),

$0.75F_y$ = 27 ksi for A36 steel.

By substituting 27 ksi for $0.75F_y$ in the foregoing expressions, we can write

$$\text{maximum end reaction} = 27 \times t \times (N + k)$$

$$\text{maximum concentrated load} = 27 \times t \times (N + 2k)$$

When these values are exceeded, the webs of the beams should be reinforced with stiffeners or the length of bearing increased.

Example. Consider the S 15 × 42.9 given in the example of Art. 5-23. Is this beam safe with respect to web crippling at the end support? The reaction transmitted by this beam was 31 kips, k = 1.375 in., and the bearing length = 8 in.

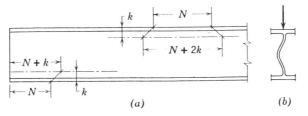

(a)   *(b)*

FIGURE 5-8

Solution: To investigate this beam for web crippling, we compute the allowable end reaction and compare this value with the 31 kips actually developed by the loading. From Table 4-2 we find that the web thickness of an S 15 × 42.9 is 0.411 in. Then

$$R_{(allowable)} = 27 \times t \times (N + k)$$
$$= 27 \times 0.411 \times (8 + 1.375) = 104 \text{ kips}$$

Because the beam reaction of 31 kips is below this allowable value, the end bearing is safe with respect to web crippling.

Problem 5-24-A*.   A W 12 × 40 ($k = 1\frac{1}{4}$ in.) has a load of 40 kips placed on its top flange at some point between the supports. The bearing length of the load is 8 in. Are stiffeners needed to prevent web crippling?

Problem 5-24-B.   The bearing length of the end reaction of a W 10 × 25 is $3\frac{1}{2}$ in. and the k dimension of this section is 1 in. Compute the magnitude of the maximum allowable end reaction with respect to web crippling.

5-25   Open Web Steel Joists

Open web steel joists are shop-fabricated, parallel chord trusses used in buildings for the direct support of floors and roof decks between main supporting beams, girders, trusses, or walls (Fig. 5-9). Floor and roof decks may consist of cast-in-place or precast concrete or gypsum, formed steel, or wood. The usual maximum spacing of joists is 24 in., but the spacing must not exceed the safe span of the wood deck or other material that is used over them.

When a wood deck is placed over the steel joists, it should be fastened to the joists by wood nailing strips (nailers) attached to the top chords of the joists. Cast-in-place concrete slabs reinforced with ribbed metal lath should not be less than 2 in. thick. The attachment of the slab or deck to the top chords of the joists must be so detailed as to brace the top chord against lateral buckling.

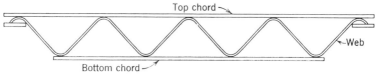

Top chord

Web

Bottom chord

FIGURE 5-9

Open web joists are manufactured in six different series but the discussion here will be limited to the J-Series, which is fabricated from A36 steel and is assigned an allowable bending stress of 22,000 psi by the Steel Joist Institute's 1965 *Standard Specifications for Open Web Steel Joists J-Series and H-Series.*[7] Information concerning the makeup and properties of specific joint types, safe load tables, and construction details are contained in the catalogs of the various manufacturers. (See also Sweet's Architectural Catalog File.)

Table 5-11 lists the safe uniformly distributed loads in pounds per linear foot of the J-Series joists. To identify the joists, the numerals that precede the letters give the nominal depth of the joist in inches the letter or letters indicate the series, and the numerals on the right designate the chord sections. The J-Series joists are made in standard depths of 8 to 24 in., in increments of 2 in., and their lengths accomodate spans up to 48 ft. The loads given in Table 5-11 are total loads: to find the *live* loads, the *dead* loads, including the weight of the joists, must be deducted.

The following example illustrates procedures that may be used in determining the size and spacing of open web steel joists.

Example. It is proposed to use open web steel joists to support a floor on which there is a live load of 100 psf. The floor construction consists of a 2-in. reinforced concrete slab, 1 in. of wood subflooring, and 1 in. of finished flooring. A ceiling of metal lath and plaster is suspended from the lower chords of the joists. The span of the joists is 16 ft and they will be spaced 18 in. on centers. What size open web steel joists should be used?

Solution: (1) Compute the design load on the floor using Table 5-8 for weights of materials.

Dead load	1-in. wood subflooring	=	3 psf
	1-in. wood finished flooring	=	3
	2-in. reinforced concrete slab	=	24
	suspended plaster ceiling	=	10
Live load			= 100
Total floor load			= 140 psf

---

[7] The chord sections of H-series joists are fabricated from steel having a yield point of 50,000 psi, and a bending stress of 30,000 psi is permitted.

TABLE 5–11. Standard Load Table for Open Web Joists, J-Series† Allowable Total Safe Loads in Pounds Per Linear Foot, Based on Allowable Stress of 22,000 psi

Joist designation	8J3	10J3	10J4	12J3	12J4	12J5	12J6	14J3	14J4	14J5	14J6	14J7	16J4	16J5	16J6	16J7	16J8
*Depth in inches	8	10	10	12	12	12	12	14	14	14	14	14	16	16	16	16	16
Resisting moment in inch kips	70	89	111	108	135	161	196	127	159	190	230	276	173	216	258	310	359
Max. end reaction in pounds	2000	2200	2400	2300	2500	2700	3000	2400	2800	3100	3400	3700	3000	3300	3600	4000	4300
**Approx. joist wgt. pounds per foot	4.8	4.8	6.0	5.1	6.0	7.0	8.1	5.2	6.4	7.3	8.4	9.7	6.6	7.6	8.5	10.1	11.3
Span in feet																	
8	400	440	480														
9	364	400	436														
10	324	367	400	383	417	450	500										
11	276	338	369	354	385	415	462										
12	238	303	343	329	357	386	429	343	400	443	486	529					
13	207	264	320	307	333	360	400	320	373	413	453	493					
14	182	232	289	281	313	338	375	300	350	388	425	463	375	413	450	500	538
15	161	205	256	249	294	318	353	282	329	365	400	435	353	388	424	471	506
16	144	183	228	222	278	300	333	261	311	344	378	411	333	367	400	444	478
17	129	164	205	199	249	284	316	235	294	326	358	389	316	347	379	421	453
18	117	148	185	180	225	268	300	212	265	310	340	370	288	330	360	400	430
19				163	204	243	286	192	240	287	324	352	262	314	343	381	410
20				149	186	222	270	175	219	262	309	336	238	298	327	364	391
21				136	170	203	247	160	200	239	290	322	218	272	313	348	374
22				125	156	186	227	147	184	220	266	308	200	250	299	333	358
23								135	170	203	245	294	185	230	275	320	344
24								125	157	187	227	272	171	213	254	306	331
25								116	145	174	210	252	158	198	236	283	319
26								108	135	162	196	235	147	184	219	264	305
27													137	171	205	246	285
28													128	160	191	230	266
29													120	150	179	215	249
30													113	141	168	202	234
31																	
32																	

Joist designation	18J5	18J6	18J7	18J8	20J5	20J6	20J7	20J8	22J6	22J7	22J8	24J6	24J7	24J8
* Depth in inches	18	18	18	18	20	20	20	20	22	22	22	24	24	24
Resisting moment in inch kips	243	293	352	406	265	316	382	455	335	420	493	367	460	540
Max. end reaction in pounds	3500	3900	4200	4500	3800	4100	4300	4600	4200	4500	4800	4400	4700	5000
** Approx. joist wgt. pounds per foot	7.9	9.0	10.2	11.3	8.1	9.2	10.6	11.9	9.6	10.5	11.9	9.9	11.1	12.4
Span in feet														
16														
17														
18	389	433	467	500										
19	368	411	442	474										
20	350	390	420	450	380	410	430	460						
21	333	371	400	429	362	390	410	438	382	409	436			
22	318	355	382	409	345	373	391	418	365	391	417			
23	304	339	365	391	330	357	374	400	350	375	400			
24	281	325	350	375	307	342	358	383	336	360	384	367	392	417
25	259	312	336	360	285	328	344	368	323	346	369	352	376	400
26	240	289	323	346	261	312	331	354	306	333	356	338	362	385
27	222	268	311	333	242	289	319	341	285	321	343	326	348	370
28	207	249	299	321	225	269	307	329	266	310	331	312	336	357
29	193	232	279	310	210	250	297	317	248	300	320	291	324	345
30	180	217	261	300	196	234	283	307	232	290	310	272	313	333
31	169	203	244	282	184	219	265	297	218	273	300	255	303	323
32	158	191	229	264	173	206	249	288	205	257	291	239	294	313
33	149	179	215	249	162	193	234	279	193	242	282	225	282	303
34	140	169	203	234	155	182	220	262	182	229	268	212	265	294
35	132	159	192	221	144	172	208	248	172	216	254	200	250	286
36	125	151	181	209	136	163	197	234	163	205	240	189	237	278
37					129	154	186	222	155	194	228	179	224	263
38					122	146	176	210	147	184	216	169	212	249
39					116	139	167	199	140	175	205	161	202	237
40					110	132	159	190	133	167	196	153	192	225
41									127	159	186	146	182	214
42									121	151	178	139	174	204
43									115	145	170	132	166	195
44												126	158	186
45												121	151	178
46												116	145	176
47												111	139	163
48												106	133	156

Loads above heavy stepped lines are governed by shear.
† Copyright 1965, Steel Joist Institute. Reprinted by permission.
* Indicates nominal depth of steel joists only.
** Approximate weight per linear foot of steel joists only. Accessories and nailer strip not included.

145

(2) Because the joists are spaced 18 in. on centers, each linear foot of joist supports a floor area of 1 × 1.5 = 1.5 sq ft. Consequently, each linear foot of joist supports a load of 140 × 1.5 = 210 lb, exclusive of the weight of the joist. We estimate that the joist weighs 5 lb per lin ft, making 210 + 5 = 215 *lb lin ft*, the load to be carried by each joist.

(3) Referring to Table 5-11, we see that a 10J3 will support a load of 232 lb per lin ft on a span of 16 ft. Since this value is in excess of our 215 lb per lin ft required, the 10J3 is accepted. The weight of this joist is given as 4.8 lb per lin ft, so our allowance of 5 lb per lin ft was ample.

Alternate Solution: Since Table 5-11 gives the resisting moment for each open web joist, another method for selecting the proper joist size is to compute the maximum bending moment. Using the information developed in Step (2) above, the total uniformly distributed load on one joist is 215 × 16 = 3440 lb or 3.44 kips. Each joist acts as a simple beam so the maximum bending moment (Case 2, Table 3-1) is

$$M = \frac{WL}{8} = \frac{3.44 \times 16}{8} = 6.88 \text{ kip-ft or } 82.5 \text{ kip-in.}$$

From Table 5-11 we find that a 10J3 provides a resisting moment of 89 kip-in. which is greater than the bending moment of 82.5 kip-in. developed by the loading. Therefore, the 10J3 can be accepted at a spacing of 18 in. on centers.

Table 5-11 also gives the maximum end reaction for each joist size; for a 10J3 it is 2200 lb. We found the total load on one joist to be 3440 lb, making the end reaction 3440 ÷ 2 = 1720 lb. Since this value is less than the allowable maximum, the 10J3 is satisfactory by this test also.

Problem 5-25-A.   Open web steel joists have a span of 18 ft and support a live load of 80 psf. The floor consists of a 2-in. reinforced concrete slab covered with 2-in. wood block flooring. A metal lath and plaster ceiling is suspended from the lower chords of the joists. What is the size of the lightest weight joists required to support this load if their spacing is 24 in. on centers?

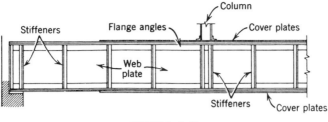

FIGURE 5-10

5-26   Plate Girders

In general, rolled beam sections are used for floor framing in building construction. It frequently happens, however, that rolled beams of adequate dimensions are not available for particularly long spans or for beams with unusually large loads. This necessitates the use of built-up sections called plate girders. The various parts of these girders are secured together by means of welding or riveting. Figure 5-10 indicates the general arrangement and elements of a riveted plate girder, and Fig. 5-11 shows the makeup of typical girder cross sections of both welded and riveted construction. Welding has largely replaced riveting in plate girder fabrication, although riveted construction is permitted by the AISC Specification.

The design of plate girders is beyond the scope of this book. However, the discussion of properties of built-up section in Art. 4-8 indicates the method of determining gross moment of inertia that forms the basis for proportioning girder cross sections to resist bending. The reader who wishes to pursue the design of plate girders

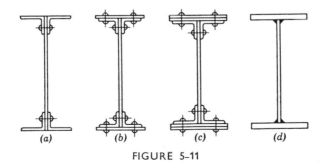

FIGURE 5-11

further should consult the 7th Edition of the AISC Manual. Part 2 of the Manual presents illustrative examples and contains extensive tables to facilitate design under the provisions of the 1969 AISC Specification.

5-27   Structural Steel Design Methods

There are two different methods employed in designing steel beams for bending stresses. The first, called *allowable stress design*, is used in this chapter and in the major portion of Section II of the book. It is the more conventional procedure based on the idea of using F_b, an allowable extreme fiber stress, as a certain fraction of the yield stress (Art. 5-5). Such allowable stresses fall below the elastic limit of the material, and we speak of them as conforming with the elastic behavior of the material. This approach to the design of steel members has been standard practice for many years.

The second method, known as *plastic design*, is a more recent development and is based on the idea of computing an ultimate load, and on utilizing a portion of the reserve strength (after initial yield stress has been reached) as part of the factor of safety (Art. 1-12). This method was introduced into the AISC Specification in 1963 and forms Part 2 of the 1969 Specification. A brief introduction to plastic design theory is given in Chapter 9.

6

Steel Columns

IIIIIIIIIIIIIIIIIIIIIIIIIIIIIIIIIIIIIIIIIIIIIIIIIIIIIIIIIIIII

6-1 Introduction

A column or strut is a compression member, the length of which is several times greater than its least lateral dimension. The term *column* usually applies to relatively heavy vertical members, while relatively light or inclined members carrying compressive stresses (such as braces and the compression members of roof trusses) are called struts.

In Art. 1-5, it was stated that the unit compressive stress in the short block shown in Fig. 1-1b could be determined by the direct stress formula $f_a = P/A$. However, it was pointed out that this expression became invalid as the length of a compression member increased. To understand why this is so, let us consider a small block of structural steel 1 in. by 1 in. in cross section and 1 or 2 in. high (Fig. 6-1a). If the allowable compressive stress is 20 ksi, the block will safely support an axial load P of 20 kips. However, if we consider a bar of the same cross section but with a length of, say, 20 in., we find that the value of P it will support is much smaller because of the tendency of this more slender bar to buckle or bend (Fig. 6-1b). Therefore, in columns the element of slenderness must be taken into account.

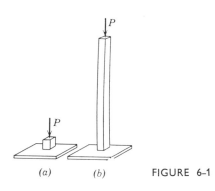

(a)   (b)   FIGURE 6-1

In timber columns the term *slenderness ratio* is defined as l/d, the unbraced length in inches divided by the dimension of the least side. In steel columns l/r is the slenderness ratio with l, as before, representing the unbraced length in inches and r being the least radius of gyration. A short column or block fails by crushing, but long slender columns fail by stresses that result from bending. In steel columns long enough to have a tendency to bend, the stresses are not uniformly distributed over the cross section and therefore the *average* unit stress must be less than 20 ksi. This average stress is dependent on the slenderness ratio, the end conditions, and the cross-sectional area of the column.

6-2 Column Sections

Because of the tendency to bend, the safe load on a column depends not only on the number of square inches in the cross section but also on the manner in which the material is distributed with respect to the axes of the section; that is, the *shape* of the column cross section is an important factor. An axially loaded column tends to bend in a plane perpendicular to the axis of the cross section about which the moment of inertia is the least. Since column cross sections are seldom symmetrical with respect to both major axes, the ideal section would be one in which the moment of inertia for each major axis is equal. Pipe columns and structural tubing meet this condition, but their use is somewhat limited because of difficulties in making beam connections. Of the two major axes of an I-beam (S shape), the moment of inertia about the axis parallel to the web is much the smaller; hence, for the amount of material in the cross section, S shapes are

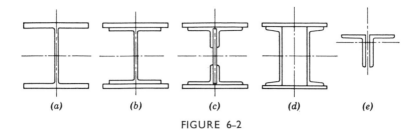

FIGURE 6-2

not economical when used as columns or struts. In the past, built-up sections such as Fig. 6-2c and d were used extensively, but wide flange sections (Fig. 6-2a) are now rolled in a large variety of sizes and are used universally because they require a minimum of fabrication. They are sometimes called H-columns. For excessive loads or unusual conditions, plates are welded or riveted to the flanges of W shapes to give added strength (Fig. 6-2b). The compression members of steel roof trusses are usually formed of two angles, as shown in Fig. 6-2e.

In steel frame buildings columns are generally in two-story lengths. In order not to conflict with the beam and girder connections, column splices, that is, the joints between column sections, are usually made about 2 ft above the floor level. In general, splices are made by means of plates bolted, welded, or riveted to the flanges of the columns, as indicated in Figs. 6-3 and 8-9. When the upper column has a smaller

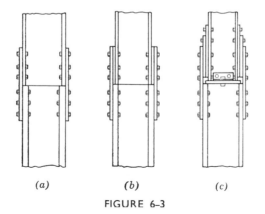

(a)   (b)   (c)

FIGURE 6-3

width than the supporting column, filler plates are used (Fig. 6-3*b*); if the difference in width is so great that full bearing between the columns is not achieved, a horizontal plate is used as in Fig. 6-3*c*.

6-3 Slenderness Ratio

In the design of steel columns it is important to remember that the *least* radius of gyration *r* is generally the one to be considered. As stated in Art. 4-9, radius of gyration is dependent on the area and shape of the cross section and its value is given by the expression $r = \sqrt{I/A}$. It is an index of the stiffness of a structural section when used as a column or strut, and hence is a measure of effectiveness in resisting buckling. For rolled sections, values of radii of gyration with respect to both major axes are given in tables of properties for designing, such as Tables 4-1 through 4-5. For built-up sections it may be necessary to compute its value. Note that in Tables 4-4 and 4-5 relating to angles, the least *r* is that with respect to axis *Z–Z*.

The slenderness ratio *l/r* for main compression members should not exceed 200. When determining the safe load on a column, one of the first steps is to compute its slenderness ratio for use in the controlling column formula. The term occurs in all steel column formulas used to determine the allowable unit compressive stress.

6-4 Effective Column Length

The AISC Specification requires that in addition to the unbraced length of a column, the effect of end conditions must be given consideration. The slenderness ratio is stated as *Kl/r*, in which *K* is a factor dependent on the restraint against rotation at the ends of a column and the means available to resist lateral motion (translation). Figure 6-4 shows diagrammatically six idealized conditions in which joint rotation and joint translation are illustrated. The term *K* is the ratio of the *effective column length* to the actual unbraced length. For average conditions in building construction the value of *K* is taken as 1 (Fig. 6-4*d*) so that the slenderness ratio *Kl/r* becomes simply *l/r*.

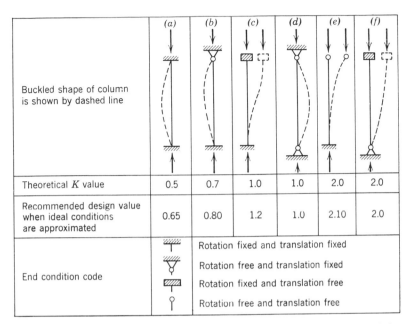

	(a)	(b)	(c)	(d)	(e)	(f)
Buckled shape of column is shown by dashed line						
Theoretical K value	0.5	0.7	1.0	1.0	2.0	2.0
Recommended design value when ideal conditions are approximated	0.65	0.80	1.2	1.0	2.10	2.0
End condition code		Rotation fixed and translation fixed				
		Rotation free and translation fixed				
		Rotation fixed and translation free				
		Rotation free and translation free				

FIGURE 6–4. Effect of end conditions. Reproduced from the 7th Edition of the *Manual of Steel Construction.* Courtesy American Institute of Steel Construction.

6-5 Column Formulas

The AISC Specification gives the following requirements for use in the design of compression members. The allowable unit stresses shall not exceed the following values:

On the gross section of axially loaded compression members when Kl/r, the largest effective slenderness ratio of any unbraced segment is less than C_c,

$$F_a = \frac{\left[1 - \dfrac{(Kl/r)^2}{2C_c^2}\right]F_y}{\text{F.S.}} \qquad \text{Formula (1)}$$

where

$$\text{F.S.} = \text{factor of safety} = \frac{5}{3} + \frac{3(Kl/r)}{8C_c} - \frac{(Kl/r)^3}{8C_c^3}$$

and

$$C_c = \sqrt{\frac{2\pi^2 E}{F_y}}$$

On the gross section of axially loaded columns when Kl/r exceeds C_c,

$$F_a = \frac{12\pi^2 E}{23(Kl/r)^2}$$ Formula (2)

On the gross section of axially loaded bracing and secondary members, when l/r exceeds 120 (for this case K is taken as unity),

$$F_{as} = \frac{F_a \,(\text{by Formula 1 or 2})}{1.6 - \dfrac{l}{200r}}$$ Formula (3)

In these formulas,

F_a = the axial compression stress permitted in the absence of bending stress in psi or ksi,

K = effective length factor (see Art. 6-4),

l = actual unbraced length in inches,

r = governing radius of gyration (usually the least) in inches,

$C_c = \sqrt{\dfrac{2\pi^2 E}{F_y}}$; for A36 steel, $C_c = 126.1$,

F_y = the minimum yield point of the steel being used (for A36 steel, $F_y = 36{,}000$) in psi,

F.S. = factor of safety (see above),

E = modulus of elasticity of structural steel, 29,000,000 psi,

F_{as} = the axial compressive stress, permitted in the absence of bending stress, for bracing and other secondary members, in psi or ksi.

To determine the allowable axial load that a main column will support, F_a, the allowable unit stress, is computed by Formula 1 or 2 and this stress is multiplied by the cross-sectional area of the column. If the column is a secondary member, or is used for bracing, Formula 3 gives the allowable unit stress; these allowable unit stresses are somewhat greater than those permitted for main members. Table 6-1 gives allowable stresses computed in accordance with these formulas. It should be examined carefully since it will be found to be of great assistance. Note particularly that this table is for use with A36 steel; tables based on other grades of steel are contained in the AISC Manual.

TABLE 6-1. Allowable Unit Stresses for Columns of A36 Steel, * in kips per square inch

Main and secondary members Kl/r not over 120						Main members Kl/r 121 to 200				Secondary members† l/r 121 to 200			
$\dfrac{Kl}{r}$	Fa (ksi)	$\dfrac{Kl}{r}$	Fa (ksi)	$\dfrac{Kl}{r}$	Fa (ksi)	$\dfrac{Kl}{r}$	Fa (ksi)	$\dfrac{Kl}{r}$	Fa (ksi)	$\dfrac{l}{r}$	Fas (ksi)	$\dfrac{l}{r}$	Fas (ksi)
1	21.56	41	19.11	81	15.24	121	10.14	161	5.76	121	10.19	161	7.25
2	21.52	42	19.03	82	15.13	122	9.99	162	5.69	122	10.09	162	7.20
3	21.48	43	18.95	83	15.02	123	9.85	163	5.62	123	10.00	163	7.16
4	21.44	44	18.86	84	14.90	124	9.70	164	5.55	124	9.90	164	7.12
5	21.39	45	18.78	85	14.79	125	9.55	165	5.49	125	9.80	165	7.08
6	21.35	46	18.70	86	14.67	126	9.41	166	5.42	126	9.70	166	7.04
7	21.30	47	18.61	87	14.56	127	9.26	167	5.35	127	9.59	167	7.00
8	21.25	48	18.53	88	14.44	128	9.11	168	5.29	128	9.49	168	6.96
9	21.21	49	18.44	89	14.32	129	8.97	169	5.23	129	9.40	169	6.93
10	21.16	50	18.35	90	14.20	130	8.84	170	5.17	130	9.30	170	6.89
11	21.10	51	18.26	91	14.09	131	8.70	171	5.11	131	9.21	171	6.85
12	21.05	52	18.17	92	13.97	132	8.57	172	5.05	132	9.12	172	6.82
13	21.00	53	18.08	93	13.84	133	8.44	173	4.99	133	9.03	173	6.79
14	20.95	54	17.99	94	13.72	134	8.32	174	4.93	134	8.94	174	6.76
15	20.89	55	17.90	95	13.60	135	8.19	175	4.88	135	8.86	175	6.73
16	20.83	56	17.81	96	13.48	136	8.07	176	4.82	136	8.78	176	6.70
17	20.78	57	17.71	97	13.35	137	7.96	177	4.77	137	8.70	177	6.67
18	20.72	58	17.62	98	13.23	138	7.84	178	4.71	138	8.62	178	6.64
19	20.66	59	17.53	99	13.10	139	7.73	179	4.66	139	8.54	179	6.61
20	20.60	60	17.43	100	12.98	140	7.62	180	4.61	140	8.47	180	6.58
21	20.54	61	17.33	101	12.85	141	7.51	181	4.56	141	8.39	181	6.56
22	20.48	62	17.24	102	12.72	142	7.41	182	4.51	142	8.32	182	6.53
23	20.41	63	17.14	103	12.59	143	7.30	183	4.46	143	8.25	183	6.51
24	20.35	64	17.04	104	12.47	144	7.20	184	4.41	144	8.18	184	6.49
25	20.28	65	16.94	105	12.33	145	7.10	185	4.36	145	8.12	185	6.46
26	20.22	66	16.84	106	12.20	146	7.01	186	4.32	146	8.05	186	6.44
27	20.15	67	16.74	107	12.07	147	6.91	187	4.27	147	7.99	187	6.42
28	20.08	68	16.64	108	11.94	148	6.82	188	4.23	148	7.93	188	6.40
29	20.01	69	16.53	109	11.81	149	6.73	189	4.18	149	7.87	189	6.38
30	19.94	70	16.43	110	11.67	150	6.64	190	4.14	150	7.81	190	6.36
31	19.87	71	16.33	111	11.54	151	6.55	191	4.09	151	7.75	191	6.35
32	19.80	72	16.22	112	11.40	152	6.46	192	4.05	152	7.69	192	6.33
33	19.73	73	16.12	113	11.26	153	6.38	193	4.01	153	7.64	193	6.31
34	19.65	74	16.01	114	11.13	154	6.30	194	3.97	154	7.59	194	6.30
35	19.58	75	15.90	115	10.99	155	6.22	195	3.93	155	7.53	195	6.28
36	19.50	76	15.79	116	10.85	156	6.14	196	3.89	156	7.48	196	6.27
37	19.42	77	15.69	117	10.71	157	6.06	197	3.85	157	7.43	197	6.26
38	19.35	78	15.58	118	10.57	158	5.98	198	3.81	158	7.39	198	6.24
39	19.27	79	15.47	119	10.43	159	5.91	199	3.77	159	7.34	199	6.23
40	19.19	80	15.36	120	10.28	160	5.83	200	3.73	160	7.29	200	6.22

Note: $C_c = 126.1$

* Reproduced from *Manual of Steel Construction*. Courtesy American Institute of Steel Construction.

† K taken as 1.0 for secondary members.

6-6   Allowable Column Loads

The allowable axial load that a steel column will support is found by multiplying the allowable unit stress by the cross-sectional area of the column. The value of Kl/r is first determined and, by referring to Table 6-1, we can establish the allowable unit stress. The area of the column cross section may be found from Table 4-1.

Example 1.   A W 10 × 49 of A36 steel is to be used as a main column in a building. Compute the maximum allowable load on this column if the unbraced height is 14 ft.

Solution: (1) Referring to Table 4-1, we find the following values for a W 10 × 49: A = 14.4 sq in., r_{X-X} = 4.35 in., r_{Y-Y} = 2.54 in. Since the column is unbraced with respect to both major axes, the least radius of gyration (2.54) will be used to determine the slenderness ratio.[1] Also, as in the average end conditions that occur in building construction (Fig. 6-4*d*), $K = 1$. The slenderness ratio is then

$$\frac{Kl}{r} = 1 \times \frac{14 \times 12}{2.54} = 66.1$$

(2) To determine F_a, the allowable unit stress, we turn to Table 6-1. There we find for Kl/r = 66 that F_a = 16.84 ksi, and for Kl/r = 67 that F_a = 16.74 ksi. The allowable stress for Kl/r = 66.1 will fall between these two values and, by interpolating, it is found to be 16.83 ksi.

(3) The allowable load on the column is then

$$P = F_a \times A = 16.83 \times 14.4 = 242 \text{ kips}$$

As noted in Art. 6-2, W shapes are commonly employed for columns in steel frame bulidings. However, where special conditions require the use of a built-up section, its area and the radius of gyration must be computed. The procedure is illustrated in the following example.

---

[1] Occasionally a column is braced by framing so that the unbraced height with respect to one axis is less than that with respect to the other. It is desirable in such cases to orient the column so that the axis having the smaller radius of gyration will be braced by the closer supports. Under these conditions, the slenderness ratio with respect to *both* major axes must be computed and the larger value used in the column formula.

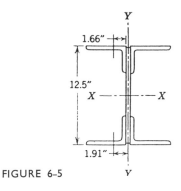

FIGURE 6-5

Example 2. Compute the allowable axial load on a built-up column section composed of one 12 × $\frac{1}{2}$ in. plate and four angles 5 × $3\frac{1}{2}$ × $\frac{1}{2}$ in. having the shorter legs against the web plate. The overall depth of the section is $\frac{1}{2}$ in. greater than the depth of the plate (Fig. 6-5). The unbraced height is 16 ft and the plate and angles are rolled from A36 steel.

Solution: (1) Referring to Table 4-5, we find the following data: area of one angle = 4 sq in., I_{X-X} = 9.99 in.^{4} (say 10), centroid is located 1.66 in. from the back of short leg. Determining the cross-sectional area,

$$\text{area of plate} = 12 \times 0.5 = 6$$
$$\text{area of 4 angles} = 4 \times 4 = 16$$
$$\text{Total area of section} = \overline{22} \text{ sq in.}$$

(2) Determine moment of inertia of the built-up section about the Y–Y axis.

$$I_{Y-Y} \text{ for plate} = \frac{bd^3}{12} = \frac{12 \times 0.5^3}{12} = 0.125 \text{ in.}^4 \quad \text{See Art. 4–5}$$

To compute I_{Y-Y} for the angles we use the equation for transferring moments of inertia from one axis to another. See Art. 4-8. Note that the centroid of the angle to the back of the short leg is 1.66 in.; therefore the distance of the centroid to Y–Y axis of the entire cross section is 1.66 + 0.25 = 1.91 in.

$$I_{Y-Y} \text{ for 1 angle} = I + Az^2 = 10 + (4 \times 1.91^2) = 24.58 \text{ in.}^4$$
$$I_{Y-Y} \text{ for 4 angles} = 4 \times 24.58 = 98.32 \text{ in.}^4$$

Therefore I_{Y-Y} for the entire area $= 98.32 + 0.125 = 98.44$ in.^{4}. By the same method I_{X-X} of the entire cross section $= 544$ in.^{4}. Because I_{Y-Y} is smaller than I_{X-X}, the former will give the smaller radius of gyration, which, of course, is the one used in computing the slenderness ratio of the column. Then

$$ r = \sqrt{\frac{I}{A}} = \sqrt{\frac{98.44}{22}} = 2.11 \text{ in.} $$

Assuming that the end conditions of the column are as indicated in Fig. 6-4*d*, the value of *K* is 1.0, and the slenderness ratio is

$$ \frac{Kl}{r} = 1 \times \frac{16 \times 12}{2.11} = 91 $$

(3) Referring to Table 6-1, we see that $F_a = 14.09$ ksi when $Kl/r = 91$. Therefore, the total allowable axial load on the built-up column is

$$ P = F_a \times A = 14.09 \times 22 = 310 \text{ kips} $$

Note: In the following problems, assume that all columns are main members, the steel is A36, and $K = 1.0$.

Problem 6-6-A*.   Compute the allowable axial load on a W 10 × 60 column with an unbraced height of 20 ft.

Problem 6-6-B.   A W 8 × 31 column has an unsupported height of 18 ft. Compute its allowable axial load.

Problem 6-6-C.   What is the maximum allowable axial load that can be supported by a W 8 × 67 column if its length is 17 ft?

Problem 6-6-D*.   Compute the allowable axial load on a built-up column section composed of two channels C 12 × 25 and two plates 14 × ½ in. The component parts are arranged as shown in Fig. 6-2 with the channels placed 6 in. back-to-back. The unsupported height of the column is 32 ft.

Problem 6-6-E.   The length of a W 14 × 87 column is 30 ft. Compute its allowable axial load.

Problem 6-6-F.   A W 14 × 320, known as a core section, is often used with cover plates to support large column loads. Properties and dimensions of the W 14 × 320 are shown in Fig. 6-6. Assuming that the cover plates are 18 × 1 in. each and that the unbraced height of the column is 24 ft, compute the allowable axial load on the built-up section.

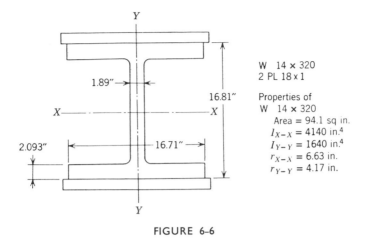

W 14 x 320
2 PL 18 x 1

Properties of
W 14 x 320
Area = 94.1 sq in.
I_{X-X} = 4140 in.⁴
I_{Y-Y} = 1640 in.⁴
r_{X-X} = 6.63 in.
r_{Y-Y} = 4.17 in.

FIGURE 6-6

6-7 Design of Steel Columns

In the absence of safe load tables the design of columns is accomplished by the trial method. Data include the load and length of the column. The designer selects a trial cross section on the basis of his experience and judgment and, by means of a column formula, computes the allowable load that it can support. If this allowable load is less than the actual load the column will be required to support, the trial section is too small and another section is tested in a similar manner.

In practice, the designer selects the proper size column section directly from tables similar to Table 6-2, which gives allowable axial loads for a number of column sections. It has been compiled from more extensive tables in the AISC Manual, and the loads are computed in accordance with the formulas in Art. 6-5. Note particularly that these allowable loads are for main members of A36 steel. Loads to the right of the heavy vertical lines are for main members with Kl/r ratios between 120 and 200. The significance of the bending factors, given at the extreme right of the table, is considered in Art. 6-10.

To illustrate the use of Table 6-2, refer to Example 1 of Art. 6-6. This problem asked that the allowable load be computed for a

TABLE 6-2. Allowable Axial Loads on Columns—

I

Nominal depth & width	Weight per ft	Area sq in.	Radius of gyration		Effective length in					
			X–X	Y–Y	6	7	8	9	10	11
4 × 4	13	3.82	1.72	0.99	62	57	51	45▮	39	32
5 × 5	16	4.70	2.13	1.26	83	79	74	69	64	58
	18.5	5.43	2.16	1.28	97	92	86	81	75	68
6 × 6	15.5	4.56	2.57	1.46	84	81	77	73	69	65
	20	5.88	2.66	1.51	109	105	100	96	91	85
	25	7.35	2.69	1.53	137	132	126	120	114	108
8 × 8	31	9.12	3.47	2.01	178	174	169	164	159	154
	35	10.3	3.50	2.03	201	197	191	186	180	174
	40	11.8	3.53	2.04	231	225	220	213	207	200
	48	14.1	3.61	2.08	276	270	263	256	249	241
	58	17.1	3.65	2.10	336	328	320	312	303	293
	67	19.7	3.71	2.12	387	379	370	360	350	339
10 × 8	33	9.71	4.20	1.94	189	184	179	173	167	161
	39	11.5	4.27	1.98	224	218	213	206	200	193
10 × 10	49	14.4	4.35	2.54	289	284	279	273	268	262
	54	15.9	4.39	2.56	319	314	308	302	296	290
	60	17.7	4.41	2.57	355	350	343	337	330	323
	66	19.4	4.44	2.58	390	383	377	369	362	354
	72	21.2	4.46	2.59	426	419	412	404	396	387
	77	22.7	4.49	2.60	456	449	441	433	424	415
	89	26.2	4.55	2.63	527	519	510	501	491	480
12 × 10	53	15.6	5.23	2.48	312	307	301	295	288	282
	58	17.1	5.28	2.51	343	337	331	324	317	310
12 × 12	65	19.1	5.28	3.02	389	384	378	373	367	361
	72	21.2	5.31	3.04	432	426	420	414	408	401
	79	23.2	5.34	3.05	473	467	460	453	446	439
	85	25.0	5.38	3.07	510	503	496	489	482	474
	92	27.1	5.40	3.08	553	546	538	530	522	514
	99	29.1	5.43	3.09	593	586	578	570	561	552
	106	31.2	5.46	3.11	637	629	620	611	602	593
14 × 10	61	17.9	5.98	2.45	358	351	345	338	330	322
	68	20.0	6.02	2.46	400	393	385	377	369	360
14 × 12	78	22.9	6.09	3.00	466	460	453	447	439	432
	84	24.7	6.13	3.02	503	496	489	482	475	467
14 × 14¼	87	25.6	6.15	3.70	528	523	518	512	506	500
	95	27.9	6.17	3.71	576	570	564	558	552	545
	103	30.3	6.21	3.72	625	619	613	606	599	592

Loads to right of heavy lines are for main members with Kl/r ratios between 120 and 200.

* Compiled from data in the 7th Edition of the *Manual of Steel Construction*. Courtesy American Institute of Steel Construction.

feet with respect to least radius of gyration												Bending factor	
12	13	14	15	16	17	18	19	20	22	24	26	B_x	B_y
27	23	20	17	15								0.701	2.065
52	46	39	34	30	27	24	21	19				0.551	1.567
62	54	47	41	36	32	28	26	23				0.547	1.534
60	55	50	45	39	35	31	28	25	21	17		0.456	1.412
80	74	68	61	54	48	43	39	35	29	24		0.439	1.328
101	94	86	78	70	62	55	49	45	37	31		0.441	1.308
148	142	136	130	123	117	110	102	95	79	66	57	0.333	0.988
168	162	155	148	141	133	125	117	109	91	76	65	0.332	0.972
193	186	178	170	162	153	144	135	125	105	88	75	0.333	0.976
233	224	215	206	196	186	176	165	154	131	110	94	0.327	0.940
283	273	263	251	240	228	216	203	190	162	136	116	0.329	0.940
328	316	304	292	279	265	251	236	221	190	159	136	0.327	0.921
155	149	142	135	127	120	112	103	95	78	66	56	0.278	1.061
186	178	170	162	154	145	136	126	116	97	81	69	0.273	1.027
255	249	242	235	228	221	213	205	197	180	161	142	0.264	0.775
283	276	268	261	253	245	236	228	219	200	180	159	0.264	0.769
315	307	299	291	282	273	264	254	244	224	202	178	0.264	0.767
346	337	328	319	310	300	290	279	269	246	222	197	0.264	0.761
378	369	359	349	339	328	317	306	294	270	244	217	0.265	0.760
406	396	385	375	364	352	341	329	316	290	263	233	0.264	0.755
470	458	447	435	422	409	396	382	368	339	308	275	0.263	0.745
275	268	260	252	244	236	227	218	209	189	169	147	0.221	0.813
302	294	286	278	269	260	251	241	231	211	189	165	0.219	0.800
354	348	341	334	326	319	311	303	294	277	259	240	0.218	0.657
394	387	379	371	363	355	346	337	328	309	289	268	0.218	0.655
431	423	415	407	398	389	379	370	360	339	317	294	0.217	0.649
465	457	448	439	430	420	410	400	389	367	344	319	0.216	0.643
505	496	486	476	466	456	445	434	422	399	374	347	0.217	0.643
543	533	523	512	501	490	478	467	454	429	402	374	0.216	0.637
583	572	561	550	539	527	514	502	489	462	433	404	0.216	0.635
314	306	297	288	278	268	258	248	237	214	190	165	0.195	0.833
351	342	332	322	311	301	289	278	266	241	214	186	0.195	0.830
424	416	408	399	390	381	371	362	352	331	309	285	0.190	0.664
458	450	441	431	422	412	402	391	381	358	335	310	0.189	0.659
493	487	480	473	465	458	450	442	434	417	399	381	0.186	0.532
538	531	523	516	508	499	491	482	473	455	436	416	0.185	0.529
585	577	569	560	552	543	534	524	515	495	474	452	0.185	0.527

W 10 × 49 of A36 steel used on an unbraced height of 14 ft. Referring to Table 6-2, we see at once that the allowable axial load is 242 kips, the same value we found by computation.

Although the designer may select the proper column section by merely referring to the safe load tables, it is well to understand the application of the formula by means of which the tables have been computed. To this end, the *design procedure* is outlined below. When the design load and length (unbraced height) have been established, the following steps are taken:

Step 1:   Assume a trial section and note from the table of properties both the cross-sectional area and the least radius of gyration.

Step 2:   Compute the slenderness ratio Kl/r, l being the unsupported length of the column in inches. For the value of K, see Art. 6-4.

Step 3:   Compute F_a, the allowable unit stress, by using a column formula or by reference to Table 6-1.

Step 4:   Multiply F_a found in Step 3 by the area of the column cross section. This gives the allowable load *on the trial column section.*

Step 5:   Compare the allowable load found in Step 4 with the design load. If the allowable load on the trial section is less than the design load (or if it is so much greater as to make use of the section uneconomical), try another section and test it in the same manner. The reader should note that, except for assuming a trial section, the above operations were carried out in Example 1 of Art. 6-6.

Problem 6-7-A.   Using Table 6-2, select a column section to support an axial load of 128 kips if the unbraced height is 15 ft. A36 steel is to be used and K, the effective length factor, is 1.0.

Problem 6-7-B*.   Same data as in Problem 6-7-A except that the load to be supported is 250 kips and the length is 24 ft.

Problem 6-7-C.   Same data as in Problem 6-7-A except that both a 10-in. and 12-in. section are to be selected for a load of 360 kips with the unbraced height of 16 ft.

6-8 Double-Angle Struts

Struts, including compression members of roof trusses, commonly consist of two angle sections separated by the thickness of a connection plate at each end, and fastened together at intervals by fillers and welds or rivets. These members are designed in accordance with the requirements and formulas for columns given in Art. 6-5. To assure that the angles act as a unit, the intervals between fillers (intermittent connections) are determined so that the slenderness ratio of either angle between fasteners does not exceed the governing slenderness ratio of the built-up section.

The AISC Manual contains a series of tables giving the allowable concentric (axial) loads for struts of two angles with $\frac{3}{8}$ in. separation back-to-back. Table 6-3, relating to unequal leg angles with long legs back-to-back, has been abstracted from this series and lists allowable loads with respect to both the $X-X$ and $Y-Y$ axes. The smaller (least) radius of gyration gives the smaller allowable load, and, unless the member is braced with respect to the weaker axis, this is the tabular load to be used. The usual practice is to assume $K = 1.0$ when using these tables. The following example shows how the loads in the table are computed.

Example. Two $4 \times 3 \times \frac{3}{8}$ angle sections spaced with their long legs $\frac{3}{8}$ in. back-to-back are used as a compression member. If the member is A36 steel and has an effective length of 8 ft, compute the allowable axial load.

Solution: At the bottom of Table 6-3 under Properties, we find that the area of the two-angle member is 4.97 sq in., and that the radii of gyration are $r_x = 1.26$ in. and $r_y = 1.31$ in. Using the smaller r, the slenderness ratio is

$$\frac{Kl}{r} = 1 \times \frac{8 \times 12}{1.26} = 76.2$$

Referring to Table 6-1 and interpolating, we find that $F_a = 15.7$ ksi for a slenderness ratio of 76.2, making the allowable load

$$P = F_a \times A = 15.7 \times 4.97 = 78 \text{ kips}$$

This value is, of course, readily verified by entering Table 6-3 under $X-X$ axis with an effective length of 8 ft, and then proceeding horizontally to the column giving loads for the $4 \times 3 \times \frac{3}{8}$ angles.

The design of double-angle struts as compression members of roof trusses is considered in Chapter 20.

TABLE 6-3. Double-Angle Struts *

Allowable concentric loads in kips. Unequal leg angles

Long legs ⅜ in. back to back of angles

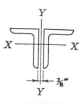

Size	5 × 3½					5 × 3			4 × 3½				
Thickness	¾	⅝	½	7⁄16		½	7⁄16		⅝	½	7⁄16	⅜	5⁄16
Weight per foot	39.6	33.6	27.2	24.0		25.6	22.6		29.4	23.8	21.2	18.2	15.4

$F_y = 36$ ksi — Effective length in feet KL with respect to indicated axis

X-X axis

KL (5×3½)	¾	⅝	½	7⁄16	KL (5×3)	½	7⁄16	KL (4×3½)	⅝	½	7⁄16	⅜	5⁄16
0	251	213	173	152	0	162	143	0	186	151	133	115	97
4	231	196	159	140	2	157	138	2	177	144	128	110	93
6	216	184	150	132	4	149	132	4	165	135	119	103	87
8	200	170	139	123	6	141	124	6	151	123	109	94	79
10	181	154	126	112	8	130	115	8	133	109	97	84	71
12	161	137	113	100	10	119	105	10	113	93	83	72	61
14	138	118	97	86	12	106	94	12	91	75	67	59	50
15	126	108	89	79	14	92	82	14	68	56	50	44	38
16	113	97	81	72	15	84	75	16	52	43	38	34	29
18	89	77	64	57	16	76	68	18	41	34	30	27	23
20	72	62	52	46	18	61	54	20	33	27	25	22	18
22	60	51	43	38	20	49	44	21					17
24	50	43	36	32	22	41	36						
25	46	40	33	30	24	34	31						
26		37	31	27	26	29	26						

Y-Y axis

KL (5×3½)	¾	⅝	½	7⁄16	KL (5×3)	½	7⁄16	KL (4×3½)	⅝	½	7⁄16	⅜	5⁄16
0	251	213	173	152	0	162	143	0	186	151	133	115	97
4	230	195	158	139	2	155	137	2	179	146	129	111	94
6	216	183	148	130	4	145	128	4	171	139	123	106	89
8	199	168	136	119	6	132	117	6	161	131	116	100	84
10	180	152	122	107	8	118	104	8	150	121	107	92	77
12	159	133	107	93	10	101	89	10	137	111	97	84	70
14	136	113	90	78	12	82	72	12	122	98	87	74	62
15	123	102	81	70	14	62	54	14	106	85	75	64	53
16	110	91	72	62	16	47	41	15	97	78	68	58	49
18	87	72	57	49	18	38	33	16	88	70	61	53	44
20	70	58	46	39	20	30	26	18	70	56	49	42	35
22	58	48	38	33				20	57	45	39	34	28
24	49	40	32	27				22	47	37	33	28	23
25	45	37						24	40	31	27	23	19
								25	36	29	25	22	18
								26	34	27	23	20	

Properties of 2 angles—⅜ in. back-to-back

	¾	⅝	½	7⁄16		½	7⁄16		⅝	½	7⁄16	⅜	5⁄16
Area A (in.²)	11.6	9.84	8.00	7.05		7.50	6.62		8.59	7.00	6.18	5.34	4.49
r_x (in.)	1.55	1.56	1.58	1.59		1.59	1.60		1.22	1.23	1.24	1.25	1.26
r_y (in.)	1.53	1.51	1.49	1.47		1.25	1.24		1.60	1.58	1.57	1.56	1.55

Heavy line indicates $Kl/r = 120$. Values omitted for $Kl/r > 200$.

* Compiled from data in the 7th Edition of the *Manual of Steel Construction*. Courtesy, American Institute of Steel Construction.

TABLE 6-3. Double-Angle Struts (*Continued*)

Allowable concentric loads in kips. Unequal leg angles

Long legs ⅜ in. back to back of angles

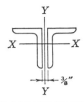

$F_y = 36$ ksi

Effective length in feet KL with respect to indicated axis

Size 4 × 3

Thickness	⅝	½	⁷⁄₁₆	⅜	⁵⁄₁₆
Weight per foot	27.2	22.2	19.6	17.0	14.4

X–X axis

0	172	140	124	107	90
2	164	134	119	103	86
4	154	126	111	96	81
6	140	115	101	88	74
8	124	102	90	78	66
10	106	88	77	67	57
12	85	71	63	55	47
14	64	54	47	42	36
16	49	41	36	32	27
18	39	33	29	25	22
20	31	26	23	20	17
21				19	16

Y–Y axis

0	172	140	124	107	90
2	165	135	119	103	86
4	156	127	112	97	81
6	144	117	103	89	75
8	130	105	93	80	67
10	115	92	81	70	58
12	97	77	68	58	48
13	88	69	61	52	43
14	78	61	53	45	37
16	60	47	41	35	29
18	47	37	32	27	23
20	38	30	26	22	18
21	35	27	24	20	17
22	32	25	21		

Size 3½ × 2½

Thickness	½	⁷⁄₁₆	⅜	⁵⁄₁₆
Weight per foot	18.8	16.6	14.4	12.2

X–X axis

0	119	105	91	77
2	113	100	86	73
4	104	92	80	67
6	93	82	71	60
8	79	70	61	52
10	64	57	50	42
11	56	49	43	37
12	47	42	37	31
14	35	31	27	23
16	26	23	21	18
17	23	21	18	16
18	21	19	16	14

Y–Y axis

0	119	105	91	77
2	113	100	87	73
4	105	92	80	67
6	94	83	72	60
8	82	72	62	52
10	68	59	50	42
11	60	52	44	37
12	51	44	37	31
14	38	32	28	23
16	29	25	21	17
18	23	20	17	14

Size 2½ × 2

Thickness	⅜	⁵⁄₁₆	¼
Weight per foot	10.6	9.0	7.2

X–X axis

0	67	57	46
2	61	52	42
3	58	49	40
4	53	45	37
5	48	41	34
6	42	36	30
7	36	31	26
8	30	26	21
9	23	20	17
10	19	16	14
11	16	14	11
12	13	11	9
13			8

Y–Y axis

0	67	57	46
2	63	53	43
3	60	51	41
4	57	48	39
5	53	45	36
6	49	41	33
7	45	38	30
8	40	34	27
9	35	29	23
10	30	24	19
11	24	20	16
12	21	17	13
13	18	14	11
14	15	12	10
15	13	11	9
16	12		

Properties of 2 angles—⅜ in. back-to-back

	4 × 3					3½ × 2½				2½ × 2		
Area A (in.²)	7.97	6.50	5.74	4.97	4.18	5.50	4.87	4.22	3.55	3.09	2.62	2.13
r_x (in.)	1.23	1.25	1.25	1.26	1.27	1.09	1.09	1.10	1.11	0.768	0.776	0.784
r_y (in.)	1.36	1.33	1.32	1.31	1.30	1.14	1.12	1.11	1.10	0.961	0.948	0.935

Heavy line indicates $Kl/r = 120$. Values omitted for $Kl/r > 200$.

Problem 6-8-A*. Using Table 6-3, select the lightest weight double-angle strut that will support a concentric load of 60 kips if the effective length is 10 ft.

Problem 6-8-B*. Using Table 6-3, find the allowable axial load on a double-angle strut composed of two angles 5 × 3½ × ¾ if the effective length is 12 ft.

Problem 6-8-C. Using Table 6-3, find the allowable load on a double-angle strut composed of two angles 5 × 3 × ½ if the effective length is 10 ft.

6-9 Steel Pipe and Structural Tubing Columns

Unfilled steel pipe columns are frequently used as main compression members in building construction. They are usually designed by the use of safe load tables in routine work.

Table 6-4 gives allowable concentric loads[2] for Standard steel pipe columns. The loads are for main members and are in accordance with the AISC Specification for steel with a yield point of 36 ksi. The diameters at the head of the table are *nominal* diameters, and the actual dimensions may vary slightly from the nominal. In computing the allowable loads, K is 1.0. In addition to the table for Standard steel pipe columns, the AISC Manual contains additional tables for Extra strong and Double-extra strong steel pipe.

Example. Using Table 6-4, select a steel pipe column to carry a load of 226 kips if the unbraced height is 13 ft. Verify the value given in the table by computing the allowable axial load.

Solution: (1) Entering Table 6-4 with an effective length of 13 ft, we find that a load of 226 kips can be supported by a 10-in. diameter Standard pipe column (Pipe 10 Std.).

(2) To verify the tabular load, we first note from the Properties listed in the table that the cross-sectional area of this column is 11.9

---

[2] The reader should be certain that he understands the significance of *axial* and *concentric* as related to column loading. As used in this book, the terms are synonymous and refer to vertical loads applied at the centroid of the cross section (or balanced about it) so that the resultant load acts along the vertical axis of the column. This is in contradistinction to *eccentric* loading which is considered in Art. 6-11.

TABLE 6-4. Typical Column Safe Load Table *

Standard steel pipe
Allowable concentric loads in kips $F_y = 36$ ksi

Nominal dia.	12	10	8	6	5	4	3½	3
Wall thickness	0.375	0.365	0.322	0.280	0.258	0.237	0.226	0.216
Weight per foot	49.56	40.48	28.55	18.97	14.62	10.79	9.11	7.58
6	303	246	171	110	83	59	48	38
7	301	243	168	108	81	57	46	36
8	299	241	166	106	78	54	44	34
9	296	238	163	103	76	52	41	31
10	293	235	161	101	73	49	38	28
11	291	232	158	98	71	46	35	25
12	288	229	155	95	68	43	32	22
13	285	226	152	92	65	40	29	19
14	282	223	149	89	61	36	25	16
15	278	220	145	86	58	33	22	14
16	275	216	142	82	55	29	19	12
17	272	213	138	79	51	26	17	11
18	268	209	135	75	47	23	15	10
19	265	205	131	71	43	21	14	9
20	261	201	127	67	39	19	12	
22	254	193	119	59	32	15	10	
24	246	185	111	51	27	13		
26	238	176	102	43	23			
28	229	167	93	37	20			
30	220	158	83	32	17			
32	211	148	73	29				
34	201	137	65	25				
36	192	127	58	23				
38	181	115	52					
40	171	104	47					

Effective length in feet KL with respect to radius of gyration

Properties

Area A (in.²)	14.6	11.9	8.40	5.58	4.30	3.17	2.68	2.23
I (in.⁴)	279.	161.	72.5	28.1	15.2	7.23	4.79	3.02
r (in.)	4.38	3.67	2.94	2.25	1.88	1.51	1.34	1.16
B (Bending factor)	0.333	0.398	0.500	0.657	0.789	0.987	1.12	1.29
a (Multiply values by 10^6)	41.7	23.9	10.8	4.21	2.26	1.08	0.717	0.447

Heavy line indicates $Kl/r = 120$. Values omitted for $Kl/r > 200$.

* Taken from a more complete set of tables in the 7th Edition of the *Manual of Steel Construction*. Courtesy American Institute of Steel Construction.

sq in., and the radius of gyration (the same about any axis through the centroid) is 3.67 in. Then, taking K as 1.0, the slenderness ratio is

$$\frac{Kl}{r} = 1 \times \frac{13 \times 12}{3.67} = 42.5$$

(3) Referring to Table 6-1, we find that the allowable unit stress F_a for a slenderness ratio of 42.5 is between 19.03 and 18.95 ksi. By interpolating, $F_a = 18.99$ ksi.

(4) The allowable axial load is

$$P = F_a \times A = 18.99 \times 11.91 = 226 \text{ kips}$$

Referring to Table 6-4, we find that this value is in agreement with the tabulated allowable load.

Steel structural tubing columns are fabricated in both square and rectangular shapes. Square tubing is available in sizes from 4 to 10 in. and in weights from approximately 12 to 74 lb per lin ft. Rectangular tubing is produced in a variety of cross sections ranging from 5 × 3 in. to 12 × 6 in., with varying wall thicknesses and weights. The AISC Manual contains safe load tables for square and rectangular structural tubing columns. Table 6-5 has been abstracted from these tables to illustrate their general arrangement. Their method of use is similar to that for the steel pipe column safe load tables.

Problem 6-9-A. Using Table 6-4, select a steel pipe column to support an axial load of 145 kips if the unbraced height is 15 ft. Verify the value given in the table by computing the allowable load.

Problem 6-9-B. Using Table 6-5, select a structural tubing column to carry an axial load of 88 kips on an effective length of 10 ft. Verify the tabular value by computing the allowable load.

6-10 Bending Factors for Columns

The columns previously discussed have been axially or concentrically loaded. It frequently happens, however, that in addition to the axial load the column maybe subjected to bending stresses that result from eccentric loads. Figure 6-7 indicates a column having both a concentric and an eccentric load. The design of eccentrically loaded columns is accomplished by testing trial sections. As an aid to design, it will be found convenient to convert the axial and eccentric loads

TABLE 6-5.   Typical Column Safe Load Table *

Square structural tubing
Allowable concentric loads in kips $F_y = 36$ ksi

Nominal size	5 × 5				4 × 4			
Wall thickness	½	⅜	5/16	¼	½	⅜	5/16	¼
Weight per foot	27.68	21.94	18.77	15.42	20.88	16.84	14.52	12.02
2	171	135	116	95	127	103	89	74
3	168	133	114	94	124	100	87	72
4	164	130	112	92	120	98	84	70
5	160	127	109	90	116	94	82	68
6	156	124	107	88	111	91	79	66
7	151	121	104	86	106	87	76	63
8	147	117	101	83	101	83	72	60
9	141	113	98	81	95	79	69	57
10	136	109	94	78	88	74	65	54
11	130	105	91	75	82	69	61	51
12	124	101	87	72	75	64	57	48
13	118	96	83	69	68	59	52	44
14	111	91	79	66	60	53	47	40
15	104	86	75	62	52	47	42	36
16	97	81	70	59	46	42	37	32
17	90	75	66	55	41	37	33	29
18	82	69	61	51	36	33	30	25
19	74	63	56	47	33	29	27	23
20	67	57	51	43	29	27	24	21
22	55	47	42	35	24	22	20	17
24	46	40	35	30		18	17	14
26	40	34	30	25				
28	34	29	26	22				
30		25	22	19				

($F_y = 36$ ksi) — Effective length in feet KL with respect to radius of gyration

Properties								
Area A (in.²)	8.14	6.45	5.52	4.54	6.14	4.95	4.27	3.54
I (in.⁴)	25.7	22.0	19.5	16.6	11.4	10.2	9.23	8.00
r (in.)	1.78	1.85	1.88	1.91	1.36	1.44	1.47	1.50
B}Bending factor	0.792	0.733	0.708	0.684	1.08	0.971	0.925	0.885
a}Multiply values by 10⁶	3.84	3.29	2.91	2.47	1.69	1.53	1.37	1.19

Heavy line indicates $Kl/r = 120$. Values omitted for $Kl/r > 200$.

* Taken from a more complete set of tables in the 7th Edition of the *Manual of Steel Construction*. Courtesy American Institute of Steel Construction.

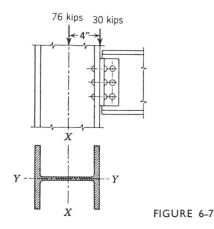

FIGURE 6-7

into single equivalent axial load. Having done this, the safe load tables may be used to select the trial section.

On the right-hand side of Table 6-2 the *bending factors* B_x and B_y are listed. The bending factor is the area of the cross section divided by its section modulus. Because there are two major section moduli there are two bending factors; B_x for the X–X axis and B_y for the Y–Y axis. For example, the area of a W 10 × 49 is 14.4 sq in. and the section moduli with respect to the X–X and Y–Y axes are 54.6 in.^{3} and 18.6 in.^{3}, respectively. Then

$$B_x = \frac{A}{S_x} = \frac{14.4}{54.6} = 0.264$$

and

$$B_y = \frac{A}{S_y} = \frac{14.4}{18.6} = 0.775$$

Note that these are the values given in Table 6-2.

Bending factors are used to convert the effect of eccentricity to an equivalent axial load. To accomplish this, the *bending moment* resulting from the eccentric load is multiplied by the appropriate bending factor. Then, *the total equivalent axial load (P') is equal to the sum of the axial load and the eccentric load plus the product of the bending moment due to the eccentric load and the appropriate bending factor.*

6-11   Trial Section for Eccentrically Loaded Columns

The trial section used in designing a column subjected to both axial and eccentric loads may be established by first finding an *approximate* equivalent axial load. This procedure is illustrated in the following example.

Example.   An 8-in. column of A36 steel with an unsupported height of 13 ft supports an axial load of 76 kips and a load of 30 kips applied 4 in. from the X–X axis. The arrangement is shown in Fig. 6-7. Determine the trial column section.

Solution: (1) The bending moment produced by the eccentric load is

$$M = 30 \times 4 = 120 \text{ kip-in}.$$

However, only the general dimensions of the section are known at this point, so we do not know the exact value of the bending factor.

(2) Referring to the 8-in. column sections in Table 6-2, select tentatively a bending factor of 0.333. This may be revised later. Then the bending moment multiplied by the bending factor is 120 × 0.333 = 39.9 or 40 kips which is an equivalent axial load for the eccentric load.

(3) Now, in accordance with the principle stated in Art. 6-10, the approximate total equivalent load on the column is

$$P' = 76 + 30 + 40 = 146 \text{ kips}$$

Referring to Table 6-2 again, we find that a W 8 × 35 will carry 162 kips with an effective length of 13 ft, and a W 8 × 31 is listed for 142 kips.

Since the foregoing procedure gives results that are approximate on the safe side, the W 8 × 35 could be the accepted section. However, if it is desired to determine the lightest weight section that can be used, the W 8 × 31 should be investigated for compliance with the AISC Specification requirements for design of columns with combined loading. This is not a simple procedure, and the diversity of factors involved makes it inadvisable to include treatment of these complex requirements in a book of this scope. Reference to the AISC Manual is recommended for readers who desire to study the complete specification requirements covering combined axial compression and bending in columns.

It should be noted that the selection of a trial section by the equivalent axial load method is always conservative and increasingly so as the ratio of the eccentric load to the axial load and the column slenderness ratio increase. Nevertheless, for many situations that arise in routine practice, the trial section determined by this method may be taken as the accepted section.

In conventional building construction it is assumed that the effect of an eccentric load disappears at each story height. Consequently in the above example, the design load that the W 8 × 35 transmits to the column in the story below, or to a base plate, is not 146 kips but 106 kips plus the column weight. In this example only one eccentric load was given. If, in addition, there is an eccentric load about the $Y-Y$ axis, its equivalent axial load plus the magnitude of the load is added to the 146 kips to determine the total approximate equivalent axial load on the column.

6-12   Reduction in Column Live Loads

In a building containing a number of floors the loads on the columns increase for each successive lower floor. These loads consist of dead floor loads, live loads, and the weight of the columns in the stories above. Minimum live loads for various classes of building are given in Table 5-9. It is improbable, however, that the full live load will occur on all floors simultaneously. With the exception of buildings used for storage, public garages, and similar types of occupancy, building codes generally permit a reduction in the live loads on columns for buildings more than 5 or 6 stories high. Although specific code requirements vary, one provision often used permits the following live load reductions on columns:

carrying one floor	0 %
carrying two floors	10 %
carrying three floors	20 %
carrying four floors	30 %
carrying five floors	40 %
carrying six floors	45 %
carrying seven or more floors	50 %

It is customary for the designer to tabulate in some convenient manner the loads on columns and footings. The following example is given to show one method that is sometimes used.

Example. Determine the column loads on an interior column of a steel frame building 10 stories high. The bay dimensions are 20 × 22 ft. and each column supports a floor area of 440 sq ft. The floor live load is 60 psf and the floor dead load, including an allowance for beams and girders, is 70 psf. The roof has a live load of 30 psf and a dead load of 50 psf. Use the schedule for live load reduction given above.

Solution: It should be noted that columns are designated by the story through which they run. For example, a column between the 7th and 8th floors, supporting the 8th floor, is known as a 7th-story column. The computations are best arranged in tabular form such as that shown in Table 6-6. In this example, the computations have been carried out for the upper stories only but the design load on the 6th-story column has been determined as 234,820 lb (say 235 kips).

Loads for columns in the stories below are determined in the same manner. A convenient method of recording final design loads is shown in Fig. 6-8.

6-13 Column Base Plates

The loads from steel columns are generally transferred to the foundation bed through reinforced concrete footings. Because the allowable compressive strength of concrete is considerably less than the unit compressive stress in the steel column, it is necessary to provide a steel base plate under the column to spread the load sufficiently to prevent crushing of the concrete. The typical arrangement is shown in Fig. 6-9a.

It is essential that the end of the column and the base plate be in absolute contact, and plates 4 in. and more in thickness are planed on the upper surface; the lower surface rests on a layer of cement grout on the concrete. Steel columns are usually secured to the footing by steel anchor bolts embedded in the concrete that pass through the base plate and angles riveted or welded to the column flanges. For light columns the angles are often omitted, and the base plate is secured to the column by fillet welds.

TABLE 6-6. Live Load Reduction Computations

Column Carrying	Roof and Floor Loads			Design Load (lb)
	Individual Floor (lb)	Accumulated (lb)	Reduction (lb)	
Roof	L.L. 440 × 30 = 13,200			
	D.L. 440 × 50 = 22,000			
	Column = 300			
	——			
	35,500	35,500	none	35,500
10TH floor	L.L. 440 × 60 = 26,400			
	D.L. 440 × 70 = 30,800			
	Column = 400			
	——			
	57,600	93,100	none	93,100
9TH floor	L.L. 440 × 60 = 26,400			
	D.L. 440 × 70 = 30,800			
	Column = 500			
	——			
	57,700	150,800	26,400 × 2 × 0.1 = 5,280	145,520
8TH floor	L.L. 440 × 60 = 26,400			
	D.L. 440 × 70 = 30,800			
	Column = 600			
	——			
	57,800	208,600	26,400 × 3 × 0.2 = 15,840	192,760
7TH floor	L.L. 440 × 60 = 26,400			
	D.L. 440 × 70 = 30,800			
	Column = 700			
	——			
	57,900	266,500	26,400 × 4 × 0.3 = 31,680	234,820

Roof	
	Load at floor level
10th floor	35.5 kips
9th floor	93.1 kips
8th floor	146 kips
7th floor	193 kips
6th floor	235 kips

FIGURE 6–8

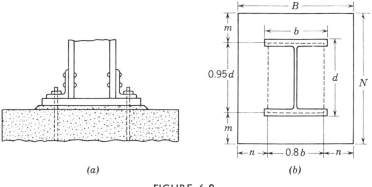

(a)   *(b)*

FIGURE 6-9

The first step in the design of a column base plate is to determine its area. This is readily accomplished by dividing the column load by the allowable unit bearing pressure on the concrete. The plate thickness is determined on the assumption that the plate acts as an inverted cantilever tending to curl around the edges of the column. The column load is assumed to be uniformly distributed over the dotted rectangular area shown in Fig. 6-9*b*, and the projection *m* or *n* (whichever is greater) is used to compute the bending moment. The design procedure follows.

Step 1:   Determine *A*, the required area of the plate.

$$A = \frac{P}{F_p}$$

in which *A* = area of the plate in square inches,

   P = total column load in pounds,

   F_p = allowable bearing value of the concrete in pounds per square inch—the AISC Specification gives the stress F_p as $0.25\, f_c'$, when the full area of the concrete is covered by the plate and $0.375\, f_c'$ when the plate is one third of the concrete area; for a concrete commonly used, $f_c' = 3000$ psi, hence, $0.25\, f_c' = 750$ psi and $0.375\, f_c' = 1125$ psi,

   A = *B* × *N* (Fig. 6-9*b*).

Step 2:   Make a diagram similar to Fig. 6-9*b*; select a plate whose area is *B* × *N* and with dimensions *m* and *n* nearly equal.

Step 3: Compute the dimensions *m* and *n*, using the values 0.95*d* and 0.80*b* as shown in Fig. 6-9*b*.

Step 4: Solve for *t*, the thickness of the plate in the following formula, using the value of *m* or *n*, whichever is greater.[3]

$$t = \sqrt{\frac{3f_p m^2}{F_b}} \quad \text{or} \quad t = \sqrt{\frac{3f_p n^2}{F_b}}$$

in which *t* = thickness of the bearing plate, in inches,

f_p = *actual* bearing pressure on the foundation, in psi,

F_b = allowable bending stress in the base plate, in psi—the AISC Specification gives the value of F_b as $0.75F_y$, F_y being the yield point of the steel plate; thus for A36 steel $F_y = 36,000$ psi and $F_b = 0.75 \times 36,000$, or $F_b = 27,000$ psi (see Table 5-2).

Example. Design a column base plate of A36 steel to support a W 12 × 58 column section with a load of 340,000 lb. The plate will rest on a concrete footing for which $F_p = 750$ psi.

Solution: (1) The required area of the plate is

$$A = \frac{P}{F_p} = \frac{340,000}{750} = 453 \text{ sq in.}$$

As a trial, let $B = 19$ in. and $N = 24$ in. Then *A*, the area of the plate, is 19 × 24 = 456 sq in.

(2) Referring to Table 4-1, we find the dimensions of a W 12 × 58 are depth = 12.19 in. and flange width = 10.014 in. Now make a diagram similar to Fig. 6-9*b*.

(3) Using Fig. 6-9*b*, compute the values on *m* and *n* as follows:

$$0.95d = 0.95 \times 12.19 = 11.6 \text{ in.} \qquad m = \frac{24.0 - 11.6}{2} = 6.2 \text{ in.}$$

$$0.80b = 0.80 \times 10.01 = 8.01 \text{ in.} \qquad n = \frac{19 - 8.01}{2} = 5.5 \text{ in.}$$

(4) Of the two projections, *m* is the greater, 6.2 in.; this dimension is used to compute the thickness of the plate. Because the plate area is 456 sq in., the *actual* bearing pressure on the concrete footing is

---

[3] For the derivation of this formula see *Simplified Design of Structural Steel*, by Harry Parker, Fourth Edition (New York: Wiley, 1974), Art 10-15.

$f_p = 34,000/456 = 746$ psi. Then

$$t = \sqrt{\frac{3f_p m^2}{F_b}} = \sqrt{\frac{3 \times 746 \times 6.2 \times 6.2}{27,000}} = \sqrt{3.19}$$

and
$$t = 1.79 \text{ in}$$

which is the required theoretical thickness of the plate. Accept a thickness of $1\frac{7}{8}$. The plate adopted is, therefore, $19 \times 24 \times 1\frac{7}{8}$ in.

Problem 6-13-A*. Design a base plate of A36 steel for a W 10 × 33 column that supports an axial load of 180 kips. The footing on which the base plate rests is concrete for which $F_p = 1125$ psi.

Problem 6-13-B. Design a base plate of A36 steel for a W 8 × 31 column section having an axial load of 175 kips. The base plate rests on a concrete footing for which the allowable bearing pressure is 750 psi.

6-14 Grillage Foundations

In order to distribute a column load over the foundation bed, steel grillage is occasionally used instead of a reinforced concrete footing.

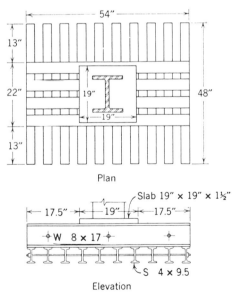

Plan

Elevation

FIGURE 6-10

Because reinforced concrete footings are more economical, they have largely supplanted steel grillages for use as column footings except in special cases where very heavy column loads are involved.

Grillage foundations consist of one or more layers of steel beams arranged as shown in Fig. 6-10, which represents a typical design for a column load of 216 kips. Space is left between the flanges for placing and rodding the concrete. The concrete fills all spaces between beams and the entire assembly is encased in concrete not less than 4 in. thick. Pipe separators hold the beams in position or welded diaphragms are used as separators.

The general structural action of a grillage foundation may be visualized from Fig. 6-10. By means of the base plate, the column load is transferred to the first tier of beams which then spreads the load over the lower tier at a reduced bearing pressure. The bearing pressure on the bottom of the grillage is limited to the bearing capacity of the concrete, inasmuch as a thin concrete mat is usually placed under the grillage even where the supporting material is bedrock.

7

Bolted and Riveted Connections

||||||||||||||||||||||||||||||||||||||||||||||||||||||||||||||

7-1 General

Bolts or welds are generally used for making connections in modern steel building construction. Riveting was once the dominant method of joining structural steel members, but it has been largely supplanted by high-strength bolting and welding. However, in our discussion of bolts and rivets as fasteners, we will consider rivets first because much of the technology related to bolted connections has developed from the long experience with rivets.

7-2 Riveting

The simplest type of riveted joint is illustrated in Fig. 7-1a. The two bars of metal to be connected are punched or drilled and held securely, one on the other, with the holes in alignment. A heated rivet is placed in the hole and a riveting tool is pressed against the head. Next, the projecting shank is covered by a power riveter that delivers rapid blows, filling the hole with rivet metal, deforming the shank, and forming a head. As the rivet cools, there is a slight shrinkage in its length which draws the two plates tightly together.

For structural steel work in buildings $\frac{3}{4}$-in. and $\frac{7}{8}$-in. rivets are the sizes commonly used.

Rivet holes are either punched or drilled. If the thickness of the material is not greater than the diameter of the rivet plus $\frac{1}{8}$ in., the holes may be punched. The holes should be $\frac{1}{16}$ in. larger than the diameter of the rivet. Punching structural steel damages the material adjacent to the hole. Consequently, in computing the effective cross-sectional area of a punched member, the diameter of the rivet hole is taken to be $\frac{1}{8}$ in. greater than the diameter of the rivet.

7-3 Gage Lines, Pitch, and Edge Distance

The lines, parallel to the length of a member, on which bolts or rivets are placed are called *gage lines*. The gage is the normal (perpendicular) distance between the gage line and the edge of a member. If there are two parallel gage lines, there are two gage distances as indicated in the figure accompanying Table 7-1. The usual gage dimensions for standard structural steel angles are given in Table 7-1. The gage dimensions for the flanges of wide flange shapes, standard I-beams, and channels may be found in the AISC Manual tables "Dimensions for detailing."

TABLE 7-1.   Usual Gage Dimensions for Angles *

Leg	8	7	6	5	4	3½	3	2½	2
g	$4\frac{1}{2}$	4	$3\frac{1}{2}$	3	$2\frac{1}{2}$	2	$1\frac{3}{4}$	$1\frac{3}{8}$	$1\frac{1}{8}$
g_1	3	$2\frac{1}{2}$	$2\frac{1}{4}$	2					
g_2	3	3	$2\frac{1}{2}$	$1\frac{3}{4}$					

* Reproduced from data in the 7th Edition of the *Manual of Steel Construction*. Courtesy American Institute of Steel Construction.

The center-to-center distance between adjacent bolts or rivets, whether they are on the same or different gage lines, is called the *pitch*. The minimum distance between centers of rivet and bolt holes should not be less than $2\frac{2}{3}$ times the nominal diameter of the fastener, but preferably not less than three diameters; the usual standard minimum is $2\frac{1}{2}$in., and the maximum is 6 in.

If a bolt or rivet is placed too close to the edge of a member, there is a tendency to tear or deform the adjacent metal. To prevent this type of failure, certain prescribed distances are maintained between the center of a hole and the edge of the member. This is called the *edge distance*. Greater edge distances are required for sheared edges than for the rolled edges of structural shapes and plates. Minimum edge distances for $\frac{3}{4}$-in. and $\frac{7}{8}$-in. bolts or rivets are, respectively, $1\frac{1}{4}$ in. and $1\frac{1}{2}$ in. at sheared edges and 1 in. and $1\frac{1}{8}$ in. at rolled edges. The maximum distance from the center of any rivet or bolt hole to the nearest edge of a plate is 12 times the thickness of the plate but not over 6 in.

7-4 Structural Action in Riveted Joints

Two plates held together by one or more rivets, as shown in Fig. 7-1a, constitute a lap joint; it is the simplest type of riveted connection. A type more common in steel building construction consists of three thicknesses of metal as indicated in Fig. 7-1e.

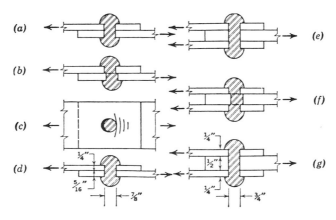

FIGURE 7-1

In order to determine the number of rivets required for a connection, we must know the load to be transmitted through the joint and the allowable working value for a single rivet. The load at the connection divided by the allowable working value of one rivet gives the number of rivets required. There are several ways in which a riveted joint may fail, but we shall be concerned with only the two major types of failure, *shear* and *bearing*.

Figure 7-1*b* represents a shear failure of the rivet shown in Fig. 7-1*a*. This rivet is in *single shear*; that is, there is only one plane through the shank of the rivet on which the shearing force acts. The allowable working value for shear for this rivet is the cross-sectional area of the rivet multiplied by the allowable unit shearing stress of the rivet steel. If there are three thicknesses of metal, as shown in Fig. 7-1*e*, the shear failure would occur as indicated in Fig. 7-1*f*; this rivet is in *double shear*. Because there are two planes through the rivet shank resisting shear failure, the allowable working value for shear is twice the cross-sectional area of the rivet multiplied by the allowable unit shearing stress of the rivet steel.

If the plates are relatively thin with respect to the diameter of the rivet, failure may occur by crushing of the plate due to the bearing pressure exerted by the rivet shank. See Fig. 7-1*c*. The area assumed to resist bearing is a rectangle, the dimensions of which are the nominal diameter of the rivet and the thickness of the plate. Therefore the allowable bearing value of a rivet is this area multiplied by the allowable unit bearing stress. When rivets are in single shear (Fig. 7-1*a*) they are said to be in single-shear bearing also, or simply in *single bearing*. If the joint is arranged as in Fig. 7-1*e*, the rivet is in double shear and the inner plate is in *enclosed bearing*. The allowable unit bearing stress is the same for either condition.

It was stated in Art. 7-2 that, as a rivet cools, there is a slight shrinkage which draws the connected parts tightly together. However, the resulting friction between the parts is neglected when computing the strength of the connection because the degree of clamping effect from rivet shrinkage cannot be measured with precision. The edges of the plates are in bearing against the rivet, and the shank of the rivet resists shear. This is called a *bearing-type connection* in contrast to the *friction-type connection* attained with high-strength bolts (Art. 7-7).

Summarizing with respect to bearing-type connections; when two plates are connected by a rivet (Fig. 7-1a), the plates are in single bearing and the rivet is in single shear; when there are three plates (Fig 7-1e), the rivets are in double shear, the outer plates are in single bearing, and the central plate is in enclosed bearing.

7-5 Allowable Stresses and Working Values for Rivets

The 1969 AISC Specification permits the following allowable unit stresses for Grade 1, hot-driven rivets of ASTM A502 steel in bearing-type connections:

unit shearing stress   15 ksi

unit bearing stress   48.6 ksi

Table 7-2 is a summary of allowable unit stresses in shear and tension for rivets and bolts. Rivets are not often used in positions where direct tensile stresses are developed, except in some types of brackets supporting eccentric loads on columns.

It follows from the discussion in Art. 7-4 that the strength of a rivet in single shear is equal to its cross-sectional area A times the allowable unit shearing stress F_v. Consequently, the allowable working value (load) in single shear for a $\frac{3}{4}$-in. diameter rivet is $(\pi D^2/4) \times F_v = 0.4418 \times 15 = 6.63$ kips. The allowable working value in double shear would, of course, be twice the single shear value.

Allowable working values for rivets of different diameters are given in Table 7-3 for both single and double shear, and for bearing in metal of different thicknesses. The bearing values are computed by multiplying the bearing area (nominal diameter of rivet $\times$ thickness of plate) by the allowable bearing stress. For a $\frac{3}{4}$-in. rivet in a $\frac{3}{8}$-in. plate, the allowable working value in bearing is $(0.75 \times 0.375) \times 48.6 = 13.7$ kips. The use of Table 7-3 is illustrated in the following examples.

Example 1. Figure 7-1d shows a $\frac{7}{8}$-in. rivet connecting $\frac{1}{4}$-in. and $\frac{5}{16}$-in. plates. Determine the *controlling value* of the rivet.

Solution: (1) The rivet is in single shear. Entering Table 7-3 at the column for $\frac{7}{8}$-in. diameter, and moving down the page to the third line, the allowable shearing value is found to be 9.02 kips.

TABLE 7-2. Allowable Tension and Shear Unit Stresses for Rivets and Bolts *
(kips per square inch)

		Shear (F_v)	
Description of fastener	Tension (F_t)	Friction-type connections	Bearing-type connections
A502, Grade 1, hot-driven rivets	20.0		15.0
A502, Grade 2, hot-driven rivets	27.0		20.0
A307 bolts	20.0[1]		10.0
Threaded parts[3] of steel meeting the requirements of Sect. 1.4.1	$0.60F_y$[1]		$0.30F_y$
A325 and A449 bolts, when threading is *not* excluded from shear planes	40.0[2]	15.0	15.0
A325 and A449 bolts, when threading is excluded from shear planes	40.0[2]	15.0	22.0
A490 bolts, when threading is *not* excluded from shear planes	54.0[2,4]	20.0	22.5
A490 bolts, when threading is excluded from shear planes	54.0[2,4]	20.0	32.0

* Reproduced from the 7th Edition of the *Manual of Steel Construction*. Courtesy American Institute of Steel Construction.

[1] Applied to tensile stress area equal to $0.7854\left(D - \dfrac{0.9743}{n}\right)^2$ where D is the major thread diameter and n is the number of threads per inch.

[2] Applied to the nominal bolt area.

[3] Since the nominal area of an upset rod is less than the stress area, the former area will govern.

[4] Static loading only.

TABLE 7-3. Allowable Loads for Rivets and Threaded Fasteners in Bearing-Type Connections in kips *

Unit Shearing Stress = 15.0 ksi

Unit Bearing Stress = 48.6 ksi

Diameter		$\frac{5}{8}$ in.	$\frac{3}{4}$ in.	$\frac{7}{8}$ in.	1 in.	$1\frac{1}{8}$ in.
Area, sq in.		0.0368	0.4418	0.6013	0.7854	0.9940
Single shear		4.60	6.63	9.02	11.78	14.19
Double shear		9.20	13.25	18.04	23.56	29.82
Bearing						
Thickness of plate in inches	$\frac{1}{4}$	7.59	9.11	10.6	12.2	13.7
	$\frac{5}{16}$	9.49	11.4	13.3	15.2	17.1
	$\frac{3}{8}$	11.4	13.7	15.9	18.2	20.5
	$\frac{7}{16}$	13.3	15.9	18.6	21.3	23.9
	$\frac{1}{2}$	15.2	18.2	21.3	24.3	27.3
	$\frac{9}{16}$	17.1	20.5	23.9	27.3	30.8
	$\frac{5}{8}$	19.0	22.8	26.6	30.4	34.2

* Compiled from data in the 7th Edition of the *Manual of Steel Construction.* Courtesy American Institute of Steel Construction.

(2) The value of the rivet in bearing will be limited by the thickness of the thinner of the two plates, in this case $\frac{1}{4}$ in. Referring to Table 7-3 again, and moving down the page under $\frac{7}{8}$-in. diameter to the line for $\frac{1}{4}$-in. bearing thickness, we find the allowable bearing value listed as 10.6 kips.

(3) Comparing the allowable working values found above, the controlling value is, of course, the smaller of the two, or 9.02 kips.

Example 2.   Figure 7-1g indicates a joint made with ¾-in. rivets. The outer plates are ¼ in. thick and the enclosed plate is ½ in. thick. Determine the controlling value of one rivet for this connection.

Solution: (1) The rivet is in double shear. Entering Table 7-3 under ¾-in. diameter and moving down the page to the fourth line, the allowable shearing value is found to be 13.25 kips.

(2) The enclosed plate is ½ in. thick and the combined thickness of the two outer plates (which act in the same direction with the load divided between them) is also ½ in. Consequently, the value of one rivet in bearing is limited by a ½-in. thickness of metal. Referring to Table 7-3, and moving down the page under ¾-in. diameter to the line for ½-in. bearing, we find the allowable working value for bearing listed as 18.2 kips.

(3) Comparing the two allowable working values found above, the controlling value of one rivet in this connection is 13.25 kips.

Example 3.   A connection similar to the one shown in Fig. 7-1e is to be made with ¾-in. rivets. The two outer plates acting to the left are each ¼ in. thick and the enclosed plate acting to the right is ⁵⁄₁₆ in. thick. If the load to be transmitted across the joint is 30 kips, how many rivets are required?

Solution: (1) The rivets are in double shear. Referring to Table 7-3, the allowable working value (allowable load) on a ¾-in. rivet in double shear is found to be 13.25 kips.

(2) The enclosed plate is ⁵⁄₁₆ in. thick and the combined thickness of the two outer plates is ¼ + ¼ = ½ in. Therefore the value of one rivet in bearing is limited by the ⁵⁄₁₆-in. plate. Referring to Table 7-3, we find the allowable load for a ¾-in. rivet in ⁵⁄₁₆-in. bearing is 11.4 kips.

(3) Comparing the allowable loads in shear and bearing, we see that the controlling value of one rivet in this connection is 11.4 kips.

(4) The number of rivets required in the connection is, then, $30 \div 11.4 = 2.63$. Therefore, 3 rivets are required.

Problem 7-5-A*.   Two ¼-in. plates are held together with ⅞-in. rivets, as in Fig. 7-1a. What is the controlling value of one rivet?

Problem 7-5-B.   What is the controlling value of a rivet in the preceding problem if the rivet is ¾ in. in diameter?

Problem 7-5-C. In a connection similar to that shown in Fig. 7-1*e*, the outer plates are ⅜ in. thick and the enclosed plate is ½ in. thick. If ⅞-in. rivets are to be used, determine the controlling value of one rivet.

Problem 7-5-D. A ⁵⁄₁₆-in. plate and a ¼-in. plate are to be connected by ¾-in. rivets, as in Fig. 7-1*a*. If the load transferred from one plate to the other is 16 kips, how many rivets are required?

Problem 7-5-E. Two ¼-in. plates are connected by ⅞-in. rivets (Fig. 7-1*a*). The load to be transferred from one plate to the other is 26 kips. How many rivets are required?

Problem 7-5-F*. Rivets 1 in. in diameter are used to connect three plates as shown in Fig. 7-1*e*. The outer plates are each ⁵⁄₁₆ in. thick and the enclosed plate is ⁷⁄₁₆ in. thick. Find the number of rivets required in the connection if the load transferred by the plates is 69 kips.

7-6 Unfinished Bolts

Under certain conditions in bearing-type connections, unfinished bolts are used in place of rivets. These are made by automatic machines from rods of A307 steel as they come from the mill. The sizes of unfinished bolts are not exact and the allowable unit shearing stress is only 10 ksi as against 15 ksi for rivets (Table 7-2). The allowable unit bearing stress, however, is 48.6 ksi for A307 bolts as well as for rivets.

Unfinished bolts are not permitted under some building codes in buildings exceeding certain specified height limits; nor should they be used for connections subjected to vibration, impact, or fatigue. Where these conditions exist, nuts may become loose and the strength of the connection impaired. The holes for unfinished bolts, as for rivets should be $\frac{1}{16}$ in. larger than the diameter of the bolt. It should be noted that although the bearing capacities of A307 unfinished bolts may be read from Table 7-3, the allowable shearing values may not. The shear values given in this table are based on $F_v = 15$ ksi (applicable to bolts made from A325 and A449 steel; see Table 7-2) whereas $F_v = 10$ ksi for A307 bolts.

7-7 High-Strength Bolts

The current wide use of bolting in structural steel framing has resulted from the development of the friction-type connection made

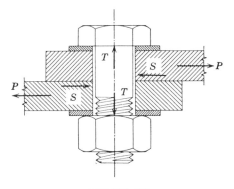

FIGURE 7-2

possible by high-strength bolts. Figure 7-2 indicates two plates held together by a high-strength bolt with a nut and two washers. This is a friction-type connection. The nut is tightened to produce a tensile stress T in the bolt that approaches its yield stress, thus clamping the plates tightly together. The frictional resistance S prevents slip between the plates. If the joint load P exceeds the magnitude of S, slip occurs and the edges of the plates are brought into contact with the shank of the bolt, thereby producing bearing and shearing stresses in the bolt. If there is *no slip*, the load P is transmitted from one plate to the other by *frictional* resistance that results from T, the high tension in the bolt. In this case, there are no shearing or bearing stresses on the bolt. Theoretically, the bolt's shank and the edges of the plates through which it passes are not in contact at all because the holes are punched slightly larger than the bolt. Even when a connection of this type is subjected to vibration, the high residual tensile stress prevents loosening.

The tensile stresses set up in the bolts must be controlled and, when all fasteners in the joint are tight, the minimum bolt tension should approximate the values given in Table 7-4. Tightening of high-strength bolts in friction-type connections is accomplished by special torque or impact wrenches. Because the efficiency of this connection depends on the control used in achieving the proper tensile stresses in the bolts, special attention must be given to supervision of their installation and tightening.

TABLE 7-4. High Strength Bolt
Tension

Nominal bolt diameter in inches	Minimum bolt tension (proof load) in kips
$\frac{5}{8}$	19.2
$\frac{3}{4}$	28.4
$\frac{7}{8}$	30.1
1	47.3

High-strength bolts are usually made of A325 or A449 steel. The design procedure is predicated on taking the allowable capacity (working value) of a bolt equal to the allowable shear value of a rivet of equivalent size, *with no consideration of bearing stresses*. In friction-type connections this hypothetical shear value is based on an allowable unit shearing stress of 15 ksi whether or not threading is excluded from the shear planes (Table 7-2). Therefore, Table 7-3 may be used in designing friction-type connections if the portion of the table relating to bearing values is disregarded.

When high-strength bolts are used in connections where tightening of the nut is done by hand wrenches, attaining only a snug fit, there is insufficient tension developed to prevent slip. Consequently, the allowable value of a bolt is determined as for any bearing-type connection, the allowable shear stress for A325 bolts being limited to 15 ksi when threading is *not* excluded from the shear planes and 22 ksi when threading is excluded (Table 7-2).

7-8 Net Sections: Angles in Tension

Single and double angles are often used in direct tension as hangers, and double angles are widely used as tension members of roof trusses. When the end connections of such tension members are made with rivets or bolts, holes have to be punched through the angles to accommodate the fasteners; hence, the full or gross area of the cross section is not available to resist the tensile force. The term *net section* is used to denote the gross cross-sectional area minus

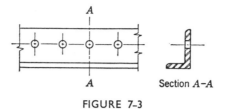

A

Section *A–A*

FIGURE 7-3

holes or openings in the plane of the normal cross section. Figure 7-3 indicates that the net section of the angle shown occurs on a line *A–A* taken through the center of a hole. As stated earlier, rivet and bolt holes are punched $\frac{1}{16}$ in. larger in diameter than the diameter of the fastener to provide clearance, and an additional $\frac{1}{16}$ in. is allowed for damage to the metal during punching. Therefore, the net area of the angle shown in Fig. 7-3 is its gross area minus the rectangular area that results from multiplying the thickness of the angle leg by the nominal fastener diameter plus $\frac{1}{8}$ in.

To illustrate the computation of net area, consider an angle 5 × $3\frac{1}{2}$ × $\frac{1}{2}$ punched for a single row of rivets as indicated in Fig. 7-3. The gross area of this angle is given in Table 4-5 as 4 sq in. The area to be deducted for the rivet hole is $(\frac{3}{4} + \frac{1}{8}) \times \frac{1}{2} = 0.4375$ sq in. Therefore the effective or net area is $4.0 - 0.4375 = 3.56$ sq in. The values listed in Table 7-5 have been computed in a similar manner. Note that the table gives effective net areas for *single* angles with one hole deducted; it will be found extremely useful in the design of tension members.

Example. A mezzanine floor in a building is partially supported by hangers from the roof construction above. Each hanger is composed of two steel angles spaced $\frac{3}{8}$ in. back-to-back, and carries a tensile force of 81 kips. The ends of the hangers are connected to steel plates by a single row of $\frac{7}{8}$-in. rivets. What size angles of A36 steel should be used?
Solution: (1) Since the total force in the member is 81 kips, each angle will carry $81 \div 2 = 40.5$ kips. Table 5-2 gives the allowable tensile stress $F_t = 22$ ksi. Therefore each angle must have a net area of $40.5 \div 22 = 1.84$ sq in.

TABLE 7-5. Effective Net Areas of Angles with One Hole Deducted

Size (inches)	Gross area (sq in.)	Net area (sq in.)		Size (inches)	Gross area (sq in.)	Net area (sq in.)	
		⅞ in. rivets	¾ in. rivets			⅞ in. rivets	¾ in. rivets
2½ × 2 × 3/16	0.81		0.64	4 × 3½ × ¼	1.81	1.56	1.59
¼	1.06		0.84	5/16	2.25	1.94	1.98
5/16	1.31		1.04	3/8	2.67	2.29	2.34
				7/16	3.09	2.65	2.70
3 × 2 × ¼	1.19	0.94	0.97	½	3.50	3.00	3.06
5/16	1.47	1.16	1.20				
3/8	1.73	1.35	1.40	5 × 3 × ¼	1.94	1.69	1.72
				5/16	2.40	2.09	2.13
3 × 2½ × ¼	1.31	1.06	1.09	3/8	2.86	2.48	2.53
5/16	1.62	1.31	1.35	7/16	3.31	2.87	2.92
3/8	1.92	1.54	1.59	½	3.75	3.25	3.31
3½ × 2½ × ¼	1.44	1.19	1.22	5 × 3½ × ¼	2.06	1.81	1.84
5/16	1.78	1.47	1.51	5/16	2.56	2.25	2.29
3/8	2.11	1.73	1.78	3/8	3.05	2.67	2.72
7/16	2.43	1.99	2.04	7/16	3.53	3.09	3.14
½	2.75	2.25	2.31	½	4.00	3.50	3.56
				5/8	4.92	4.29	4.37
3½ × 3 × ¼	1.56	1.31	1.34				
5/16	1.93	1.62	1.66	6 × 3½ × ¼	2.31	2.06	2.09
3/8	2.30	1.92	1.97	5/16	2.87	2.56	2.60
7/16	2.65	2.21	2.26	3/8	3.42	3.04	3.09
½	3.00	2.50	2.56	½	4.50	4.00	4.06
4 × 3 × ¼	1.69	1.44	1.47	6 × 4 × 3/8	3.61	3.23	3.28
5/16	2.09	1.78	1.82	7/16	4.18	3.74	3.80
3/8	2.48	2.10	2.15	½	4.75	4.25	4.31
7/16	2.87	2.43	2.48	9/16	5.31	4.74	4.82
½	3.25	2.75	2.81				

(2) Referring to Table 7-5, we find that an angle $3\frac{1}{2} \times 3 \times \frac{3}{8}$, punched for a ⅞-in. rivet, has a net area of 1.92 sq in. Consequently, we can accept a member composed of two angles $3\frac{1}{2} \times 3 \times \frac{3}{8}$.

Problem 7-8-A*. A hanger composed of two angles sustains a tensile force of 50 kips. If ¾-in. rivets are to be used in the member, determine the size of the angles.

Problem 7-8-B. A double-angle tension member sustains a load of 100 kips. If ⅞-in. rivets are used, determine the size of the angles.

Problem 7-8-C. Two angles $3\frac{1}{2} \times 2\frac{1}{2} \times 5/16$, punched for a single line of ¾-in. rivets, are to be used as a hanger. If the load to be carried is 64 kips, are the angles large enough?

7-9 Beam Framing Connections

Figure 7-4 illustrates typical framed connections for floor beams. A common type of connection consists of two connecting angles riveted to the web of the beam; the outstanding legs of the angles are then bolted or riveted to the column or girder (Fig. 7-4*a*, *b*, *c*, *d*, and *g*). The seat angle shown in Figs. 7-4*a* and *b* is not counted on to carry any of the beam reaction but is used as an aid in erection. In order to position a spandrel beam nearer to the outer face of a wall, the arragement shown in Figs. 7-4*e* and *f* is often employed.

It is not necessary for the designer to compute the proper size of connecting angles and the number of bolts or rivets for the the usual framed connection. Standard beam connections have been developed, and the AISC Manual contains extensive tables to facilitate their use. In order that the most economical framing connections can be provided by the steel fabricator, the magnitude of beam reactions should be shown on the structural working drawings. When such reactions are known, the AISC Manual tables of framed beam connections are highly effective in selecting the proper connection for a given beam size and reaction.

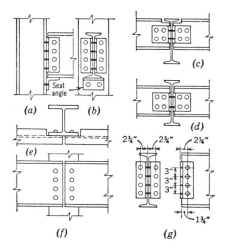

FIGURE 7-4

7-10   Free-End and Moment Connections

The framing connections discussed in Art. 7-9 fall under the category of "simple" or free-end connections. That is, insofar as gravity loading is concerned, the ends of the beams and girders are connected for shear only, and are assumed free to rotate under gravity load. This is the type of connection used in Type 2 of the three types of steel construction recognized in the AISC Specification.

Type 1 construction, commonly designated as continuous or rigid frame, assumes that beam-to-column connections possess sufficient rigidity to prevent rotation of the beam ends as the member deflects under its load. This means that the connection must transmit some bending moment between beam and column, and consequently it is called a moment-resisting connection or simply a *moment connection*. Although Type 1 continuous framing can be achieved by proper design of bolted or riveted connections, it is accomplished much more effectively in welded construction. Moment-resisting connections are used in multistory steel frame buildings to provide lateral stability against the effects of wind and earthquake forces. A fully continuous frame of Type 1 construction is statically indeterminate and its analysis and design are beyond the scope of this book.[1]

In general, the design methods and procedures for structural steel treated in this volume are applicable to Type 2 construction and follow the provisions of Section 1.12.1 of the AISC Specification which states that "Beams, girders and trusses shall ordinarily be designed on the basis of simple spans whose effective length is equal to the distance between centers of gravity of the members to which they deliver their end reactions."

---

[1] The reader who wishes to pursue further the design of continuous framing is referred to *Steel Buildings: Analysis and Design*, by S. W. Crawley and R. M. Dillon (New York: Wiley, 1970), Chapter 9.

8

Welded Connections

IIIIIIIIIIIIIIIIIIIIIIIIIIIIIIIIIIIIIIIIIIIIIIIIIIIIIIIIIIIIIII

8-1 General

In welded construction it is usually possible to attach one member directly to another without the use of additional connecting plates or angles, such as are necessary when using bolts or rivets. A welded connection requires no holes for fasteners and therefore the gross rather than the net section may be considered when determining the effective cross-sectional area of members resisting tension.

As noted in Art. 7-10, moment-resisting connections are readily achieved by welding; consequently welded connections are customary in Type 1 construction in order to develop continuity in the framing. Welding may also be used in Type 2 construction but care must be taken in design to assure that a rigid connection is not provided where free-end conditions have been assumed in the design of the framing. A combination of welding and bolting is sometimes used, generally called "shop-welded and field-bolted construction." Here, connection angles with holes in the outstanding legs are welded to a beam in the fabricating shop and then bolted to a girder or column in the field.

8-2 Electric Arc Welding

Although there are several welding processes, electric arc welding is generally used for building construction. In this type of welding an electric arc is formed between an electrode and the two pieces of metal that are to be joined. The intense heat melts a small portion of each member as well as the end of the electrode. The term *penetration* is used to indicate the depth from the original surface of the base metal to the point at which fusion ceases. The globules of melted metal from the electrode flow into the molten seat and on cooling are united with the members that are to be welded together.

8-3 Types of Welding Joints

When two members are to be joined, the ends may or may not be grooved as a preparation for welding. In general, there are three classifications of joint: *butt joints, tee joints,* and *lap joints.* The selection of the type of weld to use depends on the magnitude of the load, the manner in which it is applied, and the cost of preparation and welding. Several types of joints are shown in Fig. 8-1. The type of joint and preparation permit a number of variations. In addition, welding may be done from one or both sides.

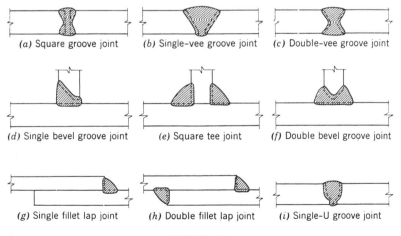

(a) Square groove joint   (b) Single-vee groove joint   (c) Double-vee groove joint

(d) Single bevel groove joint   (e) Square tee joint   (f) Double bevel groove joint

(g) Single fillet lap joint   (h) Double fillet lap joint   (i) Single-U groove joint

FIGURE 8-1

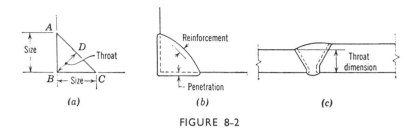

FIGURE 8-2

The weld most commonly used for structural steel in building construction is the *fillet weld*. It is approximately triangular in cross section and is formed between the two intersecting surfaces of the joined members. See Figs. 8-2a and b. The *size* of a fillet weld is the leg length of the largest inscribed isosceles right triangle, *AB* or *BC* (Fig. 8-2a). The *root* of the weld is the point at the bottom of the weld (point *B* in Fig. 8-2a). The *throat* of a fillet weld is the distance from the root to the hypotenuse of the largest isosceles right triangle that can be inscribed within the weld cross section (distance *BD* in Fig. 8-2a). The exposed surface of a weld is not the plane surface indicated in Fig. 8-2a but is usually somewhat convex. See Fig. 8-2b. Therefore the actual throat may be somewhat greater than that shown in Fig. 8-2a. This additional material is called *reinforcement*. It is not included in determining the strength of a weld.

A single-vee groove weld between two members of unequal thickness is shown in Fig. 8-2c. The *size* (throat dimension) of a butt weld is the thickness of the thinner part joined, with no allowance made for the weld reinforcement.

8-4 Stresses in Welds

If the dimension (size) of *AB* in Fig. 8-2a is 1 unit in length, $(AD)^2 + (BD)^2 = 1^2$. Because AD and BD are equal, $2(BD)^2 = 1^2$ and $BD = \sqrt{0.5}$, or 0.707. Therefore the throat of a fillet weld is equal to the *size* of the weld multiplied by 0.707. As an example, consider a $\frac{1}{2}$-in. fillet weld. This would be a weld having the dimensions *AB* or *BC* equal to $\frac{1}{2}$ in. In accordance with the foregoing, the throat would be $0.5 \times 0.707 = 0.3535$ in. Then, if the allowable unit shearing stress on the throat is 21 ksi, the allowable working strength

TABLE 8-I. Allowable Working Strength of Fillet Welds

Size of fillet weld (inches)	Allowable loads in kips per lin in.	
	E 60 XX electrodes $F_{vw} = 18$ ksi	E 70 XX electrodes $F_{vw} = 21$ ksi
$\frac{3}{16}$	2.4	2.8
$\frac{1}{4}$	3.2	3.7
$\frac{5}{16}$	4.0	4.6
$\frac{3}{8}$	4.8	5.6
$\frac{1}{2}$	6.4	7.4
$\frac{5}{8}$	8.0	9.3
$\frac{3}{4}$	9.5	11.1
1	12.7	14.8

of a $\frac{1}{2}$-in. fillet weld is $0.3535 \times 21 = 7.42$ kips *per lin in. of weld.* If the allowable unit stress is 18 ksi, the allowable working strength is $0.3535 \times 18 = 6.36$ kips *per lin in. of weld.*

The allowable unit shearing stresses used in the preceding paragraph are for welds made with E 70 XX and E 60 XX type electrodes on A36 steel. Table 8-1 gives allowable working strengths of fillet welds of various sizes, with values rounded to $\frac{1}{10}$ kip. Particular attention is called to the fact that *the stress in a fillet weld is considered as shear on the throat, regardless of the direction of the applied load.* Neither plug nor slot welds (Art. 8-7) are assigned any values other than shear in determining their resistance. For butt welds, the allowable stresses are the same as those for the metal of the connected parts, called the *base metal.*

The relation between weld size and the thickness of material in joints connected only by fillet welds is shown in Table 8-2. The maximum size of a fillet weld applied at the squared edge of a plate or section $\frac{1}{4}$ in. or more in thickness should be $\frac{1}{16}$ in. less than the nominal thickness of the edge. Along edges of material less than $\frac{1}{4}$ in. thick, the maximum size may be equal to the thickness of the material.

The effective area of butt and fillet welds is considered to be the

TABLE 8-2.   Relation Between Material Thickness and Minimum Size of Fillet Welds *

Material thickness of thicker part joined (inches)	Minimum size of fillet weld (inches)	Material thickness of thicker part joined (inches)	Minimum size of fillet weld (inches)
To $\frac{1}{4}$ inclusive	$\frac{1}{8}$	Over $1\frac{1}{2}$ to $2\frac{1}{4}$	$\frac{3}{8}$
Over $\frac{1}{4}$ to $\frac{1}{2}$	$\frac{3}{16}$	Over $2\frac{1}{4}$ to 6	$\frac{1}{2}$
Over $\frac{1}{2}$ to $\frac{3}{4}$	$\frac{1}{4}$	Over 6	$\frac{5}{8}$
Over $\frac{3}{4}$ to $1\frac{1}{2}$	$\frac{5}{16}$		

* Taken from the 7th Edition of the *Manual of Steel Construction*. Courtesy American Institute of Steel Construction.

effective length of the weld multiplied by the effective throat thickness. The minimum effective length of a fillet weld should not be less than four times the weld size. For starting and stopping the arc approximately $\frac{1}{4}$ in. should be added to the design length of fillet welds.

Figure 8-3a represents two plates connected by fillet welds. The welds marked A are longitudinal welds; B indicates a traverse weld. If a load is applied in the direction shown by the arrow, the stress distribution in the longitudinal weld is not uniform and the stress in the transverse weld is approximately 30% greater per unit of length.

Added strength is given to a transverse fillet weld that terminates at the end of a member, as shown in Fig. 8-3b, if the weld is returned around the end for a distance not less than twice the weld size. These end returns, called *boxing*, provide considerable resistance to the tendency of tearing action on the weld.

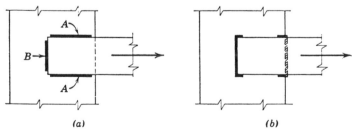

(a)   (b)

FIGURE 8-3

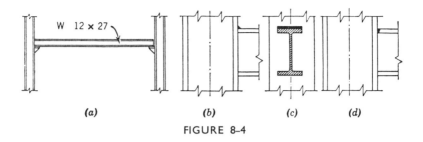

(a)   (b)   (c)   (d)

FIGURE 8-4

The $\frac{1}{4}$-in. fillet weld is considered to be the minimum practical size and a $\frac{5}{16}$-in. weld is probably the most economical size that can be obtained by one pass of the electrode. Fillet welds larger than $\frac{5}{16}$ in. must be made with two or more "passes" of the electrode.

Example. A W 12 × 27 of A36 steel is to be welded to the face of a steel column with E 70 XX electrodes. The general arrangement is shown in Fig. 8-4a. With respect to the upper flange only, compute the strength of fillet and butt welds. If we assume that the beam is to be welded to produce continuous action in a moment-resisting connection (Art. 7-10), the upper flange will be in tension at the column. See Fig. 8-7a.

Solution: (1) First assume that the left end of the beam is in contact with the column and that a $\frac{3}{8}$-in. fillet weld is run across the upper face of the beam flange as shown in Figs. 8-4b and c. Table 4-1 shows that the flange width of the W 12 × 27 is 6.5 in. and the flange thickness is 0.40 in. The length of fillet weld is, therefore, 6.5 in.

(2) Referring to Table 8-1 under E 70 XX electrodes, the strength of a $\frac{3}{8}$-in. fillet weld is given as 5.6 kips per lin in., making the allowable strength of the weld 5.6 × 6.5 = 36.4 kips. Note that this weld resists tensile stresses but, as stated earlier, the strength of a fillet weld is determined by shear at the throat of the weld.

(3) Next, instead of the fillet weld, suppose that the upper flange is beveled and that a butt weld is used as shown in Fig. 8-4d. The area of the weld resisting tension is the flange width multiplied by the flange thickness, or 6.5 × 0.40 = 2.6 sq in. Since the allowable tensile stress for A36 steel from which the beam is made is 22 ksi (Table 5-2), the allowable strength of the butt weld in tension is 22 × 2.6 = 57.2 kips.

Problem 8-4-A. A W 10 × 21 of A36 steel is to have its upper flange welded to the supporting column as in the preceding example and as shown in Figs. 8-4*b* and *c*. Determine the allowable working strength of the joint if a ⁵⁄₁₆-in. fillet weld made with E 70 XX electrodes is used.

8-5 Design of Welded Joints

The most economical choice of weld to use for a given condition depends on several factors. It should be borne in mind that·members to be connected by welding must be firmly clamped or held rigidly in position during the welding process. In riveting a beam to a column it is necessary to provide a seat angle as a support to keep the beam in position for riveting the connecting angles. The seat angle is not considered as adding strength to the connection. Similarly, with welded connections, seat angles are commonly used. The designer must have in mind the actual conditions during erection and must provide for economy and ease in working the welds. Seat angles or similar members used to facilitate erection are *shop welded* before the material is sent to the site. The welding done during erection is called *field welding*. The designer in preparing welding details indicates on the drawings which are shop and which are field welds. Conventional welding symbols indicate the type, size, and position of the various welds. Only engineers or architects experienced in the design of welded connections should design or supervise welded construction. It is apparent that a wide variety of connections is possible; experience is the best aid in determining the most economical and practical.

The following examples illustrate the basic principles on which welded connections are designed.

Example 1. Figures 8-5*a* and *b* indicate a steel bar welded to the back of a channel. The bar is made of A36 steel and has cross-sectional dimensions of 3 × ⁷⁄₁₆ in. A ³⁄₈-in. fillet weld made with E 70 XX electrodes is to be used. If it is desired to develop the full strength of the bar in tension, what is the length of weld required?
Solution: (1) The area of the bar is 3 × 0.4375 = 1.313 sq in. Since the allowable unit tensile stress in the bar is 22 ksi (Table 5-2), the tensile strength of the bar is $F_t \times A = 22 \times 1.313 = 28.9$ kips.

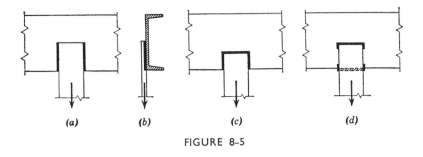

(a)   (b)   (c)   (d)

FIGURE 8–5

(2) Table 8-1 gives the allowable working strength of a $\frac{3}{8}$-in. fillet weld as 5.6 kips per lin in. Therefore, the length of weld necessary to develop the strength of the bar in tension is 28.9 ÷ 5.6 = 5.16 in. Placing half the length along each edge, the bar would have to overlap the channel 5.16 ÷ 2 = 2.58 in. However, $\frac{1}{4}$ in. should be added to this dimension to allow for starting and stopping the arc, so the length of weld on each edge becomes 2.58 + 0.25 = 2.83 in., say $2\frac{7}{8}$ in. Other possible positions for placing the weld are shown in Figs. 8-5c and d. Regardless of the pattern adopted, the minimum design length of the weld must be 5.16 in.

Example 2. A single-angle hanger $3\frac{1}{2} \times 3\frac{1}{2} \times \frac{5}{16}$ is to be connected to a plate as indicated in Fig. 8-6. Type E 70 XX electrodes will be used to make $\frac{1}{4}$-in. fillet welds. If all steel is A36, what should

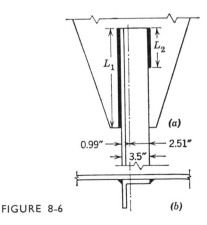

0.99″   2.51″

3.5″

(a)

(b)

FIGURE 8-6

the dimensions of the welds be to develop the full tensile strength of the hanger?

Solution: (1) The allowable working strength of the $\frac{1}{4}$-in fillet weld is 3.7 kips per lin in. (Table 8-1). The cross-sectional area of the angle is given in Table 4-4 as 2.09 sq in. Since A36 steel has an allowable unit tensile stress of 22 ksi (Table 5-2), the tensile strength of the angle is 22 × 2.09 = 46 kips. Therefore, the required length of $\frac{1}{4}$-in. fillet weld to develop the strength of the angle is 46 ÷ 3.7 = 12.4 in.

(2) A structural angle is an unsymmetrical section and the welds marked L_1 and L_2 in Fig. 8-6 will be made unequal in length so that their stresses will be proportioned in accordance with the distributed area of the angle. Referring to Table 4-4, we find that the centroid of the angle section is 0.99 in. from the back of the angle; hence the two welds are 0.99 and 2.51 in. from the centroidal axis, as recorded in Fig. 8-6a. The lengths of welds L_1 and L_2 are made inversely proportional to their distances from the axis, but the sum of their lengths is 12.4 in. Therefore

$$L_1 = \frac{2.51}{3.5} \times 12.4 = 8.9 \text{ in.}$$

and

$$L_2 = \frac{0.99}{3.5} \times 12.4 = 3.5 \text{ in.}$$

These are the design lengths required and, as noted earlier, each weld would actually be made $\frac{1}{4}$ in. longer than its required design length.

Problem 8-5-A. An angle 4 × 4 × ¼ of A36 steel is to be welded to a plate to resist the full tensile strength of the angle. Using ⅜-in. fillet welds made with E 70 XX electrodes, compute the lengths L_1 and L_2 as shown in Fig. 8-6.

8-6 Beams with Continuous Action

As noted earlier, one of the advantages of welding is that beams having continuous action at the supports (Type 1 construction) are readily provided for. The usual bolted or riveted connections of

Type 2 construction are assumed to offer no rigidity at the supports, and the bending moment throughout the length of the beam is positive. By the use of welding, however, a beam may be connected at its supports in such a manner that its ends are *fixed* or *restrained* and a negative bending moment results (Art. 3-7). For the same span and loading, the maximum bending moment for a restrained beam is smaller than for a simple beam, and a lighter beam section may be used.

When beams are rigidly connected by means of moment-resisting connections, the fibers in the upper flange *at the supports* are in tension and the lower flange is in compression. This is shown diagrammatically in Fig. 8-7a. Consequently, when designing welds for beams which have continuous action, we must provide for tension and compression in the upper and lower flanges, respectively, at the supports. The example of Art. 8-4 illustrated how this may be accomplished with either fillet or butt welds.

Figure 8-7b indicates that a beam must be fabricated so there will

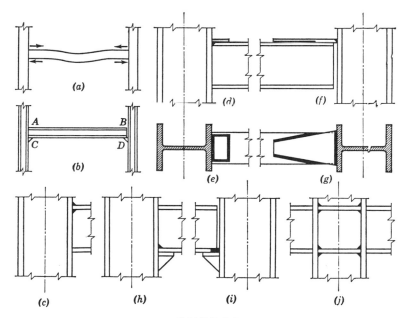

FIGURE 8-7

be clearance between its ends and the column faces. This is necessary for erection purposes. In order to provide for continuous action, a number of arrangements of the welds at the columns are possible; several of these are shown in Fig. 8-7.

8-7 Plug and Slot Welds

One method of connecting overlapping plates uses welds in holes made in one of the two plates. See Fig. 8-8. Generally, the terms plug and slot welds refer to welds in which the entire area of the hole or slot receives weld metal. A somewhat similar weld consists of a fillet weld at the circumference of a hole as shown in Fig. 8-8c. Plug or slot welds are used to transmit shear in a lap joint. The maximum and minimum diameters of the holes and slots and the maximum length of slot welds are shown in Fig. 8-8. If the plate containing the hole is not more than $\frac{5}{8}$ in. thick, the hole should be filled with weld metal. If the plate is more than $\frac{5}{8}$ in. thick, the weld metal should be at least one half the thickness of the material but not less than $\frac{5}{8}$ in.

The stress in a plug or slot weld is considered to be shear on the area of the weld at the plane of contact of the two plates to be connected. The allowable unit shearing stress when E 70 XX electrodes are used is 21 ksi.

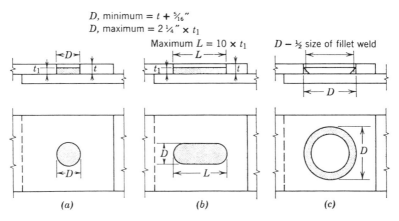

FIGURE 8-8

Example. A plug weld consists of weld metal in a $1\frac{1}{4}$-in. diameter hole through a plate $\frac{1}{2}$ in. thick, as indicated in Fig. 8-8a. Compute the load this weld will transmit.

Solution: The effective weld area is the area of the $1\frac{1}{4}$-in. diameter circle or $D^2 \times 0.7854 = 1.25^2 \times 0.7854 = 1.227$ sq in. The allowable shearing stress of the weld metal is 21 ksi; therefore, the load the plug weld will transmit is $1.227 \times 21 = 25.8$ kips.

Problem 8-7-A. A plug weld consists of a $1\frac{1}{2}$-in. diameter hole filled with weld metal in a $\frac{5}{8}$-in. plate. What load will this weld transmit if E 70 XX electrodes are used?

Problem 8-7-B*. The hole for a slot weld is 1 in. wide and 5 in. long. The ends of the slot are semicircular, as shown in Fig. 8-8b. Compute the load this weld will transmit if E 70 XX electrodes are used.

8-8 Miscellaneous Welded Connections

Part 4 of the AISC Manual contains a series of tables pertaining to the design of welded connections; they cover free-end as well as moment-resisting connections for use in Type 2 and Type 1 construction, respectively. In addition, suggested framing and column splicing details are presented.

A few miscellaneous connections are shown in Fig. 8-9. As an aid in erection, certain parts are welded together in the shop before being sent to the site. Connection angles may be shop-welded to beams and the angles field-welded to girders or columns. The beam connection shown in Fig. 8-9a shows a beam supported on a seat that has been shop-welded to the column. A small connection plate is shop-welded to the lower flange of the beam and the plate is bolted to the beam seat. After the beams have been erected and the frame plumbed, the beams are field-welded to the seat angles. This type of beam connection provides no degree of continuity, there being no restraint, hence no bending moment in the beam at the support.

Figures 8-9b and c illustrate beams connected to columns so that a continuous action results. Auxiliary plates are used to make the connection at the upper flange.

Beam seats shop-welded to columns are shown in Figs. 8-9d, e, and f. Figure 8-9d shows a short length of angle welded to the column with no stiffeners. Stiffeners (triangular-shaped plates) are welded

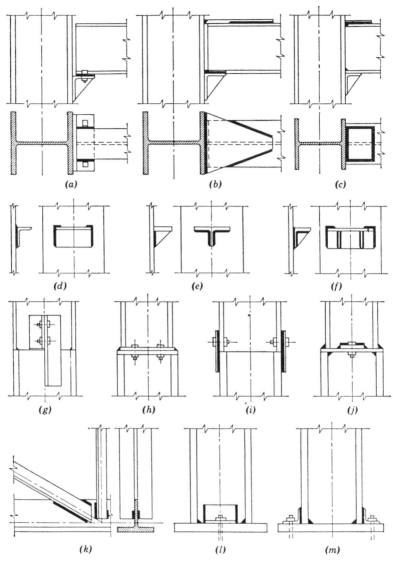

FIGURE 8-9

to the legs of the angles, as shown in Fig. 8-9*f*, and they add materially to the strength of the seat. Another method of forming a beam seat is to use a T-section or a half I-beam, as shown in Fig. 8-9*e*.

Various column splices are shown in Figs. 8-9*g*, *h*, *i*, and *j*. The auxiliary plates and angles are shop-welded to the columns to provide bolted connections in the field before making the permanent welds.

Figure 8-9*k* shows a type of welded connection used in trusses where the lower chord is composed of a structural T-section or a split I-beam. The angle sections of the truss web members are welded directly to the stem of the tee, without the use of a connecting plate.

Welded connections for columns and base plates are shown in Figs. 8-9*l* and *m*. The angles are shop-welded to the columns and the columns are field-welded to the base plates.

9

Plastic Design Theory

IIIIIIIIIIIIIIIIIIIIIIIIIIIIIIIIIIIIIIIIIIIIIIIIIIIIII

9-1 Introduction

As noted in Art. 5-27, the *plastic design method* is an alternative to *allowable stress design*. The latter is predicated on the use of working stresses less than the elastic limit of the material; plastic design procedures are based on the idea of computing an ultimate load and on utilizing a portion of the reserve strength (after initial yield stress has been reached) as part of the factor of safety (Art. 1-12). Detailed consideration of the plastic design method is beyond the scope of this book, but the following brief discussion should serve as an introduction to the theory on which it is based.

It has been found by tests that members can carry loads much higher than anticipated even when the yield point stress is reached at sections of maximum bending moment. This is particularly evident in continuous beams and structures employing rigid frames. An inherent property of structural steel is its ability to resist large deformations without failure. The deformations in these structures fall within the *plastic range*, deformations that occur without an increase in bending stress. Because of this phenomenon, the *plastic*, sometimes called the *ultimate strength design theory*, has been developed.

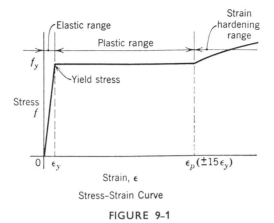

Stress–Strain Curve

FIGURE 9-1

9-2 Stress–Strain Diagram

Figure 9-1 is a graphic representation of a steel specimen as it deforms under stress. The horizontal scale is greatly exaggerated. The term *strain* is synonymous with *deformation*. The purpose of this graph is to show that up to a stress f_y, the yield point, the deformations are directly proportional to the applied stresses and that beyond this point of maximum elastic strain there is a deformation without an increase in stress. The total deformation is approximately 15 times that produced elastically. This deformation is called the *plastic range*, beyond which *strain hardening* begins when further deformation can occur only with an increase in stress.

9-3 Plastic Moment, Plastic Hinge

Article 4-7 explains the design of members in bending in accordance with the theory of elasticity. When the extreme fiber stress does not exceed the elastic limit, the bending stresses in the cross section of a beam are directly proportional to their distances from the neutral surface. In addition, the strains (deformations) in these fibers are directly proportional to their distances from the neutral surface. Both stress and strain are zero at the neutral surface and both increase to maximum magnitudes at the fibers farthest from the neutral

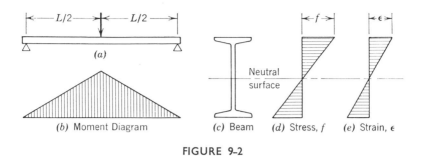

FIGURE 9-2

surface. This condition is indicated in Fig. 9-2*d* for the maximum extreme fiber stress which occurs only at the center of the span where the bending moment has its maximum value. Figure 9-2*e* shows ϵ, the deformation of the fiber farthest from the neutral surface.

When a beam is loaded to produce an extreme fiber stress in excess of the yield point, the property of *ductility* affects the distribution of the stresses. Ductility is the property of a material that permits it to undergo plastic deformation when subjected to stresses. Structural steel is a ductile material. Ductility permits a redistribution of stresses in a beam cross section. Fibers that were less stressed originally come to the assistance of the more highly stressed fibers.

Assume that the bending moment on a beam is of such magnitude that the extreme fiber stress is f_y, the yield stress. Then if M_y is the elastic bending moment at the yield stress, $M = M_y$, and the distribution of stress in the cross section is as shown in Fig. 9-3*a*; the maximum bending stress f_y is at the extreme fiber.

Next consider that the loading and the resulting bending moment have been increased; M is now greater than M_y. The stress on the extreme fiber is still f_y, but *the material has yielded* and a greater area of the cross section is also stressed to f_y. The stress distribution is shown in Fig. 9-3*b*.

Again imagine the load to be further increased. The stress on the extreme fiber is still f_y and, theoretically, *all fibers in the cross section are stressed to f_y.* This is an idealized plastic stress distribution; it is shown in Fig. 9-3*d*. The bending moment producing this condition is M_p, the plastic bending moment. In reality, 10% of the central portion of the cross section resists elastically distributed stresses as

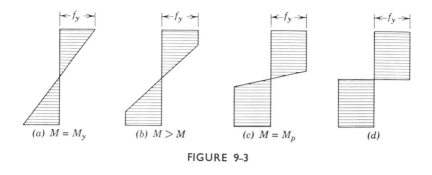

(a) $M = M_y$ (b) $M > M$ (c) $M = M_p$ (d)

FIGURE 9-3

indicated in Fig. 9-3c. This small area is considered to be negligible, and we assume that the stresses on all the fibers of the cross section are f_y, as shown in Fig. 9-3d. The section is now said to be fully plastic and a further increase in load will result in large deformations, the beam acting as if it were hinged at this section. We call this a *plastic hinge* at which free rotation is permitted only after M_p has been attained. See Fig. 9-4. At sections of the beam in which this condition prevails, the bending resistance of the cross section has been exhausted.

9-4   Plastic Section Modulus

In elastic design the moment producing the maximum allowable resisting moment may be found by the flexure formula

$$M = f \times S$$

If the extreme fiber is stressed to the yield stress,

$$M_y = f_y \times S$$

in which M_y = elastic bending moment at yield stress,
f_y = yield stress in psi or ksi,
S = section modulus.

Now let us find a similar relation between the plastic moment and its plastic resisting moment. Figure 9-5 shows the cross section of a W or S section in which the bending stress f_y, the yield stress, is

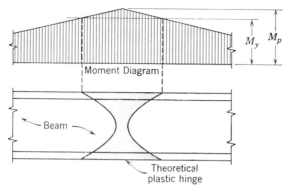

FIGURE 9-4

constant over the cross section. In the figure

A_u = upper area of the cross section above the neutral axis in square inches,

y_u = distance of the centroid of A_u from the neutral axis,

A_l = lower area of the cross section below the neutral axis in square inches,

y_l = distance of the centroid of A_l from the neutral axis.

For equilibrium the algebraic sum of the horizontal forces must be zero. Then

$$\sum H = 0$$

or

$$[A_u \times (+f_y)] + [A_l \times (-f_y)] = 0$$

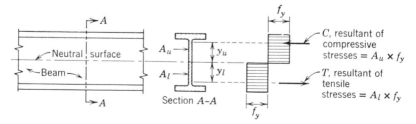

FIGURE 9-5

and

$$A_u = A_l$$

This shows that the neutral axis divides the cross section into equal areas. This is apparent in symmetrical sections but it applies to unsymmetrical sections as well. Also, the bending moment equals the sum of the moments of the stresses in the section. Thus for M_p the plastic moment,

$$M_p = (A_u \times f_y \times y_u) + (A_l \times f_y \times y_l)$$

or

$$M_p = f_y[(A_u \times y_u) + (A_l \times y_l)]$$

and

$$M_p = f_y \times Z$$

The quantity $(A_u y_u + A_l y_l)$ is called the *plastic section modulus* of the cross section and it is designated by the letter Z. Since it is an area multiplied by a distance, it is in units to the third power. If the area is in units of square inches and the distance is in linear inches, Z, the section modulus, is in units of inches to the third power, the same as for S, the elastic section modulus. Tables of plastic section moduli for W and S shapes are given in Part 2 of the AISC Manual.

The plastic section modulus is always larger than the elastic section modulus.

It is important to note that, in plastic design, the neutral axis for unsymmetrical cross sections does not pass through the centroid of the section. In plastic design the neutral axis divides the cross section into *equal* areas.

9-5 Shape Factor

As noted in the preceding article, the plastic section modulus is always larger than the elastic section modulus. The ratio of the former to the latter is called the *shape factor* and is denoted by the letter u. Then

$$u = \frac{Z}{S}$$

For the commonly used wide flange and standard I-beam sections, the shape factor is approximately 1.12 for bending about the strong axis of the section. This means that these sections can support approximately 12% more load than we expect them to carry on the basis of elastic design.

9-6 Load Factor

The term *load factor* is given to the ratio of the ultimate load to the design load on a member. Because we cannot determine loads with minute accuracy and because our materials are not always completely uniform, it would be dangerous to allow the actual load on a beam to become as large as the load that produces the plastic bending moment, M_p (Fig. 9-3c). In elastic design the allowable bending stress F_b is decreased to a fraction of F_y, the yield stress. For compact sections $F_b = 0.66F_y$. Therefore the factor of safety against yielding is $1/0.66$, or 1.5.

In plastic design it is simpler to base computations on F_y, the yield stress, and to increase the loading by making it a multiple of the actual load. For simple and continuous beams the load factor is 1.7.

Example. A simple beam of A36 steel has a 20-ft span and supports a uniformly distributed load of 4.8 kips per lin ft, including its own weight. Full lateral support is provided. Design the beam for bending (1) in accordance with elastic theory (allowable stress design), and (2) in accordance with plastic theory.

Solution 1: (1) The maximum bending moment is $wL^2/8$. Then

$$M = \frac{wL^2}{8} = \frac{4.8 \times 20 \times 20}{8} = 240 \text{ kip-ft}$$

(2) Since the allowable bending stress for a compact beam is 24 ksi, the section modulus required is

$$S = \frac{M}{F_b} = \frac{240 \times 12}{24} = 120 \text{ in.}^3$$

Referring to the second column of Table 5-4, we find that a W 21 × 62 has a section modulus of 127 in.^{3}

Solution 2: (1) Multiplying the actual load by the load factor we have

$$w_p = w \times 1.7 = 4.8 \times 1.7 = 8.16 \text{ kips per lin ft}$$

(2) The plastic bending moment is then

$$M_p = \frac{w_p L^2}{8} = \frac{8.16 \times 20 \times 20}{8} = 408 \text{ kip-ft}$$

(3) The plastic section modulus required is

$$Z = \frac{M_p}{F_y} = \frac{408 \times 12}{36} = 136 \text{ in.}^3$$

Referring to a plastic section modulus table such as that contained in the AISC Manual, it will be found that $Z = 144$ in.^{3} for a W 21 × 62. This is the same section as that found in Solution 1. The design of simple beams by either the elastic theory or the plastic theory will usually result in the same size beam; for continuous beams, however, the plastic design method generally results in smaller beam sizes.

9-7 Scope of Plastic Design

The plastic design method has applications far beyond continuous beams; particularly in the design of rigid frames where combinations of bending and direct stress are involved. Since the design is based on higher stresses adequate attention must be given to the possibility of local buckling, and care must be taken to prevent excessive deflections that might occur because of the smaller sections that usually result from use of the theory. The reader who wishes to pursue the plastic design method further is referred to *Plastic Design in Steel*, published by the American Institute of Steel Construction.

III

WOOD CONSTRUCTION

10

Wood Beams

||||||||||||||||||||||||||||||||||||||||||||||||||||||||

10-1 Structural Lumber

Unlike the metals, wood is not a processed material but an organic material generally used in its natural state. Aside from the natural properties of the species, the most important factors that influence its strength are density, natural defects (knots, checks, slope of grain, etc.), and moisture content. Because the effects of natural defects on the strength of lumber vary with the type of loading to which an individual piece is subjected, structural lumber is classified according to its *size and use*. The three major classifications are:

Joists and Planks. Rectangular cross sections with nominal dimensions 2 in. to 4 in. thick and 4 or more in. wide, graded primarily for strength in bending edgewise or flatwise.

Beams and Stringers. Rectangular cross sections with nominal dimensions 5 × 8 in. and larger, graded for strength in bending when loaded on the narrow face.

Posts and Timbers. Square or nearly square cross sections with nominal dimensions 5 × 5 in. and larger, graded primarily for use as posts or columns but adapted to other uses where bending strength is not especially important.

The two groups of trees used for building purposes are the *softwoods* and the *hardwoods*. Softwoods such as the pines or cypress are coniferous or cone bearing, whereas hardwoods have broad leaves, as exemplified by the oaks and maples. The terms softwood and hardwood are not accurate indications of the degree of hardness of the various species of trees. Certain softwoods are as hard as the medium density hardwoods, while some species of hardwoods have softer wood than some of the softwood species. The two species of trees used most extensively in the United States for structural members are Douglas Fir and Southern Pine, both of which are classified among the softwoods.

10-2 Nominal and Dressed Sizes

An individual piece of structural lumber is designated by its *nominal* cross-sectional dimensions; the size is indicated by the breadth and depth of the cross section in inches. As an example, we speak of a "6 by 12" (written 6 × 12), and by this we mean a timber with a nominal breadth of 6 in. and depth of 12 in.; the length is variable. However, after being dressed or surfaced on four sides (S4S) the actual dimensions of this piece are $5\frac{1}{2} \times 11\frac{1}{2}$ in. Since lumber used in structural design is almost exclusively dressed lumber, the sectional properties (A, I, and S) given in Table 4-6 are for standard dressed sizes conforming to those established in the American Softwood Lumber Standard, PS 20-70, promulgated through the U.S. Department of Commerce.

10-3 Allowable Stresses for Structural Lumber

Many factors are taken into account in determining the allowable unit stresses for structural lumber. Numerous tests by the Forest Products Laboratory of the U.S. Department of Agriculture made on material free from defects have resulted in a tabulation known as *clear wood strength values*. To obtain allowable design stresses, the clear wood values are reduced by factors that take into consideration the loss of strength from defects, size and position of knots, size of member, degree of density, and condition of seasoning. These adjustments are made in accordance with ASTM Designation D-245, "Methods for Establishing Structural Grades and Related Allowable Properties for Visually Graded Lumber." Grading is necessary to identify lumber quality. Individual grades are given a commercial designation such as No. 1, No. 2, Select Structural, Dense No. 2, etc., to which a schedule of allowable unit stresses is assigned.

Table 10-1, which has been compiled from more extensive data given in the 1973 Edition of the *National Design Specification for Stress-Grade Lumber and Its Fastenings*, lists some of the most commonly used stress-grade woods and their allowable working stresses. The working stresses tabulated therein are for normal loading conditions and are applicable to lumber that will be used under continuously dry conditions, such as exist in most covered structures.[1] The stresses given are: extreme fiber in bending, F_b; tension parallel to grain F_t; horizontal shear, F_v; compression perpendicular to grain, $F_{c\perp}$; compression parallel to grain, F_c; and modulus of elasticity, E. In addition, there is a column on size classification which relates to the three major classifications described in Art. 10-1, and a column at the extreme right designates the grading rules agency concerned.

It will be noted that two sets of values are given in the table for F_b, the extreme fiber stress in bending. The values listed for single-member uses apply where an individual beam or girder carries its full design load; the values given for repetitive-member uses are intended for design of members in bending such as joists, rafters, or similar members that are spaced not more than 24 in., are 3 or more in number, and are joined by floor, roof, or other load-distributing construction adequate to support the design load.

The notes in parentheses following species designations indicate whether the lumber is surfaced (dressed) when dry or green and the maximum moisture content (max. m.c.) at which it is intended to be used. Dry lumber is defined as lumber that has been seasoned to a

---

[1] Where wet conditions exist, and in situations where a member is fully stressed to the maximum allowable unit stress for many years (full load permanently applied), the allowable stress values in Table 10-1 are subject to adjustments. Methods for making such adjustments are given in the National Design Specification cited above.

TABLE 10-1. Allowable Unit Stresses for Structural Lumber—Visual Grading*

(Allowable unit stresses listed are for normal loading conditions)

Species and commercial grade	Size classification	Allowable unit stresses in pounds per square inch							
		Extreme fiber in bending F_b		Tension parallel to grain F_t	Horizontal shear F_v	Compression perpendicular to grain $F_{c\perp}$	Compression parallel to grain F_c	Modulus of elasticity E	Grading rules agency
		Single-member uses	Repetitive-member uses						
BALSAM FIR (Surfaced dry or surfaced green. Used at 19% max. m.c.)									
Select Structural	2″ to 4″ thick	1150	1350	775	60	170	925	1,200,000	Northeastern Lumber Manufacturers Association and Northern Hardwood and Pine Manufacturers Association
No. 1		1000	1150	650	60	170	825	1,200,000	
No. 2		825	950	525	60	170	700	1,100,000	
No. 3		475	550	300	60	170	450	900,000	
Appearance		1000	1150	650	60	170	1000	1,200,000	
Select Structural	Beams and Stringers	1050	—	700	55	170	725	1,000,000	
No. 1		875	—	575	55	170	625	1,000,000	
Select Structural	Posts and Timbers	975	—	650	55	170	775	1,000,000	
No. 1		800	—	525	55	170	675	1,000,000	

CALIFORNIA REDWOOD (Surfaced dry or surfaced green. Used at 19% max. m.c.)

Grade	Size							
Clear Heart Structural	4" & less thick any width	2300	2650	1550	145	425	2150	1,400,000
Clear Structural		2300	2650	1550	145	425	2150	1,400,000
Select Structural	2" to 4" thick 6" and wider	1750	2000	1150	100	425	1550	1,400,000
Select Structural Open grain		1400	1600	925	100	270	1150	1,100,000
No. 1		1500	1700	975	100	425	1400	1,400,000
No. 1, Open grain		1150	1350	775	100	270	1050	1,100,000
No. 2		1200	1400	800	80	425	1200	1,250,000
No. 2, Open grain		950	1100	625	80	270	875	1,000,000
No. 3		700	800	475	80	425	725	1,100,000
No. 3, Open grain		550	650	375	80	270	525	900,000
Clear Heart Structural or Clear Structural	5" by 5" and larger	1850	—	1250	135		1650	1,300,000
Select Structural		1400	—	950	95	425	1200	1,300,000
No. 1		1200	—	800	95	425	1050	1,300,000
No. 2		975	—	650	95	425	900	1,100,000
No. 3		550	—	375	95	425	550	1,000,000

Redwood Inspection Service

* Compiled from data in the 1973 Edition of the *National Design Specification for Stress-Grade Lumber and Its Fastenings*. Courtesy National Forest Products Association.

TABLE 10-1. Allowable Unit Stresses for Structural Lumber—Visual Grading (*Continued*)

Species and commercial grade	Size classification	Allowable unit stresses in pounds per square inch							Grading rules agency
		Extreme fiber in bending F_b		Tension parallel to grain F_t	Horizontal shear F_v	Compression perpendicular to grain $F_{c\perp}$	Compression parallel to grain F_c	Modulus of elasticity E	
		Single-member uses	Repetitive-member uses						
DOUGLAS FIR-LARCH (Surfaced dry or surfaced green. Used at 19% max. m.c.)									
Dense Select Structural	2" to 4" thick 6" and wider	2100	2400	1400	95	455	1650	1,900,000	West Coast Lumber Inspection Bureau and Western Wood Products Association
Select Structural		1800	2050	1200	95	385	1400	1,800,000	
Dense No. 1		1800	2050	1200	95	455	1450	1,900,000	
No. 1		1500	1750	1000	95	385	1250	1,800,000	
Dense No. 2		1450	1700	950	95	455	1250	1,700,000	
No. 2		1250	1450	825	95	385	1050	1,700,000	
No. 3		725	850	475	95	385	675	1,500,000	
Appearance	2" to 4" thick 6" and wider	1500	1750	1000	95	385	1500	1,800,000	
Dense Select Structural	Beams and Stringers	1900	—	1250	85	455	1300	1,700,000	Western Wood Products Association
Select Structural		1600	—	1050	85	385	1100	1,600,000	
Dense No. 1		1550	—	1050	85	455	1100	1,700,000	
No. 1		1350	—	900	85	385	925	1,600,000	
Dense Select Structural	Posts and Timbers	1750	—	1150	85	455	1350	1,700,000	
Select Structural		1500	—	1000	85	385	1150	1,600,000	
Dense No. 1		1400	—	950	85	455	1200	1,700,000	
No. 1		1200	—	825	85	385	1000	1,600,000	

EASTERN HEMLOCK—TAMARACK (Surfaced dry or surfaced green. Used at 19% max. m.c.)

Northeastern Lumber Manufacturers Association and Northern Hardwood and Pine Manufacturers Association

Grade	Size							
Select Structural	2" to 4" thick	1550	1750	1050	85	365	1200	1,300,000
No. 1		1300	1500	875	85	365	1050	1,300,000
No. 2	6" and wider	1050	1200	700	85	365	900	1,100,000
No. 3		625	725	400	85	365	575	1,000,000
Appearance		1300	1500	875	85	365	1300	1,300,000
Select Structural	Beams and Stringers	1400	—	925	80	365	950	1,200,000
No. 1		1150	—	775	80	365	800	1,200,000
Select Structural	Posts and Timbers	1300	—	875	80	365	1000	1,200,000
No. 1		1050	—	700	80	365	875	1,200,000

ENGELMANN SPRUCE—LODGEPOLE PINE (ENGELMANN SPRUCE) (Surfaced dry or surfaced green. Used at 19% max. m.c.)

Western Wood Products Association

Grade	Size							
Select Structural	2" to 4" thick	1150	1350	775	70	195	800	1,200,000
No. 1/Appearance	6" and wider	975	1150	650	70	195	725/875	1,200,000
No. 2		800	925	525	70	195	600	1,100,000
No. 3		475	550	300	70	195	375	1,000,000
Select Structural	Beams and Stringers	1050	—	700	65	195	650	1,100,000
No. 1		875	—	575	65	195	550	1,100,000
Select Structural	Posts and Timbers	950	—	650	65	195	675	1,100,000
No. 1		775	—	525	65	195	600	1,100,000

SOUTHERN PINE (Surfaced dry. Used at 19% max. m.c.)

Southern Pine Inspection Bureau

Grade	Size							
Select Structural	2" to 4" thick	1800	2050	1200	90	405	1400	1,800,000
Dense Select Structural		2100	2400	1400	90	475	1650	1,900,000
No. 1		1500	1750	1000	90	405	1250	1,800,000
No. 1 Dense		1800	2050	1200	90	475	1450	1,900,000
No. 2		1050	1200	700	75	345	900	1,400,000
No. 2 Medium grain	6" and wider	1250	1450	825	90	405	1050	1,600,000
No. 2 Dense		1450	1650	975	90	475	1250	1,700,000
No. 3		725	825	475	75	345	650	1,400,000
No. 3 Dense		850	975	575	90	475	750	1,500,000

TABLE 10-1. Allowable Unit Stresses for Structural Lumber—Visual Grading (*Continued*)

Species and commercial grade	Size classification	Allowable unit Stresses in pounds per square inches							Grading rules agency
		Extreme fiber in bending F_b		Tension parallel to grain F_t	Horizontal shear F_v	Compression perpendicular to grain $F_{c\perp}$	Compression parallel to grain F_c	Modulus of elasticity E	
		Single-member uses	Repetitive-member uses						
SOUTHERN PINE (Surfaced green. Used any condition)									
No. 1 SR	5" and thicker	1300	—	850	110	270	925	1,600,000	Southern Pine Inspection Bureau
No. 1 Dense SR		1500	—	1000	110	315	1050	1,600,000	
No. 2 SR		1100	—	725	95	270	675	1,400,000	
No. 2 Dense SR		1300	—	850	95	315	775	1,500,000	
Dense Structural 65		1650	—	1100	105	315	1000	1,600,000	
WESTERN CEDARS (NORTH) (Surfaced dry or surfaced green. Used at 19% max. m.c.)									
Select Structural	2" to 4" thick	1250	1450	850	70	285	1100	1,100,000	National Lumber Grades Authority
No. 1		1050	1250	725	70	285	975	1,100,000	
No. 2	6" and wider	875	1000	575	70	285	825	1,000,000	(A Canadian Agency. See footnotes 2 through 9 and 12)
No. 3		525	600	325	70	285	525	900,000	
Appearance		1050	1250	725	70	285	1150	1,100,000	
Select Structural	Beams and Stringers	1100	—	675	65	285	850	1,000,000	
No. 1 Structural		900	—	475	65	285	700	1,000,000	
Select Structural	Posts and Timbers	1050	—	700	65	285	900	1,000,000	
No. 1 Structural		850	—	575	65	285	800	1,000,000	
Select	Wall and Roof Plank	1200	1400	—	—	285	—	1,100,000	
Commercial		1050	1200	—	—	285	—	1,000,000	

moisture content of 19% or less; green lumber has a moisture content in excess of 19%.

The allowable unit stresses to be used in actual design practice must, of course, conform to the requirements of the local building code. As noted earlier, many municipal codes are revised only infrequently and consequently may not be in agreement with current editions of industry-recommended allowable stresses. However, the allowable stresses for wood construction used throughout this book are those given in the National Design Specification and recommended by the National Forest Products Association.

10-4 Design for Bending

The design of a wood beam for strength in bending is accomplished by use of the flexure formula (Art. 4-7). The form of this equation used in design is

$$S = \frac{M}{F_b}$$

in which M = maximum bending moment,

F_b = allowable extreme fiber (bending) stress,

S = required section modulus.

Although section moduli for standard rectangular wood beam sizes are given in Table 4-6, it is sometimes convenient to use the formula $S = bd^2/6$ which was developed in Art. 4-6.

To determine the dimensions of a wood beam as governed by bending, first compute the maximum bending moment. Next, refer to a table such as Table 10-1 and select the species and grade of lumber that is to be used, and note the corresponding allowable extreme fiber stress, F_b. Substitute these values in the flexure formula and solve for the required section modulus. The proper beam size may be determined by referring to Table 4-6 which lists S for standard dressed sizes of structural lumber. Obviously, a number of different sections may be acceptable.

Example. A simple beam has a span of 16 ft and supports a load, including its own weight, of 6500 lb. If the timber to be used is Eastern Hemlock, Select Structural grade, determine the size of the beam to withstand bending stresses.

Solution: (1) The maximum bending moment for this simple beam, Case 2 of Table 3-1, is

$$M = \frac{WL}{8} = \frac{6500 \times 16}{8} = 13,000 \text{ ft-lb}$$

(2) Referring to Table 10-1, we find under Eastern Hemlock, beams and stringers, Select Structural, that the allowable bending stress $F_b = 1400$ psi. Then, substituting in the beam formula and converting the bending moment to inch-pounds, the required section modulus is

$$S = \frac{M}{F_b} = \frac{13,000 \times 12}{1400} = 111.4 \text{ in.}^3$$

(3) From Table 4-6 select a 6 × 12 beam ($S = 121.2$ in.³). This size is adequate to withstand the bending stresses.

(4) Suppose that we are limited in headroom under the beam and that 10 in. is the maximum depth beam that can be used. To find the required width, we may solve the formula $S = bd^2/6$ for the width b. Then

$$S = \frac{bd^2}{6} \quad \text{or} \quad 111.4 = \frac{b \times 9.5^2}{6} \quad \text{and} \quad b = 7.4 \text{ in}$$

Accept an 8 × 10 ($7\frac{1}{2} \times 9\frac{1}{2}$) beam. Table 4-6 shows the section modulus of this beam to be 112.8 in.³. It should be noted that this beam could have been selected directly from Table 4-6 by moving down the page from $S = 121.229$ for the 6 × 12 section to the next value of S equal to or greater than the required 111.4 in.³.

This example covers design for bending strength only. A complete design requires checking for horizontal shear and deflection, as explained in Arts. 10-5 and 10-6, respectively.

Problem 10-4-A*.   The No. 1 grade of Douglas Fir is to be used for a series of floor beams spanning 14 ft. If the uniformly distributed load on each beam, including its own weight, is 3200 lb, design a typical beam for strength in bending.

Problem 10-4-B*.   A simple beam of California Redwood, Select Structural grade, has a span of 18 ft with two concentrated loads of 3 kips each placed at the third points of the span. Neglecting its own weight, determine the size of the beam with respect to strength in bending.

Problem 10-4-C.   A Southern Pine beam of No. 1 SR grade has a span of 15 ft with a single concentrated load of 6 kips placed 5 ft from one of its supports. Neglecting its own weight, design the beam for strength in bending.

Problem 10-4-D.   A cantilever beam projects 6 ft from the face of a masonry wall and supports a uniformly distributed load of 2 kips, including its own weight. Determine the size of the beam with respect to bending strength if Balsam Fir, Select Structural, is to be used.

10-5   Horizontal Shear

As discussed in Art. 3-1 and illustrated in Figs. 3-1*b* and *d*, a beam has a tendency to fail in shear by the fibers sliding past each other both vertically and horizontally. Also, at any point in a beam, the intensity of the vertical and horizontal shearing stresses are equal. The vertical shear strength of wood beams is seldom of concern, because the shear resistance of wood *across* the grain is much larger than it is *parallel* to the grain where the horizontal shear forces develop.

The horizontal shearing stresses are not uniformly distributed over the cross section of a beam but are greatest at the neutral surface. The maximum horizontal unit shearing stress for rectangular sections is $\frac{3}{2}$ times the average vertical unit shearing stress. This is expressed by the formula

$$v = \frac{3}{2} \times \frac{V}{bd}$$

in which v = maximum unit horizontal shearing stress in psi,

V = total vertical shear in pounds,

b = width of cross section in inches,

d = depth of cross section in inches.[2]

[2] This is the dimension called h in Table 4-6. Notation usage is not entirely consistent in wood structural design but the variations are of minor consequence only. Reference to the size and use classification of structural lumber given in Art 10-1, and to the size classification (second column) of Table 10-1, shows that *thickness* dimensions used therein correspond to b and *width* dimensions to h in the diagrams at the head of Table 4-6 The reader should examine these relationships carefully to make certain he fully understands them.

This formula applies only to rectangular cross sections. Timber is relatively weak in resistance to horizontal shear, and short spans with large loads should always be tested for this shearing tendency. Frequently a beam large enough to resist bending stresses must be made larger in order to resist horizontal shear. Table 10-1 gives allowable horizontal unit shearing stresses for several stress grades of lumber under the column headed Horizontal Shear, F_v.

Example.   A 6 × 10 beam of Southern Pine, No. 2 dense SR grade, has a total uniformly distributed load of 6000 lb. Investigate the beam for horizontal shear.

Solution: (1) Determine the value of the horizontal unit shearing stress developed by the loading. Since the beam is symetrically loaded, $R_1 = R_2 = 6000 \div 2 = 3000$ lb; this is also the value of the maximum vertical shear. Reference to Table 4-6 shows that the dressed dimensions of a 6 × 10 beam are $5\frac{1}{2} \times 9\frac{1}{2}$ in. Then

$$v = \frac{3}{2} \times \frac{V}{bd} = \frac{3}{2} \times \frac{3000}{5.5 \times 9.5} = 86.2 \text{ psi}$$

(2) Referring to Table 10-1, under size classification 5 in. and thicker for the species and grade specified, we find that the allowable horizontal unit shearing stress $F_v = 95$ psi. Since 86.2 < 95, the beam is acceptable for horizontal shear. This is the customary procedure for investigating a beam for horizontal shear; it is to be used in solving the problems stated below. However, the formula errs on the side of safety, for it indicates greater shearing stresses than actually occur. Because of this, another method frequently permitted neglects all loads within a distance equal to the depth of the beam from both supports, when calculating V for use in the formula.

Note: In solving the following problems, use Tables 4-6 and 10-1. Neglect beam weight.

Problem 10-5-A*.   A 10 × 10 beam of Douglas Fir, Select Structural grade, supports a single concentrated load of 10 kips at the center of the span. Investigate the beam for horizontal shear.

Problem 10-5-B*.   A 10 × 14 simple beam of Eastern Hemlock, Select Structural grade, is loaded symmetrically with three concentrated loads of 4300 lb each. Is the beam safe in horizontal shear?

Problem 10-5-C. A 10 × 12 beam of Southern Pine, No. 2 Dense SR grade, is 8 ft long and has a concentrated load of 8 kips located 3 ft from one end. Investigate the beam for horizontal shear.

Problem 10-5-D. What should be the nominal cross-sectional dimensions of a simple beam of California Redwood, Clear Structural grade, to resist the horizontal shearing stresses developed by a uniformly distributed load of 12 kips?

Problem 10-5-E. A 6 × 10 beam of Douglas Fir, Dense Select Structural grade, is 18 ft long. It supports a uniformly distributed load of 300 lb per lin ft over its entire length. Is the beam safe with respect to horizontal shear?

10-6   Deflection

A wood beam supporting a floor load may be large enough to carry the load safely, but the deflection may be so great that a plaster ceiling would develop cracks or the floor might vibrate noticeably. Consequently, in addition to sufficient *strength* the beam must possess adequate *stiffness*. In practice a beam is first designed for bending and then investigated for deflection. Most building codes, as well as good practice, state that the maximum deflection for beams supporting plaster ceilings must not exceed $\frac{1}{360}$ of the span. For instance, a beam 16 ft long would have an allowable deflection of $(16 \times 12) \div 360 = 0.53$ in.

A rule of thumb for determining the limiting span with respect to deflection of a timber beam uniformly loaded is the following:

The limiting span in feet (L) for a simple beam uniformly loaded is equal to 1.1 times the depth of the beam in inches.

It may be expressed thus: $L = 1.1 \times d$. A beam 10 in. in depth would have a limiting span of 10×1.1, or 11 ft.

This rule is, of course, merely approximate, and the actual deflections are found by the use of formulas given in Table 3-1. The deflection is in inches and is represented by *D*. When using these formulas, note carefully that *l, the length of the beam, must be in inches.*

Example.   Compute the deflection of a 10 × 14 Douglas Fir beam, dense select structural grade, 15 ft long and carrying a uniformly distributed load of 16 kips, including its own weight.

Solution: (1) The deflection formula[3] for this loading (Table 3-1, Case 2) is

$$D = \frac{5}{384} \times \frac{W l^3}{E I}$$

(2) Referring to Table 4-6, we find that the dressed dimensions of this beam are $9\frac{1}{2} \times 13\frac{1}{2}$ in., and that the moment of inertia of the cross section is 1948 in.[4] (rounded tabular value). Table 10-1 gives the modulus of elasticity as 1,700,000 psi. The length of the beam in inches is $15 \times 12 = 180$ in. Substituting in the deflection formula,

$$D = \frac{5}{348} \times \frac{16{,}000 \times 180^3}{1{,}700{,}000 \times 1948} = 0.36 \text{ in.}$$

(3) The allowable deflection for this beam is $(15 \times 12) \div 360 = 0.5$ in. Since 0.36 in. < 0.5 in., the deflection is not excessive.

If we apply the rule of thumb, the limiting span for a 14-in. beam will be $d \times 1.1$, or $13.5 \times 1.1 = 14.9$ ft, which is slightly less than the actual span, which is 15 ft.

Note: In solving the following problems, use Tables 4-6 and 10-1. Neglect beam weight.

Problem 10-6-A*.   A 6 × 14 Southern Pine beam, No. 1 SR grade, has a uniformly distributed load of 6000 lb on a span of 16 ft. Investigate the deflection.

Problem 10-6-B.   An 8 × 12 beam of Douglas Fir, Dense No. 1 grade, is 12 ft in length and has a concentrated load of 5 kips at the center of the span. Compute the actual deflection and compare it with the allowable.

Problem 10-6-C.   Two concentrated loads of 3500 lb each are located at the third points of a 15-ft span. The 10 × 14 beam is Douglas Fir, Select Structural grade. Investigate the deflection.

Problem 10-6-D*.   An 8 × 14 Tamarack beam, Select Structural grade, has a span of 16 ft and a uniformly distributed load of 8 kips. Investigate the deflection.

Problem 10-6-E.   An 8 × 12 beam of Southern Pine, Dense Structural 65 grade, is used as a cantilever with a projection of 5 ft from the face of

---

[3] See Art. 5-12 for definitions of terms in the formula.

the wall in which it is embedded. There is a concentrated load of 3 kips at its unsupported end. Compute the maximum deflection.

10-7 Beam Design Procedure

In general, three steps are necessary for the proper design of wood beams. Actually, however, many designers use only the first, selecting sizes which by experience they know will meet the requirements in Steps 2 and 3. The three steps of the procedure are outlined below.

Step 1: Compute the required section modulus by means of the flexure formula, $S = M/F_b$, as explained in Art. 10-4. Obviously many sizes will meet the requirements but the most practical have widths ranging from one half to one third of the depth. Beam sizes may be selected on the basis of section modulus by referring to Table 4-6. Material which is too thin tends to bend sidewise unless properly braced. Attention has been called to the difference between the nominal and dressed (S4S) sizes of structural lumber.

Step 2: Investigate the size selected under Step 1 for horizontal shear, as explained in Art. 10-5, and increase the dimensions if necessary. Allowable horizontal shearing stresses are given in Table 10-1.

Step 3: Investigate the beam for deflection, the allowable deflection usually being $\frac{1}{360}$ of the span. Formulas for deflection under different loading conditions are given in Table 3-1, and are used as explained in Art. 10-6.

Example 1. Design beam A of the floor framing shown in Fig. 10-1, using Southern Pine, No. 2 Dense SR grade lumber. A 2-in. plank floor is used on the 5-ft spans between beams, and there is a $\frac{7}{8}$-in. finished flooring laid over the planking. The live load is 90 psf.
Solution: Record the loads and design data.

Live load $= 90$	$F_b = 1,300$ psi (Table 10-1)
2-in. plank $= 6$ (Table 5-8)	$F_v = 95$ psi (Table 10-1)
$\frac{7}{8}$-in. finish $= 3$ (Table 5-8)	$E = 1,500,000$ psi (Table 10-1)
Total $= 99$ psf	

(1) Design the beam for bending. The floor load per lin ft of beam is $99 \times 5 = 495$ lb (floor area marked by diagonals in figure),

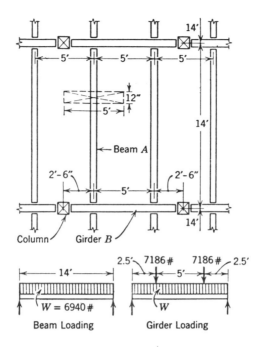

FIGURE 10-1

making the total uniform load on the beam $495 \times 14 = 6940$ lb. Then

$$M = \frac{WL}{8} = \frac{6940 \times 14}{8} = 12{,}200 \text{ ft-lb}$$

and

$$S = \frac{M}{F_b} = \frac{12{,}200 \times 12}{1300} = 112 \text{ in.}^3$$

Referring to Table 4-6, we find that a 6×12 beam has a section modulus of 121.3 (rounded tabular value) and weighs approximately 17.6 lb per lin ft. If computations are carried out to determine the revised required section modulus, taking the beam weight into account ($17.6 \times 14 = 246$ lb), the increased value of S becomes 116 in.^{3} Therefore the 6×12 beam is acceptable for bending strength. The revised total load is $6940 + 246 = 7186$ lb.

(2) Investigating for horizontal shear, we note that the beam is symmetrically loaded, the reactions are equal, and the maximum vertical shear is 7186 ÷ 2 = 3593 lb. Then

$$v = \frac{3}{2} \times \frac{V}{bd} = \frac{3}{2} \times \frac{3593}{5.5 \times 11.5} = 85.2 \text{ psi}$$

This is acceptable because the allowable unit shearing stress F_v = 95 psi.

(3) Investigating for deflection, we note that Case 2, Table 3-1 applies and that the deflection formula for this loading is

$$D = \frac{5}{384} \times \frac{Wl^3}{EI}$$

Table 4-6 shows that the moment of inertia of a 6 × 12 beam is 697.1 in.⁴. Substituting this value and the other given data in the formula,

$$D = \frac{5 \times 7186 \times 168^3}{384 \times 1,500,000 \times 697.1} = 0.41 \text{ in.}$$

Since this value is less than the allowable deflection (168 ÷ 360 = 0.46 in.), the 6 × 12 beam meets all the requirements and therefore is adopted.

Example 2. Design girder B in Fig. 10-1, using the same data given for beam A.

Solution: The loading diagram for the girder is shown in Fig. 10-1. The reaction from beam A is 3593 lb, but because similar beams frame into the girder *on each side* the concentrated loads will be 2 × 3593 = 7186 lb. The uniform load indicated on the diagram represents the weight of the girder yet to be determined.

(1) The maximum bending moment due to the symmetrically placed concentrated loads, as well as that due to the girder's weight, will occur at midspan. Insofar as the concentrated loads are concerned, $R_1 = R_2 = 7186$ lb, and the maximum moment produced by them at midspan is

$$M = (7186 \times 5) - (7186 \times 2.5) = 18,000 \text{ ft-lb}$$

This bending moment requires a section modulus of

$$S = \frac{M}{F_b} = \frac{18,000 \times 12}{1300} = 166 \text{ in.}^3$$

Referring to Table 4-6, we find that a 10 × 12 beam has $S = 209.4$ in.^{3} and that it weighs approximately 30 lb per lin ft. The bending moment at midspan produced by this weight is

$$M = \frac{wL^2}{8} = \frac{30 \times 10^2}{8} = 375 \text{ ft-lb}$$

which makes a revised maximum moment of $18,000 + 375 = 18,375$ ft-lb. The revised required section modulus is

$$S = \frac{M}{F_b} = \frac{18,375 \times 12}{1300} = 170 \text{ in.}^3$$

This value is still well within the $S = 209.4$ in.^{3} provided, so tentatively accept the 10 × 12 beam.

(2) Investigate the horizontal shear. Taking the beam weight into account, the revised reactions are 7186 plus half the beam weight $(300 \div 2)$ or $7186 + 150 = 7336$ lb, say 7340 lb. This is also the value of the maximum vertical shear. Then

$$v = \frac{3}{2} \times \frac{V}{bd} = \frac{3 \times 7340}{2 \times 109.3} = 101 \text{ psi} > 95 \text{ psi}$$

Since this is greater than the allowable stress, try a 10 × 14 beam. This section is approximately 6 lb per lin ft heavier than the 10 × 12, which results in a revised value of $V = 7370$ lb. Then

$$v = \frac{3 \times 7370}{2 \times 128.3} = 86 \text{ psi} < 95 \text{ psi}$$

The 10 × 14 beam is therefore satisfactory for horizontal shear and is adopted in place of the 10 × 12 section tentatively accepted in Step (1).

(3) Investigate the deflection. Because this is not a typical loading, there is no deflection formula for it in Table 3-1. It is necessary, therefore, to find an equivalent uniformly distributed load that would produce the same bending moment, and then use the deflection formula for a uniform load (Table 3-1, Case 2). This method

of computing the deflection of a beam gives only an approximately correct result but is useful in indicating whether the deflection is near the critical point.

The revised bending moment determined near the end of Step (1) is 18,375 ft-lb. Substituting this value in the moment equation for a uniform load, and solving for W,

$$M = \frac{WL}{8} \quad \text{or} \quad W = \frac{8M}{L}$$

$$W = \frac{8 \times 18,375}{10} = 14,700 \text{ lb}$$

which is the equivalent uniform load that produces the same bending moment as the actual loading. For this 10 × 14 beam, $l = 10 \times 12 = 120$ in., $E = 1,500,000$ psi, and $I = 1948$ in.[4] Then

$$D = \frac{5}{384} \times \frac{Wl^3}{EI} = \frac{5 \times 14,700 \times 120^3}{384 \times 1,500,000 \times 1948} = 0.11 \text{ in.}$$

which is the approximate deflection. The allowable deflection is $120 \div 360 = 0.33$ in. Therefore the 10 × 14 girder is accepted.

It should be noted that the spans used in Examples 1 and 2 were 14 ft and 10 ft, respectively. Actually these distances represent the spacing of the columns on centers and all beams and girders have slightly shorter lengths. This, however, is common practice and the small error is on the side of safety. Also, it was assumed that the members are secured against sidewise buckling. Lateral support is furnished to the girders by the floor beams framing into them; it may be provided to the beams by proper fastening of the plank floor to their top surfaces.

Problem 10-7-A. Design the beam for the floor panel shown in Fig. 10-2a. The floor construction consists of 3-in. plank underflooring and ⅞-in. top flooring. The live load is 120 psf and Southern Pine, No. 1 SR grade, is to be used.

Problem 10-7-B*. Design the girder shown in Fig. 10-2a, using the same data given for Problem 10-7-A.

Problem 10-7-C. For the floor panel shown in Fig. 10-2b, the live load is 60 psf and the flooring consists of 4-in. planking and a ⅞-in. finished floor. Design the beams using Douglas Fir, Dense No. 1 grade, lumber.

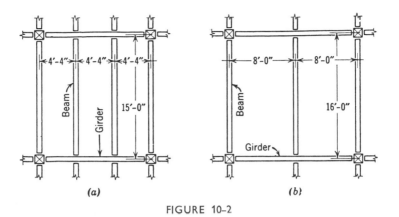

FIGURE 10-2

Problem 10-7-D. Design the girders shown in Fig. 10-2*b*, using the same data given for Problem 10-7-C.

10-8 Bearing on Supports

In addition to bending and shearing stresses in beams, beam bearings must have ample dimensions so that compressive stresses perpendicular to grain do not exceed the allowable values of $F_{c\perp}$ given in Table 10-1. The allowable stresses in the table apply to bearings of any length at the ends of beams and to all bearings 6 in. or more in length at any other location.

For bearings less than 6 in. in length and not nearer than 3 in. to the end of a member, the National Design Specification provides that the allowable stress in compression perpendicular to grain may be increased by the factor $(l_b + 0.375)/l_b$ in which l_b is the length of bearing measured along the grain of the wood. This expression yields multiplying factors as tabulated below for the bearing lengths indicated:

Bearing length in inches	$\frac{1}{2}$	1	$1\frac{1}{2}$	2	3	4	6 or more
Factor	1.75	1.38	1.25	1.19	1.13	1.10	1.0

Example. An 8 × 14 Southern Pine beam, No. 1 SR grade, has a bearing length of 6 in. at its supports. If the end reaction is 7400 lb,

is the beam adequate with respect to compressive stress perpendicular to grain?

Solution: (1) Table 4-6 shows that the dressed width of this beam is $7\frac{1}{2}$ in., making the bearing area 7.5 × 6 = 45 sq in.

(2) The bearing stess developed is equal to the reaction divided by the bearing area, or 7400 ÷ 45 = 164 psi.

(3) Referring to Table 10-1, we find that the allowable compressive stress perpendicular to grain is $F_{c\perp} = 270$ psi. Since this value is greater than the 164 psi developed, the bearing length is more than adequate.

Problem 10-8-A.   Would the beam in the above example still be safe with respect to $F_{c\perp}$ if the bearing length were reduced to 4 in.?

10-9   Joist Floors: Span Tables

Joists are the comparatively small, closely spaced beams that support floor loads. The nominal sizes commonly used are 2 × 8, 2 × 10, 2 × 12, 3 × 8, 3 × 10, and 3 × 12. The spacing of joists is determined principally by the stock lengths of metal lath and gypsum board lath used as a plaster base for plaster ceilings. The thickness of the flooring is also a consideration. The standard width of lath material is 48 in.; thus the usual spacings for joists are 16 in. and 24 in. on centers (o.c.). When the floor load or length of span is excessive, a 12 in. spacing is sometimes necessary. Figure 10-3 illustrates typical wood floor construction.

The purpose of bridging is twofold. It serves to prevent buckling by maintaining the joists in vertical position, and also aids in distributing concentrated loads to adjacent joists. This latter function is the

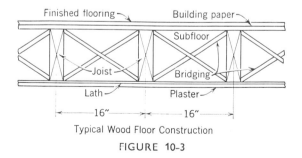

Typical Wood Floor Construction

FIGURE 10-3

basis for the higher values of F_b given in Table 10-1 under "repetitive-member uses." See Art. 10-3. Bridging should be installed at intervals of about 8 ft along the span length.

The design of joists consists of determining the load to be supported and then applying the procedures for beam design explained in Art. 10-7. However, to facilitate the selection of joist floors carrying uniformly distributed loads (by far the most common loading) extensive tables have been computed that give maximum safe spans for joists of various sizes and spacings under different loadings per square foot. Tables 10-2 and 10-3 are representative of such tables and have been compiled from data contained in *Span Tables for Joists and Rafters*, published by the National Forest Products Association. Examining the tables, we note that spans are calculated on the basis of modulus of elasticity with the required bending stress F_b listed below each span; that is, both stiffness (deflection) and bending strength have been taken into account. Maximum safe clear spans in feet are tabulated for five joist spacings[4] and selected values of E. As stated in the design criteria at the head of the tables, deflection has been limited to $\frac{1}{360}$ of the span due to live load only, whereas both live load and a dead load allowance have been used in determining the required value of F_b.

Tables 10-2 and 10-3 are based on live loads of 40 psf and 30 psf, respectively. The former value is widely accepted for apartments in multifamily housing and for the first floors of other residential construction; the 30-psf loading generally applies to upper floors of residential construction. Where heavier loadings apply, such as in commercial and industrial construction, or where local building codes call for higher values than those given here, more extensive tables should be consulted.[5] The use of joist span tables is illustrated in the following examples.

Example 1.   Using the joist span tables, select joists to carry a live load of 40 psf on a span of 15 ft 6 in. if the spacing is 16 in. on centers.

---

[4] The spacings of 13.7 in. and 19.2 in. accommodate the division of standard 8-ft length sheet material into 7 spaces and 5 spaces, respectively.

[5] A series of maximum span tables for joists based on loadings from 50 psf to 100 psf is contained in *Wood Structural Design Data*, published by the National Forest Products Association, Washington, D.C.

Solution: (1) Referring to Table 10-2, we find that 2 × 10 joists of a grade having an E value of 1,400,000 psi and an F_b value of 1150 psi may be used on a span of 15 ft 8 in. This meets the required conditions.

(2) Turning to Table 10-1, the reader should satisfy himself that, among the species and grades listed, the following selections would be satisfactory: Southern Pine No. 2, Douglas Fir No. 2, California Redwood No. 1. The values of F_b in the column headed "Repetitive member uses" apply to joist floor construction.

As with all safe load tabulations, caution must be exercised with respect to dead load allowances in joist span tables. The 10-psf allowance provided in Tables 10-2 and 10-3 is adequate to cover the weight of the joists and wood flooring. However, if a heavier type of flooring is to be used or if there will be a plaster ceiling on the underside of the joists, this additional weight must be accounted for. This is readily accomplished if the usual beam design procedure is followed as in Example 1 of Art. 10-10, but span tables may still be used for the heavier dead load condition. An adequate design may be made by consulting the next higher live load listings, using the additional 10-psf live load provided to cover the additional dead load. This procedure is illustrated below.

Example 2. Using the joist span tables, select joists to carry a live load of 30 psf on a span of 14 ft. The floor construction is similar to that shown in Fig. 10-3, with wood finished flooring and subfloor and a metal lath and plaster ceiling on the undersides of the joists.

Solution: (1) Table 10-3 is computed for a 30-psf live load but does not provide for the full weight of a plaster ceiling. The ceiling weight will vary with the thickness and type of plaster (gypsum or cement). Table 5-8 lists 10 psf for a suspended ceiling but some building codes permit an allowance of 8 psf when the lath is applied directly to the joists as indicated in Fig. 10-3. In any event we see that an 8-psf to 10-psf dead load capacity is required beyond that provided by Table 10-3. This additional dead load will be adequately covered if we make our selection from Table 10-2.

(2) Referring to Table 10-2, we find that a span of 14 ft 2 in. is safe for 2 × 8 joists 12 in. on centers, for a lumber grade having an E value of 1,600,000 psi and an F_b value of 1140 psi. Also, 2 × 10

TABLE 10-2. Maximum Spans for Floor Joists—Live Load 40 psf*
Design Criteria: Deflection for 40 psf live load limited to ⅟₃₆₀ of span
F_b determined by 40 psf live load plus 10 psf dead load

Joist size and spacing (in.)	Modulus of elasticity E in 1,000,000 psi								
	1.1	1.2	1.3	1.4	1.5	1.6	1.7	1.8	1.9
2 × 6 12.0	9-6 890	9-9 940	10-0 990	10-3 1040	10-6 1090	10-9 1140	10-11 1190	11-2 1230	11-4 1280
13.7	9-1 930	9-4 980	9-7 1040	9-10 1090	10-0 1140	10-3 1190	10-6 1240	10-8 1290	10-10 1340
16.0	8-7 980	8-10 1040	9-1 1090	9-4 1150	9-6 1200	9-9 1250	9-11 1310	10-2 1360	10-4 1410
19.2	8-1 1040	8-4 1100	8-7 1160	8-9 1220	9-0 1280	9-2 1330	9-4 1390	9-6 1440	9-8 1500
24.0	7-6 1120	7-9 1190	7-11 1250	8-2 1310	8-4 1380	8-6 1440	8-8 1500	8-10 1550	9-0 1610
32.0	6-10 1230	7-0 1300	7-3 1390	7-5 1450	7-7 1520	7-9 1590	7-11 1660	8-0 1690	8-2 1760
2 × 8 12.0	12-6 890	12-10 940	13-2 990	13-6 1040	13-10 1090	14-2 1140	14-5 1190	14-8 1230	15-0 1280
13.7	11-11 930	12-3 980	12-7 1040	12-11 1090	13-3 1140	13-6 1190	13-10 1240	14-1 1290	14-4 1340
16.0	11-4 980	11-8 1040	12-0 1090	12-3 1150	12-7 1200	12-10 1250	13-1 1310	13-4 1360	13-7 1410
19.2	10-8 1040	11-0 1100	11-3 1160	11-7 1220	11-10 1280	12-1 1330	12-4 1390	12-7 1440	12-10 1500
24.0	9-11 1120	10-2 1190	10-6 1250	10-9 1310	11-0 1380	11-3 1440	11-5 1500	11-8 1550	11-11 1610
32.0	9-0 1230	9-3 1300	9-6 1370	9-9 1450	10-0 1520	10-2 1570	10-5 1650	10-7 1700	10-10 1790

Note: Required F_b is shown below each span (psi).

* Data abstracted from *Span Tables for Joists and Rafters.* Courtesy National Forest Products Association.

TABLE 10-2. (*continued*)

Joist size and spacing (in.)		Modulus of elasticity E in 1,000,000 psi								
		1.1	1.2	1.3	1.4	1.5	1.6	1.7	1.8	1.9
2 × 10	12.0	15-11 890	16-5 940	16-10 990	17-3 1040	17-8 1090	18-0 1140	18-5 1190	18-9 1230	19-1 1280
	13.7	15-3 930	15-8 980	16-1 1040	16-6 1090	16-11 1140	17-3 1190	17-7 1240	17-11 1290	18-3 1340
	16.0	14-6 980	14-11 1040	15-3 1090	15-8 1150	16-0 1200	16-5 1250	16-9 1310	17-0 1360	17-4 1410
	19.2	13-7 1040	14-0 1100	14-5 1160	14-9 1220	15-1 1280	15-5 1330	15-9 1390	16-0 1440	16-4 1500
	24.0	12-8 1120	13-0 1190	13-4 1250	13-8 1310	14-0 1380	14-4 1440	14-7 1500	14-11 1550	15-2 1610
	32.0	11-6 1240	11-10 1310	12-2 1380	12-5 1440	12-9 1520	13-0 1580	13-3 1640	13-6 1700	13-9 1770
2 × 12	12.0	19-4 890	19-11 940	20-6 990	21-0 1040	21-6 1090	21-11 1140	22-5 1190	22-10 1230	23-3 1280
	13.7	18-6 930	19-1 980	19-7 1040	20-1 1090	20-6 1140	21-0 1190	21-5 1240	21-10 1290	22-3 1340
	16.0	17-7 980	18-1 1040	18-7 1090	19-1 1150	19-6 1200	19-11 1250	20-4 1310	20-9 1360	21-1 1410
	19.2	16-7 1040	17-0 1100	17-6 1160	17-11 1220	18-4 1280	18-9 1330	19-2 1390	19-6 1440	19-10 1500
	24.0	15-4 1120	15-10 1190	16-3 1250	16-8 1310	17-0 1380	17-5 1440	17-9 1500	18-1 1550	18-5 1610
	32.0	13-11 1220	14-4 1300	14-9 1380	15-2 1450	15-6 1520	15-10 1580	16-2 1650	16-5 1700	16-9 1770

TABLE 10-3. Maximum Spans for Floor Joists—Live Load 30 psf*
Design Criteria: Deflection for 30 psf live load limited to $\frac{1}{360}$ of span
F_b determined by 30 psf live load plus 10 psf dead load

Joist size and spacing (in.)		Modulus of elasticity E in 1,000,000 psi								
		1.1	1.2	1.3	1.4	1.5	1.6	1.7	1.8	1.9
2 × 6	12.0	10-5 860	10-9 910	11-0 960	11-3 1010	11-7 1060	11-10 1100	12-0 1150	12-3 1200	12-6 1240
	13.7	10-0 900	10-3 950	10-6 1010	10-10 1060	11-1 1110	11-3 1160	11-6 1200	11-9 1250	11-11 1300
	16.0	9-6 950	9-9 1000	10-0 1060	10-3 1110	10-6 1160	10-9 1220	10-11 1270	11-2 1320	11-4 1360
	19.2	8-11 1010	9-2 1070	9-5 1130	9-8 1180	9-10 1240	10-1 1290	10-4 1350	10-6 1400	10-8 1450
	24.0	8-3 1080	8-6 1150	8-9 1210	8-11 1270	9-2 1330	9-4 1390	9-7 1450	9-9 1510	9-11 1560
	32.0	7-6 1190	7-9 1270	7-11 1330	8-2 1410	8-4 1470	8-6 1530	8-8 1590	8-10 1650	9-0 1710
2 × 8	12.0	13-9 860	14-2 910	14-6 960	14-11 1010	15-3 1060	15-7 1100	15-10 1150	16-2 1200	16-6 1240
	13.7	13-2 900	13-6 950	13-11 1010	14-3 1060	14-7 1110	14-11 1160	15-2 1200	15-6 1250	15-9 1300
	16.0	12-6 950	12-10 1000	13-2 1060	13-6 1110	13-10 1160	14-2 1220	14-5 1270	14-8 1320	15-0 1360
	19.2	11-9 1010	12-1 1070	12-5 1130	12-9 1180	13-0 1240	13-4 1290	13-7 1350	13-10 1400	14-1 1450
	24.0	10-11 1080	11-3 1150	11-6 1210	11-10 1270	12-1 1330	12-4 1390	12-7 1450	12-10 1510	13-1 1560
	32.0	9-11 1200	10-2 1260	10-6 1340	10-9 1410	11-0 1470	11-3 1540	11-5 1590	11-8 1660	11-11 1730

Note: Required F_b is shown below each span (psi).

*Data abstracted from *Span Tables for Joists and Rafters.* Courtesy National Forest Products Association.

244

TABLE 10-3. (*continued*)

Joist size and spacing (in.)	Modulus of elasticity E in 1,000,000 psi								
	1.1	1.2	1.3	1.4	1.5	1.6	1.7	1.8	1.9
2 × 10 12.0	17-6 860	18.0 910	18.6 960	19.0 1010	19.5 1060	19-10 1100	20-3 1150	20-8 1200	21-0 1240
13-7	16-9 900	17-3 950	17-9 1010	18-2 1060	18-7 1110	19-0 1160	19-4 1200	19-9 1250	20-1 1300
16.0	15-11 950	16-5 1000	16-10 1060	17-3 1110	17-8 1160	18-0 1220	18-5 1270	18-9 1320	19-1 1360
19.2	15-0 1010	15-5 1070	15-10 1130	16-3 1180	16-7 1240	17-0 1290	17-4 1350	17-8 1400	18-0 1450
24.0	13-11 1080	14-4 1150	14-8 1210	15-1 1270	15-5 1330	15-9 1390	16-1 1450	16-5 1510	16-8 1560
32.0	12-8 1200	13-0 1260	13-4 1330	13-8 1400	14-0 1470	14-4 1540	14-7 1590	14-11 1660	15-2 1720
2 × 12 12.0	21-4 860	21-11 910	22-6 960	23-1 1010	23-7 1060	24-2 1100	24-8 1150	25-1 1200	25-7 1240
13.7	20-5 900	21-0 950	21 7 1010	22-1 1060	22-7 1110	23-1 1160	23-7 1200	24-0 1250	24-5 1300
16.0	19-4 950	19-11 1000	20-6 1060	21-0 1110	21-6 1160	21-11 1220	22-5 1270	22-10 1320	23-3 1360
19.2	18-3 1010	18-9 1070	19-3 1130	19-9 1180	20-2 1240	20-8 1290	21-1 1350	21-6 1400	21-10 1450
24.0	16-11 1060	17-5 1150	17-11 1210	18-4 1270	18-9 1330	19-2 1390	19-7 1450	19-11 1510	20-3 1560
32.0	15-4 1190	15-10 1270	16-3 1340	16-8 1400	17-0 1460	17-5 1530	17-9 1590	18-1 1650	18-5 1720

joists 16 in. on centers have a safe span of 14 ft 6 in. if the grade has E and F_b values of 1,100,000 psi and 980 psi, respectively.

(3) Among other possible selections, Table 10-1 shows that Southern Pine No. 2 Medium grain grade is satisfactory for the 2 × 8 joists on 12-in. centers, and Eastern Hemlock No. 2 for the 2 × 10 joists on 16-in. centers.

10-10 Design of Joists

This article presents two examples of joist design using the usual procedures followed for beams. Example 1 is for a uniform loading over the entire span, the condition for which span tables are invariably used in practice. Example 2 covers a special case where the joists support a partition in addition to the uniformly distributed floor load, making it necessary to compute the required section modulus and select the joist cross section accordingly.

Example 1.   Determine the size joists required at a spacing of 16 in. on centers to carry a live load of 30 psf on a span of 14ft. The floor construction consists of a double thickness of wood floor with a metal lath and plaster ceiling attached to the bottom of the joists. This is the same construction called for in Example 2 of the preceding article and illustrated in Fig. 10-3. Eastern Hemlock No. 2 grade is the lumber specified.

Solution: (1) Allowing 10 psf for the weight of joists and flooring, 10 psf for the plaster ceiling, and adding the live load of 30 psf, we obtain the total load of 50 psf. Because the joists are spaced 16 in. on centers, the load per linear foot on one joist will be $50 \times {}^{16}\!/_{12} = 66.7$ lb. The total load on one joist will be $66.7 \times 14 = 934$ lb.

(2) The maximum bending moment is

$$M = \frac{WL}{8} = \frac{934 \times 14}{8} = 1630 \text{ ft-lb}$$

(3) To determine the required section modulus, we note from Table 10-1 that F_b for Eastern Hemlock No. 2 is 1200 psi. Then

$$S = \frac{M}{F_b} = \frac{1630 \times 12}{1200} = 16.3 \text{ in.}^3$$

Referring to Table 4-6, we find that a 2 × 10 joist has a section modulus of 21.39, and therefore is acceptable for strength in bending.

(4) In this case, the deflection due to live load only is limited to $\frac{1}{360}$ of the span or (14 × 12) ÷ 360 = 168 ÷ 360 = 0.466 in. The live load per foot of span is 30 × $\frac{16}{12}$ = 40 lb, making the total live load on one joist 40 × 14 = 560 lb. The moment of inertia of a 2 × 10 is 98.9 in.[4] (Table 4-6) and the modulus of elasticity of Eastern Hemlock No. 2 is 1,100,000 psi (Table 10-1). Therefore, the computed deflection under live load (Case 2, Table 3-1) is

$$D = \frac{5Wl^3}{384EI} = \frac{5 \times 560 \times 168^3}{384 \times 1,100,000 \times 98.9} = 0.318 \text{ in.}$$

Since this value is less than the 0.466 in. allowable, the 2 × 10 joist is acceptable for deflection as well as for bending strength.

Example 2.   Floor joists spaced 16 in. on centers with a span of 14 ft support a live load of 40 psf and, in addition, a 4-in. nonbearing plastered stud partition 9 ft high. The partition runs at right angles to the joists and is located 4 ft from one end of the span as shown in Fig. 10-4a. The joists support double flooring and a plaster ceiling underneath. Design the floor joists using structural lumber for which $F_b = 1100$ psi and $E = 1,100,000$ psi.

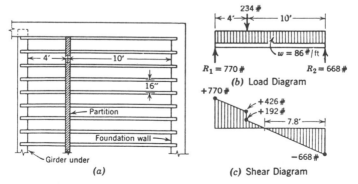

FIGURE 10-4

Solution: (1) The loading is determined as follows:

Floor loads

live load	= 40	
top floor	= 2.5	Joist spacing, 16 in. o.c.; therefore
underfloor	= 2.5	$65 \times \frac{16}{12} = 86$ lb per lin ft, the
plaster ceiling	= 10	load on one joist
joist (estimated)	= 10	
Total	= 65.0 psf	

Because the partition is 9 ft in height and the joists are 16 in. on centers, $9 \times 1.3 = 11.7$ sq ft, the area of partition supported by one joist. At 20 psf (Table 5-8), $11.7 \times 20 = 234$ lb, the partition load on one joist.

(2) To find the position of the maximum bending moment on one joist, we must draw the shear diagram. This requires drawing the load diagram of Fig. 10-4*b* and computing R_1 and R_2 by the methods explained in Arts. 2-6 and 2-7. They are found to be 770 lb and 668 lb, respectively. The shear diagram (Fig. 10-4*c*) is next drawn following the methods of Art. 3-3, and it is observed that the shear passes through zero at a point between the concentrated load and the right reaction. This is the position of the maximum bending moment. To simplify mathematics consider the forces to the right; then $V = 0 = 668 - (86 \times x)$ or $x = 7.8$ ft, the distance from R_2 to the point of maximum bending moment. See Art. 3-6.

(3) Writing the value of M at this point, again considering the forces to the right of the section,

$$M = (668 \times 7.8) - \left(86 \times 7.8 \times \frac{7.8}{2}\right) = 2594 \text{ ft-lb} = 31,128 \text{ in-lb}$$

(4) The required section modulus is

$$S = \frac{M}{F_b} = \frac{31,128}{1100} = 28.3 \text{ in.}^3$$

Table 4-6 shows that 2×12 joists have a section modulus of 31.6 in.^{3} and are therefore acceptable for strength in bending.

(5) Investigating the deflection, we first note that the allowable deflection is $(14 \times 12) \div 360 = 0.466$ in. Since the loading on the

joist is not one of the standard ones listed in Table 3-1, we will use the approximate method for computing the deflection based on an equivalent uniform load that produces the same bending moment [see Example 2, Step (3) under Art. 10-7]. The value of the bending moment determined above is 2594 ft-lb. Solving the formula $M = WL/8$ for W and making the substitutions,

$$W = \frac{8M}{L} = \frac{8 \times 2594}{14} = 1482 \text{ lb}$$

Then,

$$\text{Actual deflection } D = \frac{5Wl^3}{384EI} \qquad I = 178 \quad \text{(Table 4-6)}$$

$$D = \frac{5 \times 1482 \times (14 \times 12)^3}{384 \times 1,100,000 \times 178} = 0.463 \text{ in.}$$

It should be noted that the approximate deflection was computed for the full live and dead load rather than for live load only. Consequently there is a greater difference between the actual and computed deflections than the computations indicate. In any event, the computed deflection (0.463 in.) does not exceed the allowable value of 0.466 in. so the 2 × 12 joists 16 in. on centers are acceptable for deflection as well as for bending strength. The 2 × 12 joists are therefore adopted for the floor construction.

Problem 10-10-A*. Joists supporting a first floor live load of 40 psf in a dwelling have a span of 16 ft. The floor construction is similar to that shown in Fig. 10-3 except that there is no finished ceiling in the basement below. The lumber to be used is Balsam Fir No. 2. Determine the size and spacing of joists, using the span tables.

Problem 10-10-B. Determine the size and spacing of joists for the conditions described in Problem 10-10-A if the lumber is Douglas Fir No. 2.

Problem 10-10-C. Floor joists have a span of 12 ft 0 in., a live load of 40 psf, double wood flooring, and a plaster ceiling. In addition, at the center of the span, the joists support a 4-in. plastered stud partition 9 ft 0 in. high. If the joist lumber has $F_b = 1100$ psi and $E = 1,100,000$ psi, determine the required size and spacing of joists.

Problem 10-10-D. The floor live load in a building is 50 psf and the span is 14 ft 0 in. The joist spacing is to be 16 in. on centers and the joist lumber has $F_b = 1200$ psi with $E = 1,200,000$ psi. There is no plaster ceiling supported by the joists so the deflection is limited to $\frac{1}{240}$ of the span due to live load only. Determine the joist size.

10-11 Plank Floors

Another type of wood floor construction that may be used in residential as well as in other structures consists of planks 2 in. or more in thickness, laid with the wide faces flat, and spanning directly from beam to beam. Tongue-and-groove planking and splined planking are illustrated in Figs. 10-5a and b. Figure 10-5c shows a laminated floor; this consists of planks set on edge side by side and nailed together at frequent intervals, approximately 18 in., alternating near top and bottom. Both plank and laminated floors must, of course, be securely nailed to the supporting beams.

Four types of spans for plank floors are generally recognized: simple, two-span continuous, combination simple and two-span continuous, and controlled random. The last three types are all stiffer, in varying degree, than the simple span because of the continuity introduced by the different arrangements of the pieces of planking. The four span types are identified in Fig. 10-6. Examination of the figure shows that all planks are the same length in the simple span with end joints over each beam. The plank lengths are also equal for the two-span continuous arrangement with end joints over every other beam. For the combination span, all pieces are two spans in length except for every other piece in the end span, but the end joints

(a) Tongue–and–Groove Flooring

(b) Splined Flooring

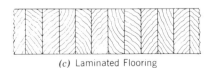

(c) Laminated Flooring

FIGURE 10-5

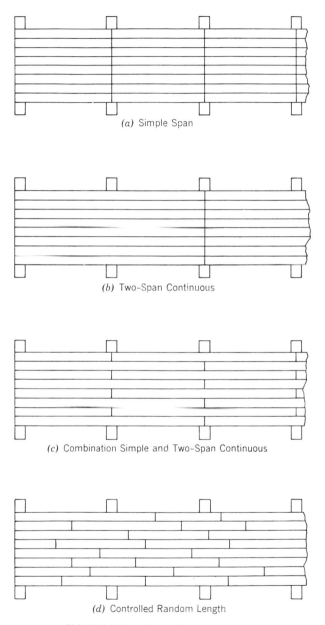

(a) Simple Span

(b) Two-Span Continuous

(c) Combination Simple and Two-Span Continuous

(d) Controlled Random Length

FIGURE 10-6. Plank floor span types.

TABLE 10-4. Safe Uniform Loads on Plank Floors* (pounds per square foot)

Nominal depth and approx. weight	Limited by bending $F_b =$ 1000 psi	Limited by deflection, $E = 1,000,000$ psi			
		Deflection ratio	Simple span	Combina-tion span	Controlled random
4-ft Span					
2 in 4.5 psf	188	1/240 1/360	117 78	168 112	153 102
3 in. 7.6 psf	521	1/240 1/360	543 362	777 518	819 546
4 in. 10.6 psf	1021	1/240 1/360	1489 992	2132 1422	2249 1499
6-ft Span					
2 in. 4.5 psf	83	1/240 1/360	35 23	50 33	45 30
3 in. 7 6 psf	231	1/240 1/360	161 107	230 153	243 162
4 in. 10.6 psf	454	1/240 1/360	441 294	632 421	666 444
8-ft Span					
2 in. 4.5 psf	47	1/240 1/360	15 10	21 14	19 13
3 in. 7.6 psf	130	1/240 1/360	68 45	97 65	102 68
4 in. 10.6 psf	255	1/240 1/360	186 124	267 178	281 187

* Data abstracted from a more complete set of tables in the *Western Woods Use Book*. Courtesy Western Wood Products Association.

TABLE 10-4. (*Continued*)

Nominal depth and approx. weight	Limited by bending $F_b = 1000$ psi	Limited by deflection, $E = 1,000,000$ psi			
		Deflection ratio	Simple span	Combination span	Controlled random
10-ft Span					
3 in. 7.6 psf	83	1/240 1/360	35 23	50 33	52 35
4 in. 10.6 psf	163	1/240 1/360	95 64	136 91	144 96
6 in. 16.6 psf	403	1/240 1/360	370 246	530 353	558 372
12-ft Span					
3 in. 7.6 psf	58	1/240 1/360	20 13	29 19	30 20
4 in. 10.6 psf	113	1/240 1/360	55 37	79 53	83 56
6 in. 16.6 psf	280	1/240 1/360	214 143	306 204	323 215
8 in. 22.7 psf	487	1/240 1/360	490 327	702 468	740 493

over intermediate beams are staggered in adjacent lines of planking. The random arrangement permits the use of economical random lengths of plank. The principal control requirement for this type of span is that end joints be well scattered and each piece of plank bear on at least one beam.

When designing plank floors, a width of 12 in. is assumed and the depth is determined by the usual methods followed in the design of beams. Although these computations are relatively simple, the proper thickness may be found directly by use of safe load tables.

Table 10-4 is an abridgment of much more extensive tables in the

Western Woods Use Book, published by the Western Wood Products Association. It gives the allowable total uniformly distributed load in pounds per square foot (for three of the four span types) for a limited number of plank thicknesses and span lengths. Two sets of values are tabulated: the maximum load as limited by bending strength, and the maximum load as limited by deflection.

With respect to bending, the tabulated loads for all span types have been determined by use of the bending moment formula $M = wL^2/8$, in which w is the total uniform load in pounds per square foot and L is the length of span in feet. The loads as limited by deflections of $\frac{1}{240}$ and $\frac{1}{360}$ of the span are different for each span type and have been calculated using the appropriate deflection formulas. Use of the table is illustrated in the following example.

Example.   Using Table 10-4, select a plank floor for the condition given below. The live load is 75 psf and a hardwood finished floor (weight allowance = 3 psf) is laid over the planking. The span is 10 ft and, since there is no plaster ceiling involved, the deflection is limited to $\frac{1}{240}$ of the span. Douglas Fir No. 2 grade lumber will be used and the planks arranged in the controlled random type of span.
Solution: (1) Since the loads given in Table 10-4 are based on $F_b = 1000$ psi and $E = 1,000,000$ psi, it is first necessary to find F_b and E for the species and grade of lumber to be used. Table 10-1 lists these values as $F_b = 1450$ psi (repetitive-member uses) and $E = 1,700,000$ psi. Because these values are higher than the corresponding ones on which Table 10-4 is based, the allowable loads for Douglas Fir No. 2 will be larger than those tabulated by the ratio 1450/1000 for loads limited by bending and 1,700,000/1,000,000 for those limited by deflection.

(2) Scanning Table 10-4 with our 78-psf superimposed load in mind, assume 3-in. plank as a trial thickness on the 10-ft span. The tabular values are

$$\text{Bending load} = 83 \text{ psf}$$

$$\text{Deflection load} = 52 \text{ psf}$$

(3) The allowable load limited by bending is

$$83 \times \frac{1450}{1000} = 120 \text{ psf}$$

and the net load that may be carried in addition to the dead load of the planking is $120 - 7.6 = 112.4$, say 112 psf.

(4) The allowable load limited by deflection is

$$52 \times \frac{1,700,000}{1,000,000} = 88.4 \text{ psf}$$

and the allowable net load in addition to the weight of the planking is $88.4 - 7.6 = 80.8$, say 81 psf.

(5) We observe that deflection controls but that the 3-in. plank is adequate to support the superimposed load of 78 psf. The trial thickness of 3 in. is, therefore, adopted.

Problem 10-11-A*.   The No. 1 grade of Balsam Fir is used for 4-in. plank flooring over several bays of a building. The combination span arrangement (simple and two-span continuous) has beam-to-beam spans of 10 ft. Determine the uniform load per square foot, in addition to its own weight, that the plank floor will safely support if deflection is limited to $\frac{1}{240}$ of the span.

Problem 10-11-B.   Three-inch planking (controlled random layup) spans 6 ft between beams. The lumber is Engelmann Spruce, No. 2 grade. Compute the safe superimposed uniform load on the floor if deflection is limited to $\frac{1}{240}$ of the span.

Problem 10-11-C.   Determine the safe superimposed load on the floor described in Problem 10-11-B if the deflection is limited to $\frac{1}{360}$ of the span.

Problem 10-11-D.   The No. 2 Medium grain grade of Southern Pine is to be used for plank floor construction in a residence. The floor beams are spaced 6 ft apart and the live load of 45 psf includes an allowance for the owner's choice of finished flooring. What thickness of planking is required if the combination simple and two-span continuous layup is used with deflection limited to $\frac{1}{360}$ of the span?

10-12   Rafter Roofs: Span Tables

Rafters are the comparatively small, closely spaced beams used to support the load on sloping roofs. The most common sizes are $2 \times 6, 2 \times 8, 2 \times 10$, and 2×12; the usual spacings are 16 in. and 24 in. on centers.

The span of a rafter is measured along the horizontal projection as indicated in Fig. 10-7. Although the dimension lines in the figure

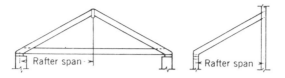

FIGURE 10-7

run between centers of supports, it is common practice to consider the span as the clear distance between supports when designing rafters spaced not more than 24 in. on centers.

The design of rafters is generally accomplished by the use of safe load tables. Tables 10-5 through 10-7 are representative of such tables and have been compiled from data in the extensive series of similar tables presented in the National Forest Products Association publication *Span Tables for Joists and Rafters*. The tables give maximum safe spans in feet for low slope (3 in 12 or less) and high slope (over 3 in 12) rafters for five spacings[6] and selected values of the allowable bending stress F_b. The modulus of elasticity E required to maintain the stated deflection limit is listed below each span.

It will be noted that the dead load allowance and the deflection limit vary among the three tables to accommodate different situations with respect to ceiling construction. The dead load allowance also provides for the rafter weight, roof sheathing, and a lightweight roofing material.

The only live load value provided for in Tables 10-5, 10-6, and 10-7 is 30 psf. Where local building codes or regional climate conditions call for a larger value, more extensive tables should be consulted The following examples illustrated the use of these tables.

Example 1. A roof with a slope of 2 in 12 has a horizontal projection rafter span of 14 ft 6 in. There is no finished ceiling and the live load on the roof is 30 psf. Select rafters from the span tables. *Solution:* (1) Referring to Table 10-5, we find that 2 × 8 rafters spaced 16 in. on centers and of a grade having an F_b value of 1300 psi and an E value of 1,180,000 psi may be used safely on a span of 14 ft 7 in. This meets the required conditions.

---

[6] The spacings of 13.7 in. and 19.2 in. accommodate the division of standard 8-ft length sheet material into 7 spaces and 5 spaces, respectively.

TABLE 10-5. Maximum Spans for Low Slope Rafters—No Finished Ceiling*
Slope = 3 in 12 or less: Live Load = 30 psf

Design Criteria: F_b determined by 10-psf dead load plus 30-psf live load
Deflection for 30-psf live load limited to $\frac{1}{240}$ of span

Rafter size and spacing (in.)		Allowable extreme fiber stress in bending, F_b (psi)								
		1100	1200	1300	1400	1500	1600	1700	2000	2400
2 × 6	12.0	11-9 1.06	12-4 1.21	12-10 1.36	13-3 1.52	13-9 1.69	14-2 1.86	14-8 2.04	15-11 2.60	
	13.7	11-0 0.99	11-6 1.13	12-0 1.27	12-5 1.42	12-10 1.58	13-3 1.74	13-8 1.90	14-10 2.43	
	16.0	10-2 0.92	10-8 1.05	11-1 1.18	11-6 1.32	11-11 1.46	12-4 1.61	12-8 1.76	13-9 2.25	
	19.2	9-4 0.84	9-9 0.95	10-1 1.08	10-6 1.20	10-10 1.33	11-3 1.47	11-7 1.61	12-7 2.05	
	24.0	8-4 0.75	8-8 0.85	9-1 0.96	9-5 1.08	9-9 1.19	10-0 1.31	10-4 1.44	11-3 1.84	12-4 2.41
2 × 8	12.0	15-6 1.06	16-3 1.21	16-10 1.36	17-6 1.52	18-2 1.69	18-9 1.86	19-4 2.04	20-11 2.60	
	13.7	14-6 0.99	15-2 1.13	15-9 1.27	16-5 1.42	16-11 1.58	17-6 1.74	18-1 1.90	19-7 2.43	
	16.0	13-5 0.92	14-0 1.05	14-7 1.18	15-2 1.32	15-8 1.46	16-3 1.61	16-9 1.76	18-2 2.25	
	19.2	12-3 0.84	12-10 0.95	13-4 1.08	13-10 1.20	14-4 1.33	14-10 1.47	15-3 1.61	16-7 2.05	
	24.0	11-0 0.75	11-6 0.85	11-11 0.96	12-5 1.08	12-10 1.19	13-3 1.31	13-8 1.44	14-10 1.84	16-3 2.41
2 × 10	12.0	19-10 1.06	20-8 1.21	21-6 1.36	22-4 1.52	23-2 1.69	23-11 1.86	24-7 2.04	26-8 2.60	
	13.7	18-6 0.99	19-4 1.13	20-2 1.27	20-11 1.42	21-8 1.58	22-4 1.74	23-0 1.90	25-0 2.43	
	16.0	17-2 0.92	17-11 1.05	18-8 1.18	19-4 1.32	20-0 1.46	20-8 1.61	21-4 1.76	23-2 2.25	
	19.2	15-8 0.84	16-4 0.95	17-0 1.08	17-8 1.20	18-3 1.33	18-11 1.47	19-6 1.61	21-1 2.05	
	24.0	14-0 0.75	14-8 0.85	15-3 0.96	15-10 1.08	16-4 1.19	16-11 1.31	17-5 1.44	18-11 1.84	20-8 2.41
2 × 12	12.0	24-1 1.06	25-2 1.21	26-2 1.36	27-2 1.52	28-2 1.69	29-1 1.86	29-11 2.04	32-6 2.60	
	13.7	22-6 0.99	23-6 1.13	24-6 1.27	25-5 1.42	26-4 1.58	27-2 1.74	28-0 1.90	30-5 2.43	
	16.0	20-10 0.92	21-9 1.05	22-8 1.18	23-6 1.32	24-4 1.46	25-2 1.61	25-11 1.76	28-2 2.25	
	19.2	19-0 0.84	19-11 0.95	20-8 1.08	21-6 1.20	22-3 1.33	23-0 1.47	23-8 1.61	25-8 2.05	
	24.0	17-0 0.75	17-9 0.85	18-6 0.96	19-3 1.08	19-11 1.19	20-6 1.31	21-2 1.44	23-0 1.84	25-2 2.41

Note: Required E is shown below each span (1,000,000 psi).

* Data abstracted from *Span Tables for Joists and Rafters.* Courtesy National Forest Products Association.

TABLE 10-6. Maximum Spans for Low or High Slope Rafters—Drywall Ceiling*
Live Load = 30 psf
Design Criteria: F_b determined by 15-psf dead load plus 30-psf live load
Deflection for 30-psf live load limited to $\frac{1}{240}$ of span

Rafter size and spacing (in.)		Allowable extreme fiber stress in bending, F_b (psi)								
		1100	1200	1300	1400	1500	1600	1700	2000	2400
2 × 6	12.0	11-1 / 0.89	11-7 / 1.01	12-1 / 1.14	12-6 / 1.28	13-0 / 1.41	13-5 / 1.56	13-10 / 1.71	15-0 / 2.18	
	13.7	10-5 / 0.83	10-10 / 0.95	11-3 / 1.07	11-9 / 1.19	12-2 / 1.32	12-6 / 1.46	12-11 / 1.60	14-0 / 2.04	
	16.0	9-7 / 0.77	10-0 / 0.88	10-5 / 0.99	10-10 / 1.10	11-3 / 1.22	11-7 / 1.35	11-11 / 1.48	13-0 / 1.89	14-2 / 2.48
	19.2	8-9 / 0.70	9-2 / 0.80	9-6 / 0.90	9-11 / 1.01	10-3 / 1.12	10-7 / 1.23	10-11 / 1.35	11-10 / 1.72	13-0 / 2.26
	24.0	7-10 / 0.63	8-2 / 0.72	8-6 / 0-81	8-10 / 0.90	9-2 / 1.00	9-6 / 1.10	9-9 / 1.21	10-7 / 1.54	11-7 / 2.02
2 × 8	12.0	14-8 / 0.89	15-3 / 1.01	15-11 / 1.14	16-6 / 1.28	17-1 / 1.41	17-8 / 1.56	18-2 / 1.71	19-9 / 2.18	
	13.7	13-8 / 0.83	14-4 / 0.95	14-11 / 1.07	15-5 / 1.19	16-0 / 1.32	16-6 / 1.46	17-0 / 1.60	18-5 / 2.04	
	16.0	12-8 / 0.77	13-3 / 0.88	13-9 / 0.99	14-4 / 1.10	14-10 / 1.22	15-3 / 1.35	15.9 / 1.48	17-1 / 1.89	18-9 / 2.48
	19.2	11-7 / 0.70	12-1 / 0.80	12-7 / 0.90	13-1 / 1.01	13-6 / 1.12	13-11 / 1.23	14-5 / 1.35	15-7 / 1.72	17-1 / 2.26
	24.0	10-4 / 0.63	10-10 / 0.72	11-3 / 0.81	11-8 / 0.90	12-1 / 1.00	12-6 / 1.10	12-10 / 1.21	13-11 / 1.54	15-3 / 2.02
2 × 10	12.0	18-8 / 0.89	19-6 / 1.01	20-4 / 1.14	21-1 / 1.28	21-10 / 1.41	22-6 / 1.56	23-3 / 1.71	25-2 / 2.18	
	13.7	17-6 / 0.83	18-3 / 0.95	19-0 / 1.07	19-8 / 1.19	20-5 / 1.32	21-1 / 1.46	21-9 / 1.60	23-7 / 2.04	
	16.0	16-2 / 0.77	16-11 / 0.88	17-7 / 0.99	18-3 / 1.10	18-11 / 1.22	19-6 / 1.35	20-1 / 1.48	21-10 / 1.89	23-11 / 2.48
	19.2	14-9 / 0.70	15-5 / 0.80	16-1 / 0.90	16-8 / 1.01	17-3 / 1.12	17-10 / 1.23	18-4 / 1.35	19-11 / 1.72	21-10 / 2.26
	24.0	13-2 / 0.63	13-9 / 0.72	14-4 / 0.81	14-11 / 0.90	15-5 / 1.00	15-11 / 1.10	16-5 / 1.21	17-10 / 1.54	19-6 / 2.02
2 × 12	12-0	22-8 / 0.89	23-9 / 1.01	24-8 / 1.14	25-7 / 1.28	26-6 / 1.41	27-5 / 1.56	28-3 / 1.71	30-7 / 2.18	
	13.7	21-3 / 0.83	22-2 / 0.95	23-1 / 1.07	24-0 / 1.19	24-10 / 1.32	25-7 / 1.46	26-5 / 1.60	28-8 / 2.04	
	16.0	19-8 / 0.77	20-6 / 0.88	21-5 / 0.99	22-2 / 1.10	23-0 / 1.22	23-9 / 1.35	24-5 / 1.48	26-6 / 1.89	29-1 / 2.48
	19.2	17-11 / 0.70	18-9 / 0.80	19-6 / 0.90	20-3 / 1.01	21-0 / 1.12	21-8 / 1.23	22-4 / 1.35	24-2 / 1.72	26-6 / 2.26
	24.0	16-1 / 0.63	16-9 / 0.72	17-5 / 0.81	18-1 / 0.90	18-9 / 1.00	19-4 / 1.10	20-0 / 1.21	21-8 / 1.54	23-9 / 2.02

Note: Required E is shown below each span (1,000,000 psi).

* Data abstracted from *Span Tables for Joists and Rafters*. Courtesy National Forest Products Association.

TABLE 10-7. Maximum Spans for Low or High Slope Rafters—Plaster Ceiling*
Live Load = 30 psf

Design Criteria: F_b determined by 15-psf dead load plus 30-psf live load
Deflection for 30-psf live load limited to ⅟₃₆₀ of span

Rafter size and spacing (in.)		Allowable extreme fiber stress in bending, F_b (psi)								
		1100	1200	1300	1400	1500	1600	1700	1800	2000
2 × 6	12.0	11-1 1.33	11-7 1.52	12-1 1.71	12-6 1.91	13-0 2.12	13-5 2.34	13-10 2.56		
	13.7	10-5 1.25	10-10 1.42	11-3 1.60	11-9 1.79	12-2 1.98	12-6 2.19	12-11 2.39		
	16.0	9-7 1.15	10-0 1.31	10-5 1.48	10-10 1.66	11-3 1.84	11-7 2.02	11-11 2.22	12-4 2.41	
	19.2	8-9 1.05	9-2 1.20	9-6 1.35	9-11 1.51	10-3 1.68	10-7 1.85	10-11 2.02	11-3 2.20	11-10 2.58
	24.0	7-10 0.94	8-2 1.07	8-6 1.21	8-10 1.35	9-2 1.50	9-6 1.65	9-9 1.81	10-0 1.97	10-7 2.31
2 × 8	12.0	14-8 1.33	15-3 1.52	15-11 1.71	16-6 1.91	17-1 2.12	17-8 2.34	18-2 2.56		
	13.7	13-8 1.25	14-4 1.42	14-11 1.60	15-5 1.79	16-0 1.98	16-6 2.19	17-0 2.39		
	16.0	12-8 1.15	13-3 1.31	13-9 1.48	14-4 1.66	14-10 1.84	15-3 2.02	15-9 2.22	16-3 2.41	
	19.2	11-7 1.05	12-1 1.20	12-7 1.35	13-1 1.51	13-6 1.68	13-11 1.85	14-5 2.02	14-10 2.20	15-7 2.58
	24.0	10-4 0.94	10-10 1.07	11-3 1.21	11-8 1.35	12-1 1.50	12-6 1.65	12-10 1.81	13-3 1.97	13-11 2.31
2 × 10	12.0	18-8 1.33	19-6 1.52	20-4 1.71	21-1 1.91	21-10 2.12	22-6 2.34	23-3 2.56		
	13.7	17-6 1.25	18-3 1.42	19-0 1.60	19-8 1.79	20-5 1.98	21-1 2.19	21-9 2.39		
	16.0	16-2 1.15	16-11 1.31	17-7 1.48	18-3 1.66	18-11 1.84	19-6 2.02	20-1 2.22	20-8 2.41	
	19.2	14-9 1.05	15-5 1.20	16-1 1.35	16-8 1.51	17-3 1.68	17-10 1.85	18-4 2.02	18-11 2.20	19-11 2.58
	24.0	13-2 0.94	13-9 1.07	14-4 1.21	14-11 1.35	15-5 1.50	15-11 1.65	16-5 1.81	16-11 1.97	17-10 2.31
2 × 12	12.0	22-8 1.33	23-9 1.52	24-8 1.71	25-7 1.91	26-6 2.12	27-5 2.34	28-3 2.56		
	13.7	21-3 1.25	22-2 1.42	23-1 1.60	24-0 1.79	24-10 1.98	25-7 2.19	26-5 2.39		
	16.0	19-8 1.15	20-6 1.31	21-5 1.48	22-2 1.66	23-0 1.84	23-9 2.02	24-5 2.22	25-2 2.41	
	19.2	17-11 1.05	18-9 1.20	19-6 1.35	20-3 1.51	21-0 1.68	21-8 1.85	22-4 2.02	23-0 2.20	24-2 2.58
	24.0	16-1 0.94	16-9 1.07	17-5 1.21	18-1 1.35	18-9 1.50	19-4 1.65	20-0 1.81	20-6 1.97	21-8 2.31

Note: Required E is shown below each span (1,000,000 psi).

* Data abstracted from *Span Tables for Joists and Rafters.* Courtesy National Forest Products Association.

(2) Turning to Table 10-1, and bearing in mind that values of F_b for repetitive-member uses apply to rafter construction, we note that the Select Structural grade of Balsam Fir and Engelmann Spruce meet the required F_b and E values quite closely. The reader should scan Table 10-1 further in order to identify other species and grades that would be satisfactory although somewhat stronger than necessary.

(3) Further examination of Table 10-5 will reveal that 2×10 rafters 24 in. on centers would be satisfactory for a grade having F_b and E values of 1200 psi and 850,000 psi, respectively; 2×8 rafters 24 in. on center could also be used with a grade having $F_b = 2000$ psi and $E = 1,840,000$ psi. Because so many combinations of size, spacing, and F_b and E values are possible, the tables should be examined carefully to determine which combination best fits a particular situation.

Example 2.   A roof of the type shown at the right of Fig. 10-7 has a slope of 3 in 12 and a horizontal projection span of 20 ft. A plaster ceiling is applied to the undersides of the rafters, and the required live load on the roof is 30 psf. Select the rafters from the span tables. *Solution:* (1) We find from scanning Table 10-7 that 2×12 rafters 16 in. on centers have an allowable span of 20 ft 6 in. if the F_b value of the grade used is 1200 psi and $E = 1,310,000$ psi. Another selection is 2×12 rafters 24 in. on centers if the grade of lumber employed has F_b and E values of 1700 psi and 1,810,000 psi, respectively.

(2) Reference to Table 10-1 shows that Southern Pine No. 2 grade meets the conditions for 2×12 rafters 16 in. on centers. For the 24-in. spacing, Southern Pine No. 1 Dense or Douglas Fir Dense No. 1 grades could be used.

Note: In the following problems the spans given are the "rafter spans," i.e., the horizontal projections of the rafter lengths.

Problem 10-12-A.   Select rafter sizes and spacings in accordance with the following data:

    (a) Span $= 16$ ft, live load $= 30$ psf, $F_b = 1500$ psi, $E = 1,200,000$ psi, and there is no finished ceiling.

    (b) Span $= 10$ ft, live load $= 30$ psf, $F_b = 1300$ psi, $E = 1,000,000$ psi, and the ceiling is drywall.

(c) Span = 22 ft, live load = 30 psf, F_b = 1400 psi, E = 1,700,000 psi, and there is a plaster ceiling.

Problem 10-12-B*.   A roof with a slope of 2½ in 12 has a rafter span of 13 ft. There is a ceiling of drywall construction, and the live load is 30 psf. Select rafters from the span tables.

Problem 10-12-C.   A roof with a slope of 3 in 12 has a rafer span of 18 ft. There is no finished ceiling, and the live load is 30 psf. Make selections from the span tables using (a) 2 × 8 rafters, (b) 2 × 10 rafters, and (c) 2 × 12 rafters. Name a satisfactory species and grade for each of your selections.

10-13   Glued Laminated Beams

Structural glued laminated lumber is composed of an assembly of wood laminations in which the grain of all laminations is approximately parallel longitudinally. The laminations are bonded by adhesives. One of the advantages of laminated lumber lies in the fact that it may be fabricated to unusually large cross sections and great lengths. The use of seasoned lumber and the dispersion of defects permits higher allowable stresses than can be employed in solid sawn lumber.

Glued laminated beams and girders are factory produced. The usual thickness of laminations is a nominal 2 in. and they may be selected free of checks or other defects found in large one-piece members. Because of this a laminated beam has greater strength than a solid member of the same grade of lumber. Allowable stresses for glued laminated structural members are given in the *National Design Specification for Stress-Grade Lumber and Its Fastenings.* Although glued laminated beams are available in a large variety of sizes, it is the ease with which very large girders can be built up that makes this type of member effective in long span construction. The reader desiring additional information on this type of construction should consult the *Timber Construction Manual,* prepared by the American Institute of Timber Construction (Second Edition, Wiley: New York, 1973).

11

Wood Columns

IIIIIIIIIIIIIIIIIIIIIIIIIIIIIIIIIIIIIIIIIIIIIIIIIIIIIIIIIIIIIII

11-1 Column Types

The type of wood column that is used most frequently is the *simple solid column*. It consists of a single piece of wood square or rectangular in cross section. Solid columns of circular cross section are also considered simple solid columns but they are used less frequently. A *spaced column* is an assembly of two pieces separated at the ends and at intermediate points along its length by blocking. Two other types are *built-up columns* with mechanical fastenings and *glued laminated columns*.

11-2 Slenderness Ratio

As stated in Art. 6-1, the slenderness ratio with respect to steel columns is defined as the ratio of the unbraced length to the least radius of gyration, or l/r. In wood construction, however, the slenderness ratio of a freestanding simple solid column is the ratio of the unbraced (unsupported) length to the dimension of its *least side*, or l/d. Both l and d are expressed in inches.

When wood compression members are braced so that the unsupported length with respect to one face is less than that with

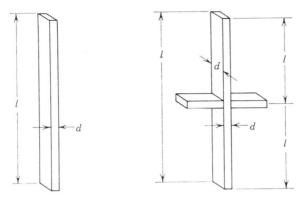

FIGURE 11-1

respect to the other, *l* is the distance between the points of support that prevent lateral movement in the direction along which the least dimension of the cross section is measured. This is illustrated in Fig. 11-1. Under these conditions, the slenderness ratio with respect to *both* faces must be computed and the larger value used in design. The slenderness ratio for simple solid columns is limited to $l/d = 50$; for spaced columns the limiting ratio is $l/d = 80$.

11-3 Column Formulas

The *National Design Specification for Stress-Grade Lumber and Its Fastenings* gives the following formula for computing the allowable unit compressive stress in axially loaded, square or rectangular, simple solid columns:

$$F'_c = \frac{0.30E}{(l/d)^2}$$

in which l/d = slenderness ratio,
$\quad\quad E$ = modulus of elasticity,
$\quad\quad F'_c$ = allowable unit stress in compression parallel to grain, adjusted for l/d ratio.

However, the maximum value of F'_c *must not exceed* F_c, the allowable unit stress in compression parallel to grain for the species and grade of lumber used (Table 10-1).

This formula applies to square-end simple solid columns as well as to the pin-end condition from which it was derived. It is appropriate for wood columns subjected to normal loading and used in dry locations. The formulas that apply to spaced columns are modifications of the one given above; they are considered in Art. 11-6.

11-4 Allowable Loads on Solid Columns

To find the safe load that a simple solid column of given cross section and length will support, the following steps may be taken.

Step 1: Consult a table of allowable unit stresses such as Table 10-1 and find the values of E and F_c for the species and grade of lumber involved.

Step 2: Referring to Table 4-6, find the dressed dimensions of the cross section and compute the slenderness ratio l/d. It should not exceed 50.

Step 3: Determine the allowable unit stress in compression parallel to grain, adjusted for l/d ratio, by substituting in the formula

$$F_c' = \frac{0.30E}{(l/d)^2}$$

Step 4: If the value of F_c' found in Step 3 does not exceed F_c, multiply F_c' by the area of the column cross section. The product will be the allowable axial load on the column.

Step 5: If the value of F_c' found in Step 3 exceeds F_c, the allowable load is determined by multiplying the area of the column by F_c.

Example 1. A 10 × 10 column of Douglas Fir, Dense Select Structural grade, has an unbraced length of 16 ft. Compute the allowable axial load.

Solution: Follow the procedure outlined above.

(1) Referring to Table 10-1, we find that for this species and grade, and size classification "Posts and Timbers," $E = 1,700,000$ psi and $F_c = 1350$ psi.

(2) The unsupported length is 16 × 12 = 192 in., and $d = 9.5$ in. (Table 4-6). Then $l/d = 192 \div 9.5 = 20.2$. Note that this value of the slenderness ratio is less than 50.

(3) Substituting in the column formula,

$$F_c' = \frac{0.30E}{(l/d)^2} = \frac{0.30 \times 1,700,000}{(20.2)^2} = 1250 \text{ psi}$$

(4) We observe that 1250 psi does not exceed the value of F_c which is 1350 psi. Consequently, the allowable axial load is F_c' times the column area (see Table 4-6) or

$$P = F_c' \times A = 1250 \times 90.25 = 112,000 \text{ lb}$$

Example 2. An 8 × 8 Southern Pine column, No. 1 Dense SR grade, has an unsupported height of 12 ft. Compute its allowable axial load.

Solution: (1) From Table 10-1 we find that $E = 1,600,000$ psi and $F_c = 1050$ psi.

(2) The unbraced height is 12 ft = 144 in. and $d = 7.5$ in. (Table 4-6). Then $l/d = 144 \div 7.5 = 19.2$. The slenderness ratio does not exceed 50.

(3) Substituting in the column formula,

$$F_c' = \frac{0.30E}{(l/d)^2} = \frac{0.30 \times 1,600,000}{(19.2)^2} = 1300 \text{ psi}$$

(4) However, we find that 1300 psi exceeds the value of F_c which is 1050 psi. Therefore we continue through Step 5 of the procedure.

(5) The allowable axial load is

$$P = F_c \times A = 1050 \times 56.25 = 59,000 \text{ lb}$$

Problem 11-4-A*. A 12 × 12 column of Southern Pine, No. 1 Dense SR grade, has an unbraced length of 20 ft. Compute its allowable axial load.

Problem 11-4-B. A column of California Redwood, Select Structural grade, has a nominal 6 × 6 cross section and is 10 ft long. Compute the allowable axial load.

Problem 11-4-C. An 8 × 8 column is 10 ft long and is made of Eastern Hemlock, Select Structural grade. Compute its allowable axial load.

Problem 11-4-D. A 10 × 12 column of Douglas Fir, Dense No. 1 grade has an unsupported length of 16 ft. Compute its allowable axial load.

11-5 Design of Solid Columns

In practice, wood columns are usually designed by the use of safe load tables. Because of the number of factors that must be considered, different forms of safe load tabulations have been developed. Table 11-1 gives the allowable unit stress (F_c') for several values of modulus of elasticity and l/d ratios, determined by the formula presented in

TABLE 11-1. Allowable Unit Stresses—Safe Unit Axial Loads for Simple Solid Wood Columns*

(F_c' from table may not exceed F_c for species and grade of lumber used)

Slenderness ratio (l/d)	Modulus of elasticity E in ksi (1000 psi)							
	1000	1100	1200	1300	1400	1500	1600	1700
10.0	3000	3300	3600	3900	4199	4499	4799	5099
10.5	2721	2993	3265	3537	3809	4081	4353	4625
11.0	2479	2727	2975	3223	3471	3719	3966	4214
11.5	2268	2495	2722	2948	3175	3402	3629	3856
12.0	2083	2291	2500	2708	2916	3125	3333	3541
12.5	1920	2112	2304	2496	2688	2880	3072	3264
13.0	1775	1952	2130	2307	2485	2662	2840	3017
13.5	1646	1810	1975	2139	2304	2469	2633	2798
14.0	1530	1683	1836	1989	2142	2295	2448	2602
14.5	1426	1569	1712	1854	1997	2140	2282	2425
15.0	1333	1466	1600	1733	1866	2000	2133	2266
15.5	1248	1373	1498	1623	1748	1873	1997	2122
16.0	1171	1289	1406	1523	1640	1757	1875	1992
16.5	1101	1212	1322	1432	1542	1652	1763	1873
17.0	1038	1141	1245	1349	1453	1557	1660	1764
17.5	979	1077	1175	1273	1371	1469	1567	1665
18.0	925	1018	1111	1203	1296	1388	1481	1574
18.5	876	964	1051	1139	1227	1314	1402	1490
19.0	831	914	997	1080	1163	1246	1329	1412
19.5	788	867	946	1025	1104	1183	1262	1341
20.0	750	825	900	975	1050	1125	1200	1275
20.5	713	785	856	928	999	1070	1142	1213
21.0	680	748	816	884	952	1020	1088	1156
21.5	649	713	778	843	908	973	1038	1103
22.0	619	681	743	805	867	929	991	1053
22.5	592	651	711	770	829	888	948	1007
23.0	567	623	680	737	793	850	907	964
23.5	543	597	651	706	760	814	869	923

* Data abstracted from *Wood Structural Design Data.* Courtesy National Forest Products Association.

TABLE 11-1   (*continued*)

Slenderness ratio (l/d)	Modulus of elasticity E in ksi (1000 psi)							
	1000	1100	1200	1300	1400	1500	1600	1700
24.0	520	572	625	677	729	781	833	885
24.5	499	549	599	649	699	749	799	849
25.0	480	528	576	624	672	720	768	816
25.5	461	507	553	599	645	692	738	784
26.0	443	488	532	576	621	665	710	754
26.5	427	469	512	555	598	640	683	726
27.0	411	452	493	534	576	617	658	699
27.5	396	436	476	515	555	595	634	674
28.0	382	420	459	497	535	573	612	650
28.5	369	406	443	480	517	554	590	627
29.0	356	392	428	463	499	535	570	606
29.5	344	379	413	448	482	517	551	586
30.0	333	366	400	433	466	500	533	566
30.5	322	354	386	419	451	483	515	548
31.0	312	343	374	405	437	468	499	530
31.5	302	332	362	393	423	453	483	513
32.0	292	322	351	380	410	439	468	498
32.5	284	312	340	369	397	426	454	482
33.0	275	303	330	358	385	413	440	468
33.5	267	294	320	347	374	400	427	454
34.0	259	285	311	337	363	389	415	441
34.5	252	277	302	327	352	378	403	428
35.0	244	269	293	318	342	367	391	416
35.5	238	261	285	309	333	357	380	404
36.0	231	254	277	300	324	347	370	393
36.5	225	247	270	292	315	337	360	382
37.0	219	241	262	284	306	328	350	372
37.5	213	234	256	277	298	320	341	362

TABLE 11-1 (continued)

Slenderness ratio (l/d)	Modulus of elasticity E in ksi (1000 psi)							
	1000	1100	1200	1300	1400	1500	1600	1700
38.0	207	228	249	270	290	311	332	353
38.5	202	222	242	263	283	303	323	344
39.0	197	216	236	256	276	295	315	335
39.5	192	211	230	249	269	288	307	326
40.0	187	206	225	243	262	281	300	318
40.5	182	201	219	237	256	274	292	310
41.0	178	196	214	232	249	267	285	303
41.5	174	191	209	226	243	261	278	296
42.0	170	187	204	221	238	255	272	289
42.5	166	182	199	215	232	249	265	282
43.0	162	178	194	210	227	243	259	275
43.5	158	174	190	206	221	237	253	269
44.0	154	170	185	201	216	232	247	263
44.5	151	166	181	196	212	227.	242	257
45.0	148	162	177	192	207	222	237	251
45.5	144	159	173	188	202	217	231	246
46.0	141	155	170	184	198	212	226	241
46.5	138	152	166	180	194	208	221	235
47.0	135	149	162	176	190	203	217	230
47.5	132	146	159	172	186	199	212	226
48.0	130	143	156	169	182	195	208	221
48.5	127	140	153	165	178	191	204	216
49.0	124	137	149	162	174	187	199	212
49.5	122	134	146	159	171	183	195	208
50.0	120	132	144	156	168	180	192	204

Art. 11-3. It will be noted that the value of F_c, allowable unit stress in compression parallel to grain, does not appear in this table. Consequently, when computing the total safe load on a column, the F_c' value obtained from the table must not exceed F_c for the species and grade of lumber used.

Table 11-1 lists values of l/d at 0.5 intervals. For most practical purposes, the tabular l/d value nearest the one calculated in Step 2 of the procedure outlined in Art. 11-4 may be used. Thus in Example 1 of Art 11-4 where $l/d = 20.2$, the value of F_c' given in the table for $l/d = 20.0$ and $E = 1,700,000$ psi is 1275 psi, slightly higher than the caluclated 1250 psi.

This table greatly facilitates the testing of trial sections in design since it eliminates the necessity of solving the column formula. The area of the trial section is multiplied directly by the tabular value of F_c' to obtain the total safe load on the column. In wood structural design, this type of table is often called Safe Unit Axial Loads because the listed values of F_c' (although stated in psi) may be thought of as loads in pounds on a 1 sq in. unit of the column cross section.

Example 1. A wood column of Douglas Fir, Select Structural grade, has an unsupported length of 14 ft and sustains an axial load of 50 kips. Design the column.
Solution: (1) For this species and grade of lumber, Table 10-1 shows that $E = 1,600,000$ psi and $F_c = 1150$ psi. Since the actual unit stress in the column must not exceed 1150 psi, let us estimate that F_c' will be about 1000 psi for the purpose of selecting a trial section. Then $50,000 \div 1000 = 50$ sq in., the estimated required area. Referring to Table 4-6, we find that a nominal 8×8 has an area of 56.25 sq in., and we select an 8×8 column for a trial section.

(2) The slenderness ratio of the trial section is $l/d = (14 \times 12) \div 7.5 = 168 \div 7.5 = 22.4$. We note that this value is less than 50, the limiting value for solid columns.

(3) Entering Table 11-1 at $l/d = 22.5$ and proceeding horizontally to the listings for $E = 1,600,000$ psi, we find that $F_c' = 948$ psi.

(3) We observe that 948 psi does not exceed $F_c = 1150$ psi; consequently the allowable axial load is F_c' times the column area or

$$P = F_c' \times A = 948 \times 56.25 = 53,300 \text{ lb} \quad \text{or} \quad 53.3 \text{ kips}$$

Since the load to be carried is 50 kips, accept the 8×8 column.

TABLE 11-2. Allowable Axial Loads on Solid Wood Columns

(F_c' may not exceed F_c for species and grade of lumber used)

Size (in.)	Area (sq in.)	Length (ft)	E = 1,200,000		E = 1,600,000		E = 1,700,000	
			F_c' (psi)	P (kips)	F_c' (psi)	P (kips)	F_c' (psi)	P (kips)
6 × 6	30.25	8	1180	35.7	1570	47.5	1670	50.5
(5½ × 5½)		10	756	22.9	1010	30.5	1070	32.4
		12	525	15.9	700	21.2	744	22.5
8 × 8	56.25	10	1410	79.2	1880	106.0	1990	112.0
(7½ × 7½)		12	977	55.0	1300	73.2	1380	77.6
		14	717	40.3	957	53.8	1020	57.3
		16	549	30.8	732	41.2	778	43.6
		18	434	24.4	579	32.5	615	34.6
10 × 10	90.25	12	1570	142	2090	188	2220	200
(9½ × 9½)		14	1150	104	1540	139	1630	147
		16	881	79.5	1175	106	1250	113
		18	696	62.8	929	83.8	987	89.0
		20	564	50.9	752	67.9	799	72.0
12 × 12	132.25	14	1690	223	2250	298	2390	316
(11½ × 11½)		15	1470	194	1960	259	2080	275
		16	1290	171	1720	227	1830	242
		17	1140	151	1530	202	1620	214
		18	1020	135	1360	180	1450	191
		20	827	109	1100	145	1170	155
14 × 14	182.25	16	1780	324	2370	432	2520	458
(13½ × 13½)		17	1580	288	2100	383	2230	406
		18	1410	256	1880	342	1990	362
		19	1260	229	1680	306	1790	326
		20	1140	208	1520	277	1610	293
		22	941	171	1260	229	1330	242

Another type of safe load table gives directly the total allowable load on a column of given cross section and length; that is, the results of multiplying F_c' values by the cross-sectional areas are tabulated. Table 11-2 gives safe total loads for a few combinations of column size, length, and E values. In using the table, care must again be exercised to see that the tabular value of F_c' does not exceed F_c, the allowable unit stress in compression parallel to grain for the species and grade of lumber used. If F_c' exceeds F_c, the tabulated value of P does not apply and the safe load on the column becomes $P = F_c \times A$.

Example 2.   Eastern Hemlock, Select Structural grade, is to be used for a column 12 ft high which has an axial design load of 50 kips. By use of Table 11-2 determine the size of the column.

Solution: (1) Table 10-1 shows that $E = 1,200,000$ psi and $F_c = 1000$ psi for this species and grade.

(2) Entering Table 11-2 under the proper E value, we find that a load $P = 55$ kips is listed for an 8 × 8 column 12 ft long.

(3) Checking the value of the unit stress given in the table, we note that $F'_c = 977$ psi under the 55-kip loading. Since this magnitude does not exceed $F_c = 1000$ psi, the 8 × 8 column is satisfactory.

Because extensive safe load tables are not always readily available, the reader should make certain that he understands thoroughly the procedure for column design presented in Example 1 of this article. With the aid of Table 11-1 and a slide rule or desk calculator, the process of assuming and testing a trial section is not a laborious one. Also, as one gains experience, his choice of initial trial section approaches more closely the one finally adopted.

Problem 11-5-A*.   By use of Table 11-2, determine the size of a column of suitable species and grade to support an axial load of 110 kips if the unbraced height is 16 ft.

Problem 11-5-B*.   An 8 × 8 column of Southern Pine, No. 1 Dense SR grade, is used for a column with a length of 12 ft. If the column load is 58 kips, is the 8 × 8 large enough? What load will it support?

Problem 11-5-C.   The Select Structural grade of Eastern Hemlock is to be used for a column 12 ft in length. If the load is 15,800 lb, what should the column size be?

Problem 11-5-D.   A column of California Redwood, No. 1 grade, has an unsupported length of 14 ft and sustains an axial load of 40 kips. Design the column.

Problem 11-5-E.   A column of Engelmann Spruce, Select Structural grade, is 16 ft long and supports an axial load of 60 kips. Design the column.

11-6   Spaced Columns

Spaced columns are used for direct support of axial loads or as compression members in wood trusses. They are formed of two or more individual solid members having their longitudinal axes parallel and separated at the ends and middle points of their length by blocking.

They are joined at the ends by connectors capable of developing the required shear resistance. The typical arrangement is shown in Fig. 11-2.

The connectors and their bolts at each end of the assembly restrain differential movement between the ends of the individual pieces, thereby introducing some end rigidity. This restraint, however, is effective only in the thickness direction. For the *individual members* of a spaced column the slenderness ratio of the narrow face l/d shall not exceed 80 but that of the wide face l/d_1 is limited to 50, the same as specified for simple solid columns. A further criterion with respect to the narrow face limits l_2/d to 40. See Fig. 11-2 for these relationships.

Because of the effect of end rigidity on the individual members, a higher unit stress is permitted for spaced columns than for simple

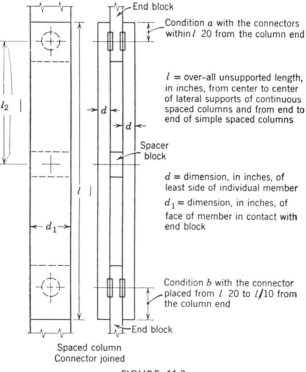

End block

Condition a with the connectors within l 20 from the column end

l = over–all unsupported length, in inches, from center to center of lateral supports of continuous spaced columns and from end to end of simple spaced columns

Spacer block

d = dimension, in inches, of least side of individual member

d_1 = dimension, in inches, of face of member in contact with end block

Condition b with the connector placed from l 20 to $l/10$ from the column end

End block

Spaced column
Connector joined

FIGURE 11-2

solid columns. The degree of restraint provided is related to the location of the connectors at the ends of the column; when they are placed in accordance with condition *a* as defined in Fig. 11-2, the allowable unit stress is determined by the formula

$$F'_c = 2.5 \times \frac{0.30E}{(l/d)^2} = \frac{0.75E}{(l/d)^2}$$

When the connectors are farther from the ends as defined for condition *b*, the allowable stress is given by

$$F'_c = 3.0 \times \frac{0.30E}{(l/d)^2} = \frac{0.90E}{(l/d)^2}$$

In either case, F'_c must not exceed F_c for the species and grade of lumber used. It will be noted that the above equations are the same as the basic column formula discussed in Art. 11-3, modified for different degrees of end rigidity. The total allowable load for a spaced column is the sum of the allowable loads for each of its individual members.

To understand how the allowable load on a spaced column is computed, consider the following example.

Example. A spaced column is 7 ft long and is composed of two 3 × 8 individual members separated by 3-in. blocks and secured together by connectors and bolts in accordance with applicable specifications. The lumber to be used is Douglas Fir, for which $E = 1,700,000$ psi and $F_c = 1250$ psi. If the positions of the connectors in end blocks are in agreement with condition *a* (see Fig. 11-2), compute the allowable axial load on the spaced column.

Solution: (1) Table 4-6 shows that the dressed size of a 3 × 8 member is $2\frac{1}{2} \times 7\frac{1}{4}$ and that its cross-sectional area is 18.13 sq in. The slenderness ratio of an individual member is then $l/d = (7 \times 12) \div 2.5 = 33.6$. We note that this value is less than 80, the limiting value for spaced columns.

(2) For condition *a* the allowable unit stress is

$$F'_c = \frac{0.75E}{(l/d)^2} = \frac{0.75 \times 1,700,000}{33.6 \times 33.6} = 1130 \text{ psi}$$

(3) We observe that 1130 psi does not exceed $F_c = 1250$ psi; consequently the allowable load on one 3×8 member is

$$P = F'_c \times A = 1130 \times 18.13 = 20{,}500 \text{ lb}$$

Because there are two 3×8 members, the total allowable load on the spaced column is $2 \times 20.5 = 41$ kips.

Tables of allowable unit stresses or safe unit axial loads for spaced columns (similar to Table 11-1 for solid columns) are contained in *Wood Structural Design Data*[1] and are, of course, a great aid when there is much designing to be done.

Problem 11-6-A*. A spaced column composed of two 2×6 members is 8 ft long. The lumber is California Redwood, Select Structural grade. If the end conditions of the spaced column agree with condition b, compute its allowable load.

Problem 11-6-B. Compute the allowable load on a spaced column of two 3×10 members having a length of 10 ft. The column ends comply with condition a and the wood is Southern Pine of a grade rated $E = 1{,}600{,}000$ psi and $F_c = 925$ psi.

11-7 Glued Laminated Columns

The allowable axial loads on glued laminated columns are determined by the column formula discussed in Art. 11-3. However, the value of F_c, the allowable compressive stress parallel to grain, may be higher than is permitted for solid wood columns. As mentioned in Art. 10-13, information on the design of glued laminated members may be found in the *Timber Construction Manual*, prepared by the American Institute of Timber Construction (Second Edition, Wiley: New York, 1973).

---

[1] Published by the National Forest Products Association, Washington, D.C. A somewhat differently arranged series of safe unit axial load tables is presented in the *Western Woods Use Book*, published by the Western Wood Products Association, Portland, Oregon.

12

Timber Connectors

IIIIIIIIIIIIIIIIIIIIIIIIIIIIIIIIIIIIIIIIIIIIIIIIIIIIIIIIII

12-1 General

A full explanation of the different types of fasteners used in wood construction is beyond the scope of this book. However, the *split ring connector* is so frequently encountered and is such an ingenious device that a brief discussion of its action and the factors that determine its strength in a joint is presented in this chapter.

12-2 Split Ring Connectors

The split ring is a most efficient device for making joints in timber construction. It is available in two sizes, the $2\frac{1}{2}$ in. and the 4 in. The smaller size is used widely in trussed-rafter construction and in wood framing for lumber having a minimum width of $3\frac{1}{2}$ in. The 4-in. split ring is used in heavier timber construction, particularly in moderate and long span roof trusses. Figure 12-1*h* shows a ring.

Split rings are placed in precut circular grooves made in the contacting faces of the pieces of wood to be joined with half the depth of the ring in the groove of each member. A $\frac{3}{4}$-in. bolt, inserted in holes concentric with the rings, holds the members together. A power-driven tool is employed for cutting the grooves. A tongue-and-groove

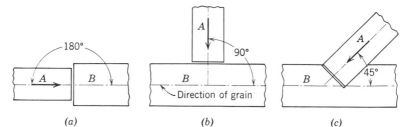

(a)   (b)   (c)

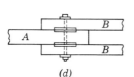

(d)   (e)   (f)

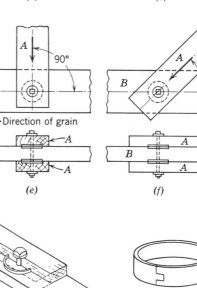

(g) Split Ring Joint
Portion of one member
cut away to show
position of rings

(h) Split Ring

FIGURE 12-1

split in the ring permits a slight variation in the ring diameter and permits a simultaneous bearing of the ring on its outer and inner surfaces. Joints employing split rings are illustrated in Figs. 12-1d, e, f, and g.

The 4-in. split ring connector is a band of metal that has a 4-in. inside diameter at the center when it is closed. The thickness of the ring material is approximately 0.2 in. at its center, and its depth is

1 in. The minimum lumber dimensions for the 4-in. split ring area $5\frac{1}{2}$-in. width, a 1-in. thickness when the ring is in only one face, and a $1\frac{1}{2}$-in. thickness when the rings are opposite in both faces. The 4-in. split ring requires a $\frac{3}{4}$-in. diameter bolt in a $\frac{13}{16}$-in. diameter hole. The washers should be 3 in. in diameter, cast or malleable iron, or a 3-in square steel plate $\frac{3}{16}$ in. thick.

12-3 Strength of Connector Joints

In general, the strength of a connector joint depends on the type and size of the connector, the species of wood, the thickness and width of the member, the distance of the connector from the end or side of the member, the spacing of the connectors, the direction of the applied load with respect to the grain of the wood, the period of load duration, and the moisture content of the wood. Because of these many factors, the design of joints in which connectors are employed is most readily accomplished by load tables and charts.

Extensive research and testing by the U.S. Forest Products Laboratory, the Timber Engineering Company, and others has resulted in valuable technical data that are employed by structural engineers. These data relate to allowable working loads, connector spacings, end distances, and edge distances. Figures 12-2, 12-4, and 12-6 are examples of such data.

12-4 Timber Species Groups

The species and density of structural lumber are important factors in determining allowable connector loads. Four species groups are identified in Table 12-1. Load charts for connectors are divided vertically into three species groups (A, B, and C) which include the species most often used with timber connectors (Fig. 12-2). As noted in the figure, Group D connector values may be found by multiplying Group C values by the factor 0.86.

12-5 Connector Loads

Wood has a much greater resistance to compressive loads when the line of action of the load is parallel to the grain than when it is

TABLE 12-1. Connector Load Grouping of Species When Stress-Graded*

Connector load grouping	Species	
Group *A*	Ash, Commercial White Beech Birch, Sweet and Yellow Douglas Fir–Larch (Dense)	Hickory and Pecan Oak, Red and White Maple, Black and Sugar Southern Pine (Dense)
Group *B*	Douglas Fir–Larch Southern Pine	Sweetgum and Tupelo
Group *C*	California Redwood (Close grain) Douglas Fir, South Eastern Hemlock– Tamarack Eastern Spruce Hem–Fir Idaho White Pine Lodgepole Pine	Mountain Hemlock Northern Pine Ponderosa Pine–Sugar Pine Red Pine Sitka Spruce Southern Cypress Spruce–Pine–Fir Yellow Poplar
Group *D*	Balsam Fir California Redwood (Open grain) Coast Sitka Spruce Cottonwood, Eastern Eastern White Pine	Engelmann Spruce Northern White Cedar Subalpine Fir Western Cedars Western White Pine

* Taken from *Western Woods Use Book*. Courtesy Western Wood Products Association.

perpendicular to it. In Figs. 12-1*a*, *b*, and *c* a force (indicated by the arrow) from one piece of timber marked *A* acts on another member marked *B*. These are solid pieces of wood; no connectors are used. Suppose that the select structural grade of Eastern Hemlock is used and that the member marked *A* is an 8 × 8 piece. Table 4-6 shows that this member has dressed dimensions of $7\frac{1}{2} \times 7\frac{1}{2}$ in. and a cross-sectional area of 56.25 sq in., which is the bearing area of piece *A* on piece *B*. Table 10-1 shows that 1000 psi and 365 psi are the allowable compressive stresses parallel to the grain and perpendicular to the grain, respectively. Therefore, in Fig. 12-1*a*, where the load is parallel to the grain of the wood, the allowable compressive load is $1000 \times 56.25 = 56,250$ lb. In Fig. 12-1*b* the load is perpendicular to the grain of the wood; hence the allowable load that *A* can exert on *B* is $365 \times 56.25 = 20,500$ lb.

An example of a load making an angle of 45° to the grain is shown in Fig. 12-1*c*. When a load acts on a surface inclined to the grain, the allowable unit stress is determined by the Hankinson formula.[1] In this case, the formula yields an allowable unit stress of 535 psi for the angle of 45°; hence the allowable load that may be transmitted from piece *A* to piece *B* is $535 \times 56.25 = 30,100$ lb.

The chart shown in Fig. 12-2 is used to determine the allowable normal load on one 4-in. split ring connector and bolt. Note that the chart is dividied into three parts according to the wood species, Groups *A*, *B*, and *C*. In each group there are several curves by means of which we can establish the allowable connector load in accordance with specific lumber thicknesses, the number of loaded faces, one or two, and the angle of load to grain. In Fig. 12-1*d* piece *A* has rings in two faces, whereas each piece marked *B* has a ring in only one face.

The curves in the chart shown in Fig. 12-2 are plotted in accordance with the Hankinson formula, with the load in pounds and the angle of load to grain as the variables. To find the allowable connector load, select the proper angle of load to grain, indicated at the top and bottom of the chart, and proceed vertically to the point of intersection with the appropriate curve. From this point proceed horizontally to the right or left to read the normal connector load.

Example. A Group *A* species of lumber is $1\frac{1}{2}$ in. thick and has 4-in. split ring connectors in two faces. If the angle of load to grain is 15°, determine the allowable normal connector load.

Solution: For Group *A* species and material $1\frac{1}{2}$ in. thick, with connectors in two faces, the appropriate curve in the chart of Fig.

[1] This formula involves the allowable compressive stresses parallel and perpendicular to the grain, and the squares of the sine and cosine of the angle. Explanations of the Hankinson formula may be found in *Simplified Design of Structural Timber*, Second Edition, by Harry Parker (New York: Wiley, 1963), Art. 3-6; and in *Wood Structural Design Data* (National Forest Products Association, 1970).

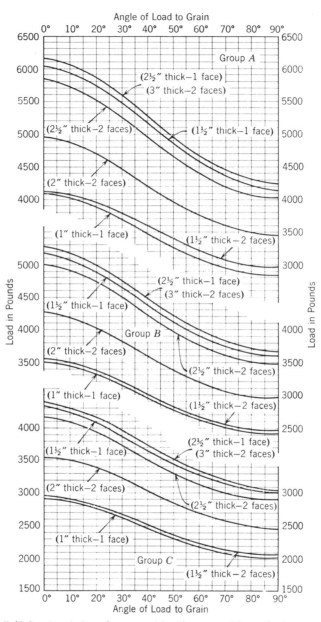

FIGURE 12-2. Load chart for normal loading—one 4-in. split ring and bolt in single shear. [*Note:* For Group *D* connector values, multiply Group *C* values by 0.86.] Courtesy Timber Engineering Company.

12-2 is the fifth from the top. This curve shows that if the load is parallel to the grain of the wood (an angle of 0°), the allowable connector load is 4140 lb. If the load makes an angle of 90° with the grain, the connector load is 2960 lb. For a 15° angle of load to grain we find that the connector load is 4030 lb.

Connectors bearing on wood at various angles to the grain are illustrated in Figs. 12-1*d*, *e*, and *f*.

12-6   End Distance

Among the placing requirements for split ring connectors is the *end distance*. This is the distance measured parallel to the grain from the center of a connector to the square-cut end of a member. Two overlapping 2 × 6 pieces of timber are secured with a 4-in. split ring connector, as shown in Fig. 12-3. The two members are in

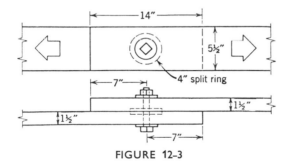

FIGURE 12-3

tension, and the distance marked 7 in. is the end distance. Figure 12-4 is the end distance chart for 4-in. split ring connectors. From this chart we see that if the end distance for a member in tension is 7 in. the connector may be counted on to support 100% of the tabulated allowable load. If, however, the end dimension is only $3\frac{1}{2}$ in., the connector load is only 62.5% of the tabulated load. This end-distance chart gives data for members in both tension and compression. For the joint shown in Fig. 12-3 imagine the two members to be in compression instead of tension. In the chart shown in Fig. 12-4 we find, for this condition, that an end distance of only $5\frac{1}{2}$ in. is required to provide 100% of the full allowable connector load.

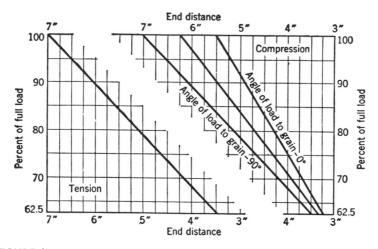

FIGURE 12-4.   End distance chart for 4-in. TECO split rings. Courtesy Timber Engineering Company.

12-7 Allowable Connector Loads

The following examples illustrate the procedure required to find the allowable connector loads under certain conditions.

Example 1.   Four 4-in. split ring connectors and two ¾-in. bolts are used to construct the joint in Fig. 12-5. The two 4 × 6 members are in tension, and the side plates are 2 × 6 pieces, 28 in. long. The lumber is Southern Pine (not dense). Compute the allowable load the connectors in this joint can transmit.

Solution: Referring to Table 12-1, we find that Southern Pine is classified as a Group *B* species. The allowable load given in the chart of Fig. 12-2 for one 4-in. split ring connector, when used in one face of a Group *B* species 1½ in. thick, is 5180 lb; when used as one of two connectors in opposite faces of a member 3 in. thick, the allowable value is 5280 lb. Therefore the safe load for the two connectors in one side of the joint is twice the smaller of these values or 2 × 5180 = 10,360 lb. Reference to the end-distance chart of Fig. 12-4 shows that a 7-in. end distance is adequate to develop the full design load of 10,360 lb in tension.

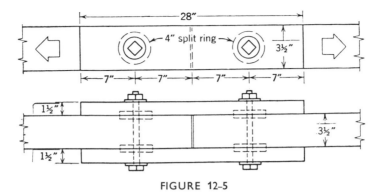

FIGURE 12-5

Example 2. Compute the allowable load the connectors in the tension joint shown in Fig. 12-5 will transmit if, instead of 28 in., the side plates have total lengths of only 26 in.

Solution: If the four connectors are so placed that the end distances are $6\frac{1}{2}$ in., the side plates will have 26-in. lengths. Then, referring to the end-distance chart (Fig. 12-4), we find that an end distance of $6\frac{1}{2}$ in. permits only 95% of the full connector load. In the preceding example the full connector load was found to be 10,360 lb. Consequently, the allowable load the connectors in the joint can transmit with the shorter side plates is 0.95 × 10,360 = 9840 lb.

12-8 Edge Distance

The *edge distance* is the distance from the edge of a member to the center of the connector closest to the edge, measured perpendicular to the edge. The *loaded edge distance* is the edge distance measured from the edge toward which the load induced by the connector acts. In Fig. 12-7 the 4 × 6 member is the loaded member; the connector bears *toward its lower* edge, and this is sometimes called the compression side of the loaded member. In this figure the loaded edge distance is $2\frac{3}{4}$ in.

Figure 12-6 is the edge distance chart for 4-in. split ring connectors. The minimum edge distance is at the right side of the chart, $2\frac{3}{4}$ in. Note that there is a variation in the percentage of full load according to the angle of load to grain. In Fig. 12-7 the angle of load to grain in the loaded member is 90°.

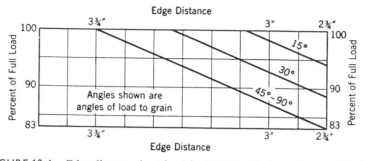

FIGURE 12-6.   Edge distance chart for 4-in. TECO split rings. Courtesy Timber Engineering Company.

Example.   In Fig. 12-7 the two tension side members are 2 × 6 pieces, and they are joined at right angles to opposite faces of a 4 × 6 center member by means of two 4-in. split ring connectors and a $\frac{3}{4}$-in. bolt. The timber is Douglas Fir (not dense). The 2 × 6 and 4 × 6 members have actual dimensions of $1\frac{1}{2} \times 5\frac{1}{2}$ in. and $3\frac{1}{2} \times 5\frac{1}{2}$ in., respectively, as shown in the figure. Calculate the allowable load the connectors can transmit from the two tension side members to the 4 × 6 horizontal center member.

Solution: (1) Consider first a connector in one of the 2 × 6 side members. Since Douglas Fir is in Group *B* species, the load chart

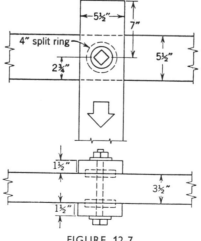

FIGURE 12-7

(Fig. 12-2) shows an allowable connector load of 5180 lb in one face of a 1½-in. thick member with a zero angle of load to grain.

(2) Figure 12-7 shows that the end distance for the side members is 7 in. Referring to Fig. 12-4, we note that this end distance permits 100% of the full load, 5180 lb. For the two connectors the allowable load is 2 × 5180 = 10,360 lb.

(3) Next, consider one of the connectors in the center member. In this member the angle of load to grain is 90°, and from the chart (Fig. 12-2) the connector load for a connector in two faces of a member 3 in. thick is 3680 lb. Now refer to the edge distance chart. See Fig. 12-6. Because the edge distance for the connector in the center member is 2¾ in. and the angle of load to grain is 90°, the allowable connector load is only 83% of the full load. Thus 3680 × 0.83 = 3050 lb, the allowable connector load for one connector. Since there are two connectors, 2 × 3050 = 6100 lb, the allowable load for the joint.

(4) The allowable connector load for the center member (6100 lb) is smaller than that for the side pieces (10,360 lb) and therefore determines the magnitude of the load that may be transmitted to it, 6100 lb.

IV

REINFORCED CONCRETE

13

Stresses in Reinforced Concrete

IIIIIIIIIIIIIIIIIIIIIIIIIIIIIIIIIIIIIIIIIIIIIIIIIIIIIIIIIIII

13-1 Introduction

Concrete is made by mixing a paste of cement and water with sand and crushed stone, gravel, or other inert material. The sand and crushed stone or gravel are called *aggregates*. After this plastic mixture is placed in forms, chemical reactions called hydration take place and the mass hardens. Concrete, although strong in compression, is relatively weak in resisting tensile and shearing stresses which develop in structural members. To overcome this lack of resistance, steel bars are placed in the concrete at the proper positions; the result is *reinforced concrete*. In beams and slabs the principal function of the concrete is to resist compressive stresses, whereas the steel bars resist tensile stresses.

13-2 Design Methods

The design of reinforced concrete structural members may be accomplished by two different methods. The first, called *working*

289

stress design, is the principal method used in this section of the book; the second method is known as *ultimate strength design*. The 1963 Code of the American Concrete Institute contained separate sections covering each of these methods but the 1971 Code, promulgated as *Building Code Requirements for Reinforced Concrete* (ACI 318-71), is built primarily around ultimate strength design, referred to in the Code as the *strength design method*. An *alternative design method* that is similar to working stress design procedures of the 1963 Code is also permitted; for ordinary beams and girders, flexural computations are identical with those of the 1963 Code. However, there are significant differences in other areas such as design for shear, anchorage length of reinforcement, and design of columns. The discussion of reinforced concrete design in Chapters 13 through 17 is keyed primarily to the working stress design method of the 1963 ACI Code, but draws also on the alternate method of the 1971 Code. A brief introduction to the theory of ultimate strength design is presented in Chapter 18.

The design of reinforced concrete members must, of course, be carried out in compliance with the building code that has jurisdiction in your locality. The 1963 ACI Code was widely adopted by building code authorities, and the 1971 standard is expected to follow the same pattern. Nevertheless, in actual design work, you should check the local code since the frequency with which it is amended may control whether or not the new provisions have been incorporated. The ACI Code is extensive, and those who desire more complete information should examine it in detail. In this elementary book space permits discussions of only the basic structural members, and many of the items referred to in the Code must necessarily be omitted.

13-3 Strength of Concrete

The designer of a reinforced concrete structure bases his computations on the use of concrete having a specified compressive strength (2500, 3000, 3500, etc., psi) at the end of a 28-day curing period. The symbol for this specified compressive strength is f'_c. Concretes of different strengths are produced by varying the proportions of cement, fine aggregate (sand), coarse aggregate, and water in the mix.

The general theory in establishing the proportions of fine and coarse aggregates is that the voids in the coarse aggregate should be filled with the cement paste and fine aggregate. The proportioning of concrete mixes and attendant procedures for strength verification are not discussed in this book. Readers interested in studying this aspect of concrete manufacture should consult *Recommended Practice for Selecting Proportions for Normal Weight Concrete* (ACI 211.1-70) and Section 4.2 of the 1971 ACI Code.

Very little concrete is proportioned and mixed at the building site today. Central or ready-mixed concrete is used whenever it is available. The use of a concrete mixed under ideal controlled conditions at a central plant affords many advantages. It is delivered to the building site in a revolving mixer, the proportions of cement, aggregate, and water are maintained accurately, any desired strength may be ordered, and the concrete thus provided is uniform in quality.

Table 13-1 gives allowable stresses in flexure (bending), shear, and bearing for concretes of four different specified compressive strengths. Note that the individual values are functions of f_c'. For example, suppose we are using a concrete for which $f_c' = 3000$ psi and we wish to know f_c, the allowable extreme fiber stress in compression. In the table we find that $f_c = 0.45 f_c'$. Then $f_c = 0.45 \times 3000 = 1350$ psi. Note that this stress is given in the table for "3000-lb concrete," as we sometimes call it.

13-4   Water–Cement Ratio

The most important factor affecting the strength of concrete is the *water–cement ratio*. This is expressed as the number of pounds of water per pound of cement used in the mix (or the number of gallons of water for each 94-lb bag of cement). The relationship is an inverse one, i.e., lower values of the ratio produce higher strengths. However, in order to produce freshly mixed concrete that possesses *workability*—the property that controls the ease with which it can be placed in the forms and around the reinforcing bars—more water must be used than the amount required for hydration of the cement. The use of too much water may cause segregation of the mix components, thereby producing nonuniform concrete. To control this

TABLE 13-1. Allowable Stresses in Concrete*

Description All values are given for normal weight concrete, 145 lb per cu ft		Based on specified concrete strength	Allowable stresses			
			For strength of concrete shown below			
			$f_c' =$ 2500 psi	$f_c' =$ 3000 psi	$f_c' =$ 4000 psi	$f_c' =$ 5000 psi
Modulus of elasticity	E_c	$57{,}000\sqrt{f_c'}$				
Modular ratio: $n = E_s/E_c$	n	$\dfrac{29{,}000{,}000}{57{,}000\sqrt{f_c'}}$	10	9	8	7
Flexure: f_c Extreme fiber stress in compression	f_c	$0.45f_c'$	1125	1350	1800	2250
Extreme fiber stress in tension in plain concrete footings and walls	f_c	$1.6\sqrt{f_c'}$	80	88	102	113
Shear: v_c (nominal permissible shear stress carried by concrete) Beams, walls, and one-way slabs	v_c	$1.1\sqrt{f_c'}$	55	60	70	78
Joist floor construction	v_c	$1.2\sqrt{f_c'}$	61	66	77	86
Two-way slabs and footings (peripheral shear)	v_c	$2\sqrt{f_c'}$	100	110	126	141
Bearing: f_c On full area		$0.297f_c'$	743	891	1190	1486
When supporting surface A_2 is wider on all sides than loaded area A_1 (but not more that $2 \times 0.297 f_c'$)		$0.297f_c'$ $\times \sqrt{A_2/A_1}$				

* Tabulated stresses are the same as those specified for the Alternate Design Method in *Building Code Requirements for Reinforced Concrete* (ACI 318-71). Data abstracted from ACI 318-71 with permission of the American Concrete Institute.

situation, building codes specify the maximum permissible water–cement ratios for concretes of specified design strengths. The ratios given in Table 13-2 are typical of those used as guides for manufacturing concretes of various strengths.

Referring to Table 13-2, we note that for nonair-entrained concrete specified to have an f_c' value of 3000 psi the maximum water–cement ratio is 0.58 by weight or 6.6 by volume.

TABLE 13-2. Maximum Permissible Water–Cement Ratios for Concrete*

| Specified compressive strength, f'_c (psi) | Maximum permissible water–cement ratio | | | |
| | Nonair-entrained concrete | | Air-entrained concrete | |
	Absolute ratio by weight	U.S. gal. per 94-lb bag of cement	Absolute ratio by weight	U.S. gal. per 94-lb bag of cement
2500	0.65	7.3	0.54	6.1
3000	0.58	6.6	0.46	5.2
3500	0.51	5.8	0.40	4.5
4000	0.44	5.0	0.35	4.0
4500	0.38	4.3	0.30	3.4

* For use when strength data from trial batches or field experience is not available. Data abstracted from *Building Code Requirements for Reinforced Concrete* (ACI 318-71) with permission of the American Concrete Institute.

13-5 Cement

The cement used most extensively in building construction is *portland cement*. Of the five types of standard portland cement generally available in the United States, and for which the American Society for Testing and Materials has established specifications, two types account for most of the cement used in buildings. These are ASTM Type I, a general purpose cement for use in concrete designed to reach its required strength in about 28 days, and ASTM Type III, a high-early-strength cement for use in concrete that attains its design strength in a period of a week or less. All portland cements set and harden by reacting with water, and this hydration process is accompanied by generation of heat. In massive concrete structures such as dams the resulting temperature rise of the materials becomes a critical factor in both design and construction, but the problem is not usually significant in building construction. (ASTM Type IV, a low-heat cement, is designed for use where the heat rise during hydration is a critical factor.)

The controlling specifications for portland cement, as called for in the ACI Code are:

(a) *Specification for Portland Cement*, ASTM C 150.

(b) *Specification for Air-Entraining Portland Cement*, ASTM C 175.

It is, or course, essential that the cement actually used in construction correspond to that employed in designing the mix to produce the specified compressive strength of the concrete.

13-6   Air-Entrained Concrete

Air-entrained concrete is produced by using an air-entraining portland cement or by introducing an air-entraining admixture during mixing of the concrete. In addition to improving workability, entrained air permits lower water–cement ratios (see Table 13-2) and significantly improves the durability of hardened concrete. Air-entraining agents produce billions of microscopic air cells per cubic foot; they are distributed uniformly throughout the mass. These minute voids prevent the accumulation of water in larger voids which, on freezing, would permit the water to expand and result in spalling of the exposed surface under frost action.

The use of entrained air is common in concrete for building construction. Although air-entraining cements reduce the strength of concrete somewhat, requiring slightly richer mixtures to obtain the same strengths produced by normal portland cement, some reduction in strength is acceptable in view of the other favorable characteristics of air-entrained concrete. When admixtures are used to produce air-entrainment, they should conform to *Specification for Air-Entraining Admixtures for Concrete*, ASTM C 260.

13-7   Steel Reinforcement

The steel used in reinforced concrete consists of round bars, mostly of the deformed type with lugs or projections on their surfaces. The purpose of the surface deformations is to develop a greater bond between the concrete and the steel. The bars used for reinforcement are made from billet steel, rail steel, and axle steel. The most common *grades* of reinforcing steel are Grade 60 and Grade 40, having yield strengths of 60,000 psi and 40,000 psi, respectively. In working stress design, the allowable tensile stress in the reinforcement is 24,000 psi for Grade 60 steel and 20,000 psi for Grade 40.

TABLE 13-3.   Areas and Perimeters of Standard Deformed Bars

Bar designation number	Nominal dimensions		
	Diameter (in.)	Cross-sectional area (sq in.)	Perimeter (in.)
3	0.375	0.11	1.178
4	0.500	0.20	1.571
5	0.625	0.31	1.963
6	0.750	0.44	2.356
7	0.875	0.60	2.749
8	1.000	0.79	3.142
9	1.128	1.00	3.544
10	1.270	1.27	3.990
11	1.410	1.56	4.430
14	1.693	2.25	5.32
18	2.257	4.00	7.09

Table 13-3 gives the areas and perimeters of standard deformed bars; the designer should confine himself to these sizes. The bar designation numbers are based on the number of eighths of an inch included in the nominal diameter of the bars.

The clear distance between parallel bars shall not be less than the nominal diameter of the bars, nor 1 in. The clear distance between layers of bars shall not be less than 1 in. and the bars in the upper layers shall be placed directly above those in the bottom layer.

Ample concrete protection for the reinforcing bars must be provided. For reinforcement near surfaces not exposed to the ground or weather the concrete covering should be not less than $\frac{3}{4}$ in. for slabs, walls, and joists reinforced with #11 bars and smaller, and $1\frac{1}{2}$ in. for beams, girders, and columns. Where the surfaces are exposed to earth or weather the minimum cover should be $1\frac{1}{2}$ in. for #5 bars and smaller, and 2 in. for #6 through #18 bars. For footings in contact with the ground the minimum covering should be 3 in.

13-8 Notation Used in Reinforced Concrete

The most most common type of reinforced concrete beam is designed on the assumption that the steel resists all the tensile stresses and the concrete resists the compressive stresses. This is indicated graphically in Fig. 13-1, the concrete area resisting compression being the

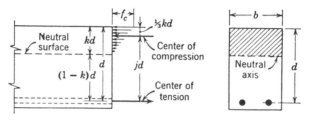

FIGURE 13-1

portion of the beam above the neutral axis indicated by cross hatching. Note particularly that d is the distance from the top of the beam to *the center line of the steel.* The following is the standard notation used in formulas of the working stress design method:

$d =$ *effective* depth of the beam, the distance from the extreme surface in compression to the center of the longitudinal reinforcement in inches,

$b =$ width of a rectangular beam in inches,

$k =$ ratio of the distance of the neutral axis of the cross section from the extreme fibers in compression to the effective depth of the beam,

$j =$ ratio of the distance from the center of compression to the center of tension and the depth, d,

$f_c =$ unit compressive stress in the extreme fiber of the concrete in pounds per square inch,

$f_s =$ unit tensile stress in the steel in pounds per square inch,

$A_s =$ area of the cross section of tensile reinforcement in square inches,

p = ratio of the area of tensile reinforcement to the effective area of concrete or $p = A_s/bd$,

n = modular ratio, modulus of elasticity of steel divided by the modulus of elasticity of concrete or $n = E_s/E_0$.

M = bending moment in inch-pounds.

13-9   Modulus of Elasticity

The modulus of elasticity E_c of hardened concrete is a measure of its resistance to deformation. The magnitude of E_c depends on w, the weight of the concrete, and on f'_c, its strength. Its value may be determined from the expression $E_c = w^{1.533}\sqrt{f'_c}$ for values of w between 90 and 155 lb per cu ft. For normal-weight concrete (145 lb per cu ft), E_c may be considered as equal to $57{,}000\sqrt{f'_c}$.

In the design of reinforced concrete members we employ the term n. As stated in the notation, this is the ratio of the modulus of elasticity of steel to that of concrete, or $n = E_s/E_c$. This ratio may be taken as the nearest whole number, but not less than 6. The modulus of elasticity of the steel reinforcement, E_s, is 29,000,000 psi.

Consider a concrete for which $f'_c = 4000$ psi and $w = 145$ lb per cu ft. Then $E_c = 57{,}000\sqrt{f'_c} = 57{,}000\sqrt{4000} = 3{,}600{,}000$ psi. Thus $n = E_s/E_c = 29{,}000{,}000 \div 3{,}600{,}000 = 8$. The values of n for four different strength concretes are given in Table 13-1.

13-10   Flexural Design Formulas for Rectangular Beams

The following formulas, with the aid of Table 13-4, may be used directly in determining the depth and longitudinal reinforcement of rectangular beams and slabs:

$$d = \sqrt{\frac{M}{Rb}} \qquad\qquad \text{Formula (1)}$$

$$A_s = pbd \qquad\qquad \text{Formula (2)}$$

$$A_s = \frac{M}{f_s j d} \qquad\qquad \text{Formula (3)}$$

$$f_c = \frac{2M}{jkbd^2} \qquad\qquad \text{Formula (4)}$$

$$f_s = \frac{M}{A_s j d} \qquad\qquad \text{Formula (5)}$$

in which d = effective depth of the beam in inches,

b = width of the beam in inches,

A_s = area of the longitudinal tensile steel reinforcement in square inches,

M = maximum bending moment in inch-pounds,

$k, j, p,$ and R are constants corresponding to unit stresses as given in Table 13-4,

$$k = \frac{n}{n + (f_s/f_c)},$$

$$j = 1 - \frac{k}{3},$$

$$p = \frac{kf_c}{2f_s},$$

$$R = \tfrac{1}{2}f_c jk,$$

f_s = unit stress in tensile steel reinforcement in pounds per square inch,

f_c = unit compressive stress in extreme fiber of concrete in pounds per square inch.

The use of these fundamental formulas is illustrated in the examples given below.

Before designing a beam, the designer must decide on the allowable stresses in both the concrete and steel reinforcement. Having done this, he refers to Table 13-4 to obtain the proper formula coefficients. Many different stresses may be employed; their magnitudes will determine the formula coefficients to be used in the design computations.

Formula (1), $d = \sqrt{M/Rb}$, is used to compute the effective depth of a rectangular beam after a width has been assumed. The clear distance between supports should not exceed 50 times the width of the beam.

TABLE 13-4. Formula Coefficients for Rectangular Beams

f_s	f_c	$n = 10(f_c' = 2500 \text{ psi})$ p	k	j	R
20,000	1,125	0.01019	0.3623	0.8792	179.2
24,000	1,125	0.00753	0.3213	0.8929	161.4

f_s	f_c	$n = 9(f_c' = 3000 \text{ psi})$ p	k	j	R
20,000	1,350	0.01293	0.3831	0.8723	225.6
24,000	1,350	0.00959	0.3410	0.8863	204.0

f_s	f_c	$n = 8(f_c' = 4000 \text{ psi})$ p	k	j	R
20,000	1,800	0.01884	0.4186	0.8605	324.2
24,000	1,800	0.01406	0.3750	0.8750	295.3

Example 1. A rectangular reinforced concrete beam has a maximum bending moment of 610,000 in-lb. Given the data $f_s = 20,000$ psi, $f_c' = 3000$ psi, $f_c = 1350$ psi, and $n = 9$, compute the depth of the beam.

Solution: Assume that b, the width of the beam, is 10 in. Referring to Table 13-4, we find that the value of R corresponding to the stresses given as data is 225.6. Then, employing Formula (1),

$$d = \sqrt{\frac{M}{Rb}} \quad \text{and} \quad d = \sqrt{\frac{610,000}{225.6 \times 10}} \quad \text{or} \quad d = 16.5 \text{ in.}$$

This is the *effective depth.* If we allow 0.5 in. for one half the diameter of a reinforcing bar plus 2 in. for fireproofing and web reinforcement, $16.5 + 0.5 + 2.0 = 19.0$ in., which is the total depth of the beam as shown in Fig. 13-2.

Example 2. Determine the cross-sectional area of the longitudinal tensile reinforcement for the beam given in the preceding example. Data: $b = 10$ in., $d = 16.5$ in.

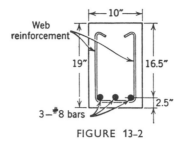

FIGURE 13-2

Solution: Since we will use Formula (2) for this determination, we consult Table 13-4 and find that $p = 0.01293$ for the applicable stress data. Then, the area of steel required is

$$A_s = pbd = 0.01293 \times 10 \times 16.5 = 2.14 \text{ sq in.}$$

Because the beam is 10 in. wide, it is advisable to use three bars, and referring to Table 13-3, we find that 3-#8 bars are adequate, $3 \times 0.79 = 2.37$ sq in.

Formula (3), $A_s = M/f_s jd$, is a general formula for finding the area of the steel tensile reinforcement in beams. If it had been used in the preceding example, it would have given the same result. Suppose, however, that the width and effective depth of a beam, limited by certain building conditions, are similar to those of the preceding example but that the bending moment is smaller. It would be uneconomical to use the formula $A_s = pbd$, for this would result in using a steel area that would be larger than necessary. Formula (3), $A_s = M/f_s jd$ is appropriate for this condition

Example 3.   A beam has a width of 10 in., an effective depth of 16.5 in., and is loaded to produce a maximum bending moment of 266,00 in-lb. Data: $f_s = 20,000$ psi, $f_y = 40,000$ psi, $f_c = 1350$ psi, and $n = 9$. Find the required area of the longitudinal tensile reinforcement.

Solution: Note that this problem is similar to the preceding one with the exception that the bending moment is smaller. It was seen that, for a load producing a moment of 610,000 in-lb, $A_s = 2.14$ sq in. Because in this example the bending moment is smaller (though the beam dimensions are the same), the required steel area may be found from Formula (3), or

$$A_s = \frac{M}{f_s jd} = \frac{266,000}{20,000 \times 0.8723 \times 16.5} = 0.925 \text{ sq in.}$$

Three #5 bars will supply this required area, $3 \times 0.31 = 0.93$ sq in. See Table 13-3.

However, the ACI Code requires a minimum steel ratio p of $200/f_y$ or, for our data, a minimum p of $200 \div 40,000 = 0.005$. Testing the reinforcement selected above, $p = A_s/bd = 0.93 \div 165 = 0.00563$. Since this value is greater than the 0.005 minimum, we accept the three #5 bars.

When we wish to compute the unit stresses in concrete and steel for beams previously designed, we use Formulas (4) and (5),

$$f_c = \frac{2M}{jkbd^2} \quad \text{and} \quad f_s = \frac{M}{A_s jd}$$

Problem 13-10-A*. A rectangular reinforced concrete beam has a maximum bending moment of 1,080,000 in-lb. Assuming the width to be 12 in., determine the effective depth and the tensile reinforcement required. Specification data: $f_s = 20,000$ psi, $f_c = 1350$ psi, and $n = 9$.

Problem 13-10-B. Determine the effective depth and the steel area required for the conditions given in the preceding problem if the specification data is: $f_s = 24,000$ psi, $f_c' = 4000$ psi, and $n = 8$.

13-11 Diagonal Tension: Web Reinforcement

A concrete beam without tensile reinforcement will fail because of excessive tensile stresses in the concrete, as indicated in Fig. 13-3a. The area of the required tensile reinforcement is computed by means of Formulas (2) or (3) as explained in the examples of Art. 13-10. The beam has a further tendency to fail by the tensile stresses that produce the inclined cracks shown in Fig. 13-3b. These cracks are due to a combination of tension and vertical shear, and the stress is known as *diagonal tension* commonly called *shear*. To prevent failure due to shearing stresses, vertical steel bars called *stirrups* are used; they are known as *web reinforcement*. Stirrups are placed in rectangular and T-beams as shown in Figs. 13-3c and d. They are generally

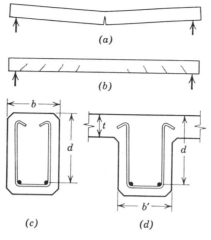

FIGURE 13-3

made of #3 or #4 bars bent in the shape of the letter U or for very wide beams, W-shaped.

Turning to Table 13-1 we find that v_c, the allowable unit shear stress for 3000-psi concrete, is 60 psi in beams. This means that if the beam loading produces a design shear stress exceeding this value, it is necessary to provide stirrups to carry the excess shear. However, even with web reinforcement, the 1963 ACI Code limits the design shear stress to $5\sqrt{f_c'}$ for the working stress design method.[1] If this limit is exceeded, the beam cross section must be made larger.

13-12 Notation and Formulas for Web Reinforcement

The following notation is used when designing web reinforcement for beams by the working stress design method:

v = actual unit shearing stress in pounds per square inch,

v_c = allowable unit shearing stress for concrete in pounds per square inch,

---

[1] This restriction is handled somewhat differently in the 1971 ACI Code, but the 1963 working stress design provision will be used here. An additional provision introduced by the 1971 Code requires that a minimum area of shear reinforcement be provided in beams wherever the design shear stress exceeds one half of v_c.

b = width of the rectangular beam in inches,
d = effective depth of the beam in inches,
V = total vertical shear at face of the support (the reaction) in pounds,
V_d = total vertical shear at d distance from the face of the support in pounds,
$V_c = v_c bd$,
$V' = V_d - V_c$,
s = spacing of stirrups in inches,
A_v = total cross-sectional area of one stirrup (both legs) in inches,
f_v = allowable tensile unit stress in the stirrups in pounds per square inch,
a = distance, parallel to the length of the beam, in which stirrups are required in inches,
L = span length of the beam in feet.

Formulas.   The first three formulas listed below deal with the intensity of the shearing stress and the amounts of shear to be carried by the concrete and by the web reinforcement, respectively. The last two expressions concern placement of the stirrups.

$$v = \frac{V_d}{bd}$$ Formula (6)

$$V_o = v_o bd$$ Formula (7)

$$V' = V_d - V_c$$ Formula (8)

$$s = \frac{A_v f_v d}{V'}$$ Formula (9)

$$a = \frac{L}{2} \times \frac{V'}{V}$$ Formula (10)

13-13   Web Reinforcement for Uniform Loads

The shear diagram for a beam with a uniformly distributed load consists of two triangles, as shown in Fig. 3-20. One of these triangles, again shown in Fig. 13-4a, represents one half the shear diagram. Concrete without reinforcement is capable of resisting v_c psi, a certain degree of shear. If the shearing stress in a beam exceeds this

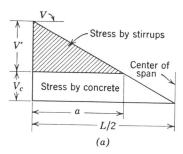

(a)

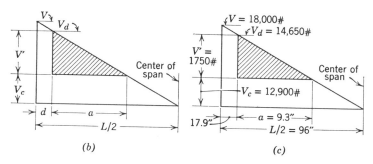

(b)   (c)

FIGURE 13-4

magnitude, web reinforcement (stirrups) is required. If v_c is multiplied by bd, the concrete area, we have $V_c = v_c bd$, Formula (7); V_c is the amount of the vertical shear that can be resisted by the concrete, and V' is the excess shear that must be resisted by the stirrups. Both V_c and V' are shown in Fig. 13-4a. If V is the maximum vertical shear at the end support, $V' = V - V_c$. Because the magnitude of the shear in a uniformly loaded beam decreases in magnitude as we leave the support, stirrups are required in only a limited portion of the beam, a certain distance from the support; in Fig. 13-4a this distance is designated as a. Note that L is the span of the beam; consequently $L/2$ is the distance from the support to the midpoint of the span because only one half the shear diagram is shown.

The current ACI specification requires that v, the nominal shear stress be computed by Formula (6), $v = V/bd$, *where the maximum shear is taken at d distance from the face of the support.* The shear at d distance is represented by V_d; hence $v = V_d/bd$. Figure 13-4b indicates V_d but, with this exception, the diagram is similar to Fig. 13-4a.

The term s represents the spacing (in inches) of the stirrups in a direction parallel to the span length of the beam. It is computed by Formula (9), $s = A_v f_v d / V'$. The allowable tensile stress in the stirrups, f_v, is 20,000 psi. It is customary to use standard hooks on the ends of stirrups, as shown in Figs. 13-3c and d.

Consider the larger triangle in Fig. 13-4a in which the base is $L/2$ and V is the height. The smaller hatched-area triangle of base a and height V' is similar. Therefore, because the two triangles are similar,

$$\frac{a}{L/2} = \frac{V'}{V} \quad \text{and} \quad a = \frac{L}{2} \times \frac{V'}{V}$$

This is Formula (10) and its use enables us to compute a, the theoretical length of the beam in which stirrups are required. Remember that this formula is appropriate for beams loaded only with uniformly distributed loads.

The discussion so far has related to stirrups. If horizontal tensile reinforcement bars are bent up, as indicated in Fig. 14-1, the bent portion is approximately at right angles to the diagonal tension cracks shown in Fig. 13-3b; the bent portion of these bars also serves as web reinforcement. Stirrups are sometimes placed at an angle with the vertical, but it is more convenient, and therefore more economical, to place them in a vertical position.

There are some special requirements relating to stirrups. Web reinforcement is required between the face of the support and the section at distance d therefrom; it must also be provided for a distance equal to d *beyond* the point where stirrups are theoretically required. The maximum spacing of stirrups is $d/2$, and the cross-sectional area of one stirrup shall not be less than 0.15% of the area bs, or $0.0015 \times b \times s$.

Formula (10) is to be used only with beams uniformly loaded. Suppose we have a beam loaded with concentrated loads, as shown in Table 3-1, Case 4. Usually the weight of the beam (a distributed load) is relatively small compared with the concentrated load or loads and the shear diagram is represented by two rectangles; the shear between the concentrated loads is comparatively small. Consequently, for a beam with two concentrated loads as shown in Table

3-1, Case 4, we use the basic formula $v = V_d/bd$ to determine whether stirrups are required; if they are, they are placed only in those portions of the beam that extend from the supports up to the concentrated loads.

13-14 Design of Beams with Web Reinforcement

The following example illustrates the procedure for design of simply supported rectangular beams when web reinforcement is required.

Example. A reinforced concrete simple beam has a span of 16 ft and supports a uniformly distributed load of 32,000 lb, not including the weight of the beam. Design the horizontal tensile reinforcement and the web reinforcement. The design will be based on the following data:

$$f'_c = 3000 \text{ psi} \qquad \text{Grade 40 bars}$$
$$f_c = 1350 \text{ psi} \qquad f_y = 40,000 \text{ psi}$$
$$v_c = \text{limited to 60 psi} \quad f_s = 20,000 \text{ psi}$$
$$n = 9 \qquad f_v = 20,000 \text{ psi}$$

Design coefficients for these data are found in Table 13-4:

$$p = 0.01293 \qquad k = 0.3831 \qquad j = 0.8723 \qquad R = 225.6$$

Solution: (1) To begin, we must estimate the beam weight. Allowing 1 in. of depth for each foot of span, $16 \times 1 = 16$ in. Adding 2.5 in. for one half the diameter of a bar, plus the fireproofing below, $16 + 2.5 = 18.5$ in., say 20 in. For a beam 20 in. in depth let us assume its width to be 12 in. Then its estimated weight at 150 lb per cu ft $= (12 \times 20)/144 \times 150 \times 16 = 4000$ lb. Hence W, the total uniformly distributed load, is $32,000 + 4000$, or 36,000 lb, and w, the distributed load per linear foot, is $36,000 \div 16$, or 2250 lb.

(2) The maximum bending moment is at the center of the span.

$$M = \frac{WL}{8} = \frac{36,000 \times 16 \times 12}{8} = 864,000 \text{ in-lb}$$

(3) Next we compute the effective depth, having assumed that b, the width, is 12 in.

$$d = \sqrt{\frac{M}{Rb}}$$ Formula (1), Art. 13-1

or

$$d = \sqrt{\frac{864,000}{225.6 \times 12}} = \sqrt{320} \quad \text{and} \quad d = 17.9 \text{ in.}$$

Consequently, a beam 12 in. wide will have a total depth of 17.9 + 2.5, or 20.4 in. and the estimated weight of 2250 lb per lin ft is sufficiently accurate.

(4) The area of the longitudinal tensile reinforcement is found by use of Formula (2) of Art. 13-10, or

$$A_s = pbd = 0.01293 \times 12 \times 17.9 = 2.78 \text{ sq in.}$$

Referring to Table 13-3, we find that one #8 bar has a cross-sectional area of 0.79 sq in. Thus $4 \times 0.79 = 3.16$ sq in. and we accept 4-#8 bars.

(5) Turning our attention to the design of web reinforcement, we note that each reaction, and hence the end shear, is equal to half the total load, or $V = 36,000 \div 2 = 18,000$ lb. The shear that determines the necessity for using web reinforcement is found at d distance from the face of the support; it is designated as V_d. Then

$$V_d = 18,000 - \left(\frac{17.9}{12} \times 2250\right) = 14,650 \text{ lb}$$

To find v at d distance from the face of the support, we use Formula (6) from Art. 13-12:

$$v = \frac{V_d}{bd} \quad \text{or} \quad v = \frac{14,650}{12 \times 17.9} \quad \text{and} \quad v = 68 \text{ psi}$$

Since this stress exceeds v_c, which by data is limited to 60 psi, stirrups are required.

(6) It is helpful in distinguishing the various stresses to construct a shear diagram similar to those shown in Figs. 13-4a and b. Such a

diagram is presented in Fig. 13-4c. The base of the larger triangle represents one half the length of the beam or $(16 \times 12) \div 2 = 96$ in. The height of the larger triangle is V, or 18,000 lb, and V_d, the shear at d distance (17.9 in.) from the face of the support, is 14,650 lb.

To find V_c we use Formula (7), $V_c = v_c bd$. Then $V_c = 60 \times 12 \times 17.9$, or 12,900 lb. This dimension is laid off on the shear diagram, and, because $V' = V_d - V_c$, $V' = 14,650 - 12,900$, and $V' = 1750$ lb. This magnitude is also laid off on the shear diagram (Fig. 13-4c), and we can observe the length of the beam in which stirrups must be placed. If the shear diagram had been laid off to scale, the distance a could have been measured directly. Since Fig. 13-4c obviously is not drawn to scale, we may compute the distance from Formula (10) of Art. 13-12. Then

$$a = \frac{L}{2} \times \frac{V'}{V} = \frac{16 \times 12}{2} \times \frac{1750}{18,000} = 9.3 \text{ in.}$$

(7) The next step is to select the size of stirrups to be used and to determine the spacing. Let us assume that the stirrups are made of #3 bars. In Table 13-3 we find that the cross-sectional area of a #3 bar is 0.11 sq in., and because a U-shaped stirrup has two vertical legs, $A_v = (2 \times 0.11)$ sq in. To find the stirrup spacing we use Formula (9) of Art. 13-12, or

$$s = \frac{A_v f_v d}{V'}$$

or

$$s = \frac{2 \times 0.11 \times 20,000 \times 17.9}{1750} \quad \text{and} \quad s = 45 \text{ in.}$$

This value is excessive since the maximum spacing permitted is $d/2$, or in this case, $17.9 \div 2 = 8.9$ in. Accept a spacing of 8 in.

Stirrups are placed between the face of the support and the section at d distance therefrom, and also through a distance equal to d beyond the point at which they are theoretically required. Then, because a was computed to be 9.3 in., the total distance requiring web reinforcement is $17.9 + 9.3 + 17.9 = 45.1$ in. It is customary to place the first stirrup at $s/2$ distance from the face of the support. Making the layout on this basis, there will be six stirrups at each end of the beam, as shown in Fig. 13-5.

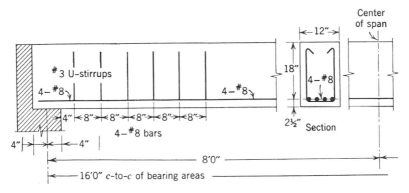

FIGURE 13-5

Checking for minimum area of web reinforcement, we know that the cross-sectional area of one stirrup shall not be less than 0.0015 × $b × s$ (Art. 13-13). Thus 0.0015 × 12 × 8 = 0.144 sq in. The total cross-sectional area of one #3 U-stirrup is 2 × 0.11 = 0.22 sq in.; hence the #3 stirrup selected above is satisfactory.

(8) As noted in Art. 13-13, the shear stress must not exceed $5\sqrt{f'_c}$ in sections of the beam with web reinforcement. For the data controlling this example, the limiting value is 5 × $\sqrt{3000}$ = 274 psi. Computing the maximum value developed by the loading,

$$v = \frac{V}{bd} = \frac{18,000}{12 \times 17.9} = 84 \text{ psi}$$

This value is well within the allowable 274 psi. In practical design work, this check is made concurrently with Step (5) above if the computed value of the shear stress at distance d from the face of the support approaches the maximum allowable value. If the shear stress is excessive, the cross section of the beam must be increased.

Problem 13-14-A*. A reinforced concrete simple beam has a span of 18 ft between faces of supports and carries a uniformly distributed load, including its own weight, of 2 kips per lin ft. Compute the width and effective depth of the beam, its longitudinal tensile reinforcement, and its web reinforcement if required. Use the design data given for the foregoing example.

13-15   Bond Stress

The tensile stress in reinforcing bars is generated by bond stresses developed between the bars and the adjacent concrete. Deformed bars are used to increase resistance to slippage and, in cases where a sufficient length of embedment to develop the required resistance is not available, the ends are formed into standard hooks. The procedure for handling bond in the 1971 ACI Code is based on the *development length* concept and is quite different from the *flexural bond* concept of the 1963 Code. Both procedures will be discussed here; flexural bond in this article, and development length of reinforcement in Art. 13-16.

The unit bond stress is computed from the expression

$$u = \frac{V}{\Sigma_0 jd} \qquad \text{Formula (11)}$$

in which u = unit bond stress in pounds per square inch,

$\quad V$ = total maximum vertical shear in pounds,

$\quad \Sigma_0$ = sum of the perimeters of all the bars crossing the section on the tension side in inches,

$\quad j$ = ratio of distance between centroid of compression and centroid of tension to the depth, d (see Fig. 13-1),

$\quad d$ = effective depth of the beam in inches.

In connection with establishing values for allowable bond stress, a distinction is made between "top bars" and other horizontal bars in the beam. Top bars are defined as horizontal reinforcement so placed that more than 12 in. of concrete is cast in the member below the bar. Air and water that rise during hydration tend to accumulate under such bars, creating conditions that result in lowered bond strength. The allowable bond stress for top bars is computed from the expression

$$u = \frac{3.4\sqrt{f_c'}}{D} \qquad \text{(but not} > 350 \text{ psi)}$$

and for all other bars,

$$u = \frac{4.8\sqrt{f_c'}}{D} \qquad \text{(but not} > 500 \text{ psi)}$$

In these formulas D is the nominal diameter of the bar in inches. These allowable stresses are for bars conforming to the ASTM A305 specification for deformed bars. For example, the allowable bond stress for a #5 top bar in 3000-psi concrete is

$$u = \frac{3.4\sqrt{f_c'}}{D} = \frac{3.4\sqrt{3000}}{0.625} \quad \text{and} \quad u = 298 \text{ psi}$$

Table 13-5 gives the allowable bond stresses for other bar sizes and concrete strengths.

TABLE 13-5.   Allowable Bond Stresses

Bar number	Top bars $u = \frac{3.4\sqrt{f_c'}}{D}$ (not to exceed 350 psi) f_c'			Other than top bars $u = \frac{4.8\sqrt{f_c'}}{D}$ (not to exceed 500 psi) f_c'		
	2500	3000	4000	2500	3000	4000
3	350	350	350	500	500	500
4	340	350	350	480	500	500
5	272	298	341	384	421	486
6	227	248	287	320	351	405
7	194	213	246	274	300	347
8	170	186	215	240	263	304
9	151	165	190	213	233	269
10	134	147	169	189	207	239
11	121	132	153	170	186	215

Consider a reinforcing bar in a beam that is subjected to a tensile stress. If D is the diameter of the bar, the cross-sectional area of the bar is $\pi D^2/4$ and, if f_s is the tensile unit stress, $\pi D^2 f_s/4$ is the tensile stress in the bar. The perimeter of a bar is πD, and if l is the length of the bar resisting bond stress, πDl is the surface area of the bar resisting bond. Thus, if u is the bond unit stress, πDlu is the total resistance to bond. For equilibrium the tensile stress in the bar and the anchorage due to bond must be equal. Hence

$$\frac{\pi D^2 f_s}{4} = \pi D l u \qquad \text{or} \qquad l = \frac{D f_s}{4u}$$

which is the length of embedment or anchorage required to develop the tensile stress in the bar.

Example 1.   A cantilever beam, as indicated in Fig. 13-6, has tensile reinforcement consisting of #6 deformed bars. The unit tensile stress

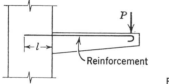

FIGURE 13-6

developed in the bars by the loading is 19,500 psi and f_c' for the concrete is 3000 psi. Compute l, the minimum required length of embedment for the bars.

Solution: Table 13-3 shows the diameter of #6 bars to be 0.75 in. From the arrangement of reinforcing shown in the figure, these are top bars. Referring to Table 13-5, we find that the allowable bond stress is 248 psi. Then

$$l = \frac{D f_s}{4u} \qquad \text{or} \qquad l = \frac{0.75 \times 19,500}{4 \times 248} \qquad \text{and} \qquad l = 14.8 \text{ in.}$$

which is the minimum length of embedment.

Example 2.   Compute the bond stress for the 4-#8 bars that constitute the longitudinal tensile reinforcement for the example in Art. 13-14.

Solution: (1) Assuming that all bars run the full length of the beam, as indicated in Fig. 13-5, there are four bars that cross the section of maximum shear ($V = 18,000$ lb) at the face of the support. Table 13-3 shows the perimeter of a #8 bar to be 3.142 in., so $\Sigma_0 = (4 \times 3.142)$ in. Referring to Table 13-4, we obtain $j = 0.8723$ for our controlling data. Then

$$u = \frac{V}{\Sigma_0 jd}$$

or

$$u = \frac{18,000}{4 \times 3.142 \times 0.8723 \times 17.9} \quad \text{and} \quad u = 92 \text{ psi}$$

This value is well within the allowable bond stress of 186 psi for #8 top bars given in Table 13-5.

(2) It will be shown later that all of the bottom bars in a beam do not have to run the entire length of the span. This is true because the full area of the tensile reinforcement (in this example $A_s = 3.16$ sq in.) is required only at the section of maximum bending moment. Consequently, at some point between midspan and either reaction some of the bars may be cut off. Assuming that only two of the four #8 bars used in this beam extend the full length, we will check to see whether the bond stress is still acceptable.

The only term that is affected in the equation of Step 1 above is Σ_0; this will become (2×3.142) in. or half the original value. Therefore u will now be equal to twice the 92 psi computed above, or 184 psi. This is still within the allowable 186 psi so the bond stress is acceptable.

Problem 13-15-A*. The reinforced concrete cantilever beam shown in Fig. 13-6 has #7 deformed tensile reinforcing bars in which the stress is 18,450 psi. If f'_c for the concrete is 2500 psi, compute the minimum length of embedment required to provide adequate anchorage.

13-16 Development Length of Reinforcement

The 1971 ACI Code defines *development length* as the length of embedded reinforcement required to develop the design strength of the reinforcement at a critical section. Critical sections occur at points of maximum stress and at points within the span where adjacent reinforcement terminates or is bent up into the top of the beam. For a uniformly loaded simple beam, one critical section is at midspan where the bending moment is maximum. This will be a point of maximum tensile stress in the reinforcement (peak bar stress), and some length of bar will be required over which the stress can be developed. Other critical sections occur between midspan and the reactions at points where some bars are cut off because they are no longer needed to resist the bending moment; such terminations create peak stress in the remaining bars that extend the full length of the beam. In either case, the necessary basic development length for #11 or smaller bars may be computed from the expression

$$l_d = \frac{0.04 A_b f_y}{\sqrt{f'_c}} \qquad \text{but not less than } 0.0004 d_b f_y$$

in which l_d = development length in inches,
  A_b = area of an individual bar in square inches,
  f_y = yield strength of reinforcement in psi,
  f'_c = specified compressive strength of concrete in psi,
  $d.$ = nominal diameter of a bar in inches.

For top bars, resulting values of l_d are multiplied by the factor 1.4, and for bar spacings 6 in. and over, by 0.8. Additional formulas applying to larger bars are given in the ACI Code. The limitation of $0.0004 d_b f_y$ becomes significant only for small bars, such as #5, or very high strength concrete.

Example 1.   Compute the development length of the reinforcement for the uniformly loaded simple beam in the example of Art. 13-14. All bars extend over the entire beam length as indicated in Fig. 13-5. *Solution:*   The reinforcement consists of four-#8 bars in the bottom of the beam. The area of one of these bars, A_b, is 0.79 sq in. According to data, $f_y = 40,000$ psi and $f'_c = 3000$ psi. Then

$$l_d = \frac{0.04 A_b f_y}{\sqrt{f'_c}} = \frac{0.04 \times 0.79 \times 40,000}{\sqrt{3000}} = 23.1 \text{ in.}$$

Since all bars extend over the entire beam length, an embedment length of $16 \div 2 = 8$ ft or 96 in. is provided. This greatly exceeds the required development length of 23.1 in., and consequently l_d is not a critical factor in this case.

Example 2.   If two of the four bars in the beam of Example 1 are cut off short of each end, at points where they are no longer needed

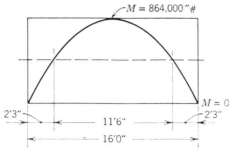

FIGURE 13-7

for bending moment, compute the required development length of the other two bars that continue into the supports.

Solution: (1) The parabolic curve in Fig. 13-7 represents the bending moment diagram for this uniformly loaded simple beam. Reference to Art. 13-14 will show that a slightly greater A_s (3.16 sq in.) is provided by the 4-#8 bars than the area theoretically required (2.78 sq in.). However, we will neglect this difference and assume that the potential resisting moment of the bars just matches the 864,000 in-lb developed by the loading. Then, the horizontal line drawn tangent to the apex of the bending moment curve represents the resisting moment of the four bars.

(2) The dashed line in the figure indicates the resisting moment of half the total reinforcement. Where this line intersects the bending moment curve, two of the four bars are no longer required and may be cut off. For a uniformly loaded simple beam, this intersection falls between ⅙ and ⅟₇ of the span length from the support. Taking $L/7$ as the approximate position, the distance in this case is $16 \div 7 = 2.29$ ft or 2 ft 3 in.

(3) At this cutoff point, the peak stress in the *continuing* bars is the same as at midspan so a development length of 23.1 in. (see Example 1 above or Table 13-6) is required beyond this critical section. Since the section occurs 2 ft 3 in. or 27 in. from the end of beam, l_d is not critical by this test. However, at simple supports, an additional condition must be satisfied as explained below.

(4) At simple supports, the diameter of the reinforcement must be small enough so that the computed development length of the bar

TABLE 13-6. Development Lengths for Tension Bars

l_d (in.) for $f_y = 40{,}000$ psi, $f'_c = 3000$ psi

Bar number	Spacing			
	Less than 6 in.		6 in. and over	
	Other	Top	Other	Top
#3	12.0	12.0	12.0	12.0
#4	12.0	12.0	12.0	12.0
#5	12.0	14.0	12.0	12.0
#6	12.9	18.0	12.0	14.3
#7	17.6	24.6	14.0	19.7
#8	23.1	32.2	18.5	25.9
#9	29.3	41.0	23.3	32.7
#10	37.1	51.8	29.7	41.6
#11	45.6	63.7	36.4	51.0
#14	62.0	87.0	50.0	70.0
#18	81.0	113.0	65.0	90.0

l_d does not exceed $(M_t/2V) + l_a$. This may be expressed as

$$l_d \lesseqgtr \frac{M_t}{2V} + l_a$$

in which M_t = theoretical moment strength in inch-pounds of a section = $0.85dA_s f_y$,

V = vertical shear at the section,

l_a = additional embedment length at support.

Figure 13-8 shows the relationship of the several terms.

For this example, $V = 18{,}000$ lb and l_a may be taken as 3 in., assuming that the end of the bar clears the end of the beam by 1 in. (Fig. 13-5). Also, $A_s = 3.16 \div 2$ and $f_y = 40{,}000$ psi. Then

$$M_t = 0.85dA_s f_y = 0.85 \times 17.9 \times 1.58 \times 40{,}000 = 960{,}000 \text{ in-lb}$$

and

$$l_d = \frac{M_t}{2V} + l_a = \frac{960{,}000}{36{,}000} + 3 = 26.6 + 3 = 29.6 \text{ in.}$$

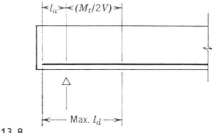

FIGURE 13-8

Since the required development length of 23.1 in. [Step (3) above] is less than 29.6 in., the $(M_t/2V) + l_a$ limitation is satisfied.

(5) Reference to Fig. 13-7 will show that, if the *terminated* bars were cut off at the points of intersection with the moment curve, they would have an overall length of 11 ft 6 in. However, the ACI Code requires that reinforcement shall extend beyond the point at which it is no longer needed in bending for a distance equal to the effective depth of the beam or 12 bar diameters, whichever is greater. In our example the effective depth is 17.9, say 18 in., and 12 times the 1-in. nominal diameter of a #8 bar is 12 in. Therefore the actual length of the terminated bars will be 11 ft 6 in. plus twice 18 in. or 14 ft 6 in.

Problem 13-16-A*.   A simply supported rectangular beam has a width of 10 in., an effective depth of 15 in., and tensile reinforcement consisting of 4-#7 bars. The span is 14 ft 0 in. center-to-center of bearing areas. Two of the four bars are cut off at the 1/7 points of the span and the remaining bars continue into the supports, extending 3 in. beyond the center of bearing at each end. The beam has been designed on the basis of $f'_c = 3000$ psi, $f_c = 1350$ psi, $f_s = 20,000$ psi, and $f_y = 40,000$ psi. It supports a total uniform load of 24.2 kips. Check the required development length of reinforcement with that provided; is it adequate?

14

Reinforced Concrete Beams

||||||||||||||||||||||||||||||||||||||||||||||||||||||||||

14-1 Typical Beams

In the preceding chapter we dealt exclusively with simply supported beams as the basis for our discussion of flexure, shear, and diagonal tension, and development length of reinforcement. In practice, however, reinforced concrete floor systems are poured simultaneously for several adjacent spans and, if the reinforcement is arranged properly, continuity is developed between the spans and we have *continuous* or *restrained* beams (Art. 2-4).

In the design of reinforced concrete beams *with uniformly distributed loads*, three span conditions occur repeatedly. First, there is the simple beam, a *single span* with no restraint at the ends. Next, there is the *end span* of a continuous beam, with no restraint at the noncontinuous end. Finally, there is the *interior span* of a fully continuous beam. These three beams are shown diagrammatically in Figs. 14-1*a*, *b*, and *c*, respectively. These figures contain much valuable information. The total uniformly distributed load is represented by W in each instance. For each of the three beams the magnitude as well as the position of M, the maximum bending moment, is given. For each beam the value of V, the maximum vertical shear, is shown; note

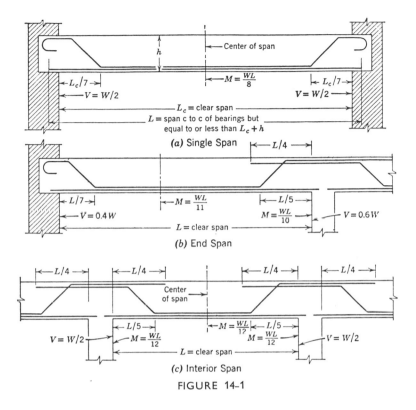

FIGURE 14-1

that different values are given for the two ends of the end span. When bent bars are to be used, the positions of the bends, the positions of the inflection points, are shown. In order to have a sufficient area of contact between the tensile reinforcement and the concrete, there must be an ample length of embedment; these lengths are shown in the figure as fractions of L. Hooks on tensile reinforcement are used only at beam terminations where there is insufficient concrete to provide an adequate length of bar.

Figure 14-1 shows a combination of bent and straight bars for the longitudinal tensile reinforcement. However, if separate systems of straight bars are used for top as well as for bottom reinforcement, the lengths of bars will be exactly as shown for the straight and bent bars in the figure.

14-2 Length of Span

In computations for reinforced concrete the span length of freely supported beams (simple beams) is generally taken as the distance between centers of supports or bearing areas; it should not exceed the clear span plus the depth of beam or slab. The span length for continuous or restrained beams is taken as the clear distance between faces of supports. These lengths are shown in Fig. 14-1.

The single span condition illustrated in Fig. 14-1a is similar to that shown in Fig. 13-5, except that some of the bottom rods have been bent up into the top of the beam. Although the bending moment is theoretically zero at this simple support, some restraint may be developed by the weight of the wall above the beam, thereby creating a negative moment in the top of the beam at the face of the wall when the member is loaded. For this reason it is common practice to place some tensile reinforcement in the top at a support of this nature; either by bending up some of the bottom steel or using short straight bars. Under these conditions, the bend is usually made at the $L_c/7$ point rather than at $L/7$.

14-3 Bending Moments

For a simple beam, that is, a single span having no restraint at the supports, the maximum bending moment for a uniformly distributed load is at the center of the span and its magnitude is $M = WL/8$ (Art. 3-11). The moment is zero at the supports and is positive over the entire span length. In continuous beams, however, negative bending moments (Art. 3-7) are developed at the supports and positive moments at or near midspan. This may be readily observed from the exaggerated deformation curve of Fig. 14-2a. The exact values of the bending moments depend on several factors but in the case of approximately equal spans supporting uniform loads, when the live load does not exceed three times the dead load, the bending moment values given in Fig. 14-2 may be used for design. It will be noted from Fig. 14-2b that the maximum positive moment for an interior (fully continuous) span is $WL/12$, and that for an end (semicontinuous) span is $WL/10$. Both of these values are, of course, smaller than the $WL/8$ of the simple span condition.

The bending moment values shown in Fig. 14-2 are more conservative than the theoretical moment factors and are used throughout the reinforced concrete section of this book. However, for continuous beams with uniformly distributed loads, a variety of conditions may occur and greater accuracy may be attained by using the factors for moments and shears given in Section 8.4 of the 1971 ACI Code.

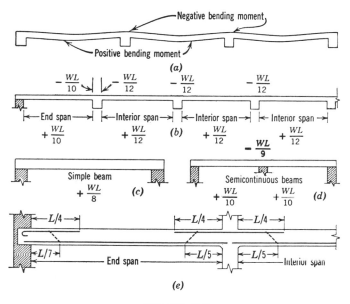

FIGURE 14-2

Readers wishing to explore this matter further should consult Chapter 8 of ACI 318-71.

The positive and negative bending moments are not necessarily equal for a given span and loading but for fully continuous beams of nearly equal span, uniformly loaded, Fig. 14-2*b* indicates that the positive and negative *design moments* may both be taken as $WL/12$. This means that the required areas of positive reinforcement (bottom steel) and negative reinforcement (top steel) will be the same. For continuous beams of equal spans, uniformly loaded, it is customary to assume that the inflection point[1] is one fifth the span length from the support. At this point, if bent bars are used, the tensile steel no longer required for positive moment is raised from the lower to the upper part of the beam to provide for the negative bending moment; otherwise, a separate system of straight bars is used at the top. The dotted diagonal lines in Fig. 14-2e indicate the positions of bends when bent bars are employed; Fig. 14-1 also shows these relationships, and Fig. 14-4 illustrates both schemes.

14-4 Design of Rectangular Reinforced Concrete Beams

The procedure outlined below may be used in the design of rectangular beams.

Step 1:   Estimate the approximate size and weight of the beam. It is customary to assume a width and to allow about 1 in. of depth for each foot of span; 150 lb per cu ft is usually taken for the weight of reinforced concrete.

Step 2:   Compute the effective depth of the beam.

$$d = \sqrt{\frac{M}{Rb}} \qquad \text{Formula (1), Art. 13-10}$$

Step 3:   Compute the area of the tensile reinforcement.

$$A_s = pbd \qquad \text{Formula (2), Art 13-10}$$

or

$$A_s = \frac{M}{f_s j d} \qquad \text{Formula (3), Art. 13-10}$$

Step 4:   Compute v, the unit shearing stress.

$$v = \frac{V}{bd} \quad \text{and} \quad v = \frac{V_d}{bd} \qquad \text{Formula (6), Art. 13-12}$$

If v exceeds v_c, stirrups are required.

---

[1] *Inflection point* was defined in Art. 3-7 as the point along the span where the curvature of the deformation diagram changes from concave to convex, and at which the bending moment is zero. In a continuous beam, the portion between the inflection point and midspan is under positive bending moment, and that between inflection point and support is under negative moment.

Step 5:   Design the web reinforcement. If the beam is loaded only with a uniformly distributed load, the procedure explained in Art. 13-14 may be followed. If the beam loading consists of a concentrated load or loads, it is necessary to construct a shear diagram to determine which portions of the beam require stirrups.

Step 6:   Check the bond stress (or the development length of reinforcement).

Step 7:   Make a design drawing that shows the dimensions of the beam, the number and size of reinforcing bars, size and placing of stirrups, lengths and points of cutoff of principal reinforcement, etc. See Fig. 14-4.

Example.   A fully continuous reinforced concrete beam has a span of 22 ft 0 in. between faces of supports and carries a uniformly distributed load of 52 kips, not including its own weight. Design the beam in accordance with the following specification data:

$$f_c' = 3000 \text{ psi} \qquad \text{Grade 40 bars}$$
$$f_c = 1350 \text{ psi} \qquad f_y = 40,000 \text{ psi}$$
$$v_c = \text{not} > 60 \text{ psi} \qquad f_s = 20,000 \text{ psi}$$
$$n = 9 \qquad f_v = 20,000 \text{ psi}$$

Table 13-4 gives the following design coefficients for these data: $p = 0.01293, k = 0.3831, j = 0.8723, R = 225.6$.

Solution: (1) Since the clear span is 22 ft, let us assume that the effective depth will be 22 in. Adding an allowance of $2\frac{1}{2}$ in. for fireproofing, we have $22 + 2.5 = 24.5$ in. for the estimated total depth. For this depth assume that $b = 12$ in., the beam width. With these dimensions, the estimated weight of the beam is

$$\frac{12 \times 24.5}{144} \times 150 \times 22 = 6750 \text{ lb}$$

and the total uniformly distributed load is

$$W = 52,000 + 6750 = 58,750 \qquad \text{say } 59,000 \text{ lb}$$

(2) This type of load and span corresponds with the beam shown in Fig. 14-1c, an interior span. Here we see that the maximum bending

moment, positive at midspan and negative over the supports, is

$$M = \frac{WL}{12} \quad \text{or} \quad M = \frac{59,000 \times 22 \times 12}{12}$$

and

$$M = 1,298,000 \text{ in.-lb}$$

Then

$$d = \sqrt{\frac{M}{Rb}} \qquad \text{Formula (1), Art. 13-10}$$

or

$$d = \sqrt{\frac{1,298,000}{225.6 \times 12}} = 21.9 \text{ in.}$$

say 22 in., which is the required effective depth. Adding $2\frac{1}{2}$ in. for one half a bar diameter plus fireproofing, the total depth is 22 + 2.5 = 24.5 in.

(3) Computing the area of the tensile reinforcement,

$$A_s = pbd \qquad \text{Formula (2), Art 13-10}$$

or

$$A_s = 0.01293 \times 12 \times 22 = 3.44 \text{ sq in.}$$

Referring to Table 13-3, we see that 2-#8 plus 2-#9 bars have a total cross-sectional area of 1.58 + 2 = 3.58 sq in. This reinforcement is tentatively accepted, pending later check for bond stress (or development length).

(4) Computing the unit shearing stress at the face of supports, $V = 59,000/2 = 29,500$ lb and

$$v = \frac{V}{bd} = \frac{29,500}{12 \times 24} = 112 \text{ psi}$$

The magnitude of V_d, the shear at distance d from the face of the support is

$$V_d = 29,500 - \left(\frac{22}{12} \times \frac{59,000}{22}\right) \quad \text{and} \quad V_d = 24,580 \text{ lb}$$

and the unit shear stress at this section is

$$v = \frac{V_d}{bd} = \frac{24,580}{12 \times 22} = 93 \text{ psi}$$

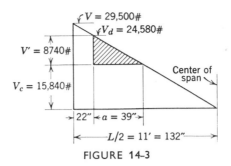

FIGURE 14-3

This magnitude exceeds the allowable 60 psi for v_c but is well within $5\sqrt{f'_c}$ which, for the data of this example, is $5\sqrt{3000} = 274$ psi; hence stirrups are required.

(5) Since the beam is uniformly loaded, we may design the web reinforcement by following the procedure explained in Steps (6) and (7) of Art. 13-14. First, we compute the values of V_c and V':

$$V_c = v_c bd \qquad \text{Formula (7), Art. 13-12}$$

or

$$V_c = 60 \times 12 \times 22 \qquad \text{and} \qquad V_c = 15{,}840 \text{ lb}$$

$$V' = V_d - V_c \qquad \text{Formula (8), Art. 13-12}$$

or

$$V' - 24{,}580 - 15{,}840 \qquad \text{and} \qquad V' = 8740 \text{ lb}$$

Now that V' and V_c have been determined we can construct one half the shear diagram (Fig. 14-3):

$$a = \frac{L}{2} \times \frac{V'}{V} \qquad \text{Formula (10), Art. 13-12}$$

or

$$a = 132 \times \frac{8740}{29{,}500} \qquad \text{and} \qquad a = 39.1 \text{ in.}$$

The stirrup spacing, assuming that #3 bars will be used ($A_v = 2 \times 0.11$ sq in.), is found by use of Formula (4), Art. 13-12.

$$s = \frac{A_v f_v d}{V'} = \frac{2 \times 0.11 \times 20{,}000 \times 22}{8740} \qquad \text{and} \qquad s = 11 \text{ in.}$$

Note that the maximum allowable spacing is also 11 in. ($d/2$ is $22/2 = 11$ in.).

The distance from the face of the support to the point beyond which stirrups are not required is $(d + a + d)$ or $(22 + 39.1 + 22) = 83.1$ in. The adopted spacing is shown in Fig. 14-4.

Checking for minimum area of web reinforcement (Art. 13-13), we note that the minimum area of one stirrup shall not be less than $0.0015 \times b \times s$ or, for this problem, $0.0015 \times 12 \times 11 = 0.198$ sq in. Since the total cross-sectional area of one #3 U-stirrup is $2 \times 0.11 = 0.22$ sq in., this requirement is adequately met.

(6) Bond stress will be checked using the procedure explained in Art. 13-15. Because this is a fully continuous beam, the tensile reinforcing bars will be placed in the top of the beam over the supports as well as in the bottom at the center of the span. To compute the bond stress we use

$$u = \frac{V}{\Sigma_0 jd} \qquad \text{Formula (11), Art. 13-15}$$

In Table 13-3 we find that the perimeters of #8 and #9 bars are 3.142 in. and 3.544 in., respectively. Then

$$u = \frac{29,500}{[(2 \times 3.142) + (2 \times 3.544)] \times 0.8723 \times 22}$$

and

$$u = 115 \text{ psi}$$

Referring to Table 13-5, we find the allowable bond stress for #8 and #9 top bars to be 186 psi and 165 psi, respectively; therefore the bond stress is acceptable and the tensile reinforcement will consist of the 2-#8 and 2-#9 bars tentatively selected in Step (3) above.

Figure 14-4a shows the dimensions of the beam as well as the size of reinforcement and its position. In the bottom of the beam, at the center of each of the continuous spans, there will be two #8 and two #9 bars to provide for the positive bending moment. One bar of each size will be bent up at the fifth point of the span (inflection point) and extended over the supports to the quarter point of the adjacent spans. This, with two bars bent up from the adjacent span, will provide two #8 plus two #9 bars for the negative bending moments

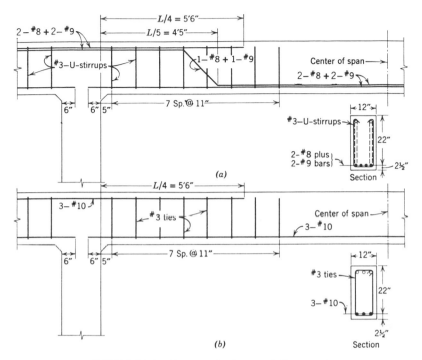

FIGURE 14-4. Rectangular beam reinforcement.

over the supports. The straight bars at the bottom of each beam will extend a distance of at least 6 in. into the supports.

The bar lengths indicated in Fig. 14-4a were determined in accordance with the fractions of L given in Fig. 14-1c. These factors have been in common use for many years and are intended to provide adequate length of embedment; they are sufficiently accurate when applied to the span, loading, and bending moment considerations discussed in Art. 14-3. When these conditions are not fulfilled, a more precise analysis must be made, and this may best be accomplished by following the development length provisions of ACI 318-71.[2]

---

[2] The development length concept as applied to simple beams was considered in Art. 13-16. Detailed provisions covering its use for continuous beams are contained in Chapter 12 of ACI 318–71. The reader desiring to investigate this matter further should consult the code and Chapter 12 of the *Commentary* on the code, both available from the American Concrete Institute, Detroit, Michigan,

(7) The design drawing should contain all of the information shown in Fig. 14-4 except the $L/4$ and $L/5$ factors; the dimensions only should be recorded. Two additional dimension lines should indicate the full clear span and the widths of the supports.

14-5 Alternate Design: Straight Bars Only

In order to satisfy the trend toward simplification in designing, detailing, fabrication, and placing of reinforcement, many designers use separate straight bars at both the top and bottom of beams instead of bent bars. The slight cost in excess weight in this arrangement over the combination of straight and bent bars is probably balanced by the ease of preparing design and shop drawings, bills of materials, and fabrication and placement of reinforcement. Bars not fabricated in accordance with the drawings, or those lost or mislaid, are more easily replaced if no bending is involved.

In a system of only straight bars, the tensile reinforcement in the preceding example might consist of 3-#10 straight bars, as shown in Fig. 14-4b. Instead of stirrups, #3 ties are used. These ties completely encircle the tensile reinforcement and serve as web reinforcement and as an aid in placing and holding the top bars in position.

Problem 14-5-A. An interior span of a fully continuous beam is 20 ft 0 in. between faces of supports. It carries a uniformly distributed load, not including its own weight, of 30 kips. Design the beam in accordance with the following specification data: $f'_c = 3$ ksi, $n = 9$, $f_s = 20$ ksi, $f_v = 20$ ksi, $f_c = 1350$ psi, $v_c = $ limited to 60 psi, $v = $ limited to 274 psi.

14-6 Web Reinforcement for Girders

Our consideration of reinforced concrete beam design thus far has been limited to beams carrying uniformly distributed loads. The principal loads on girders, however, are concentrated loads, only the weight of the girder constituting a distributed load. Formula (10), Art. 13-12, which gives the length of beam in which stirrups are required, is for use with uniformly distributed loads only; for beams or girders unsymmetrically loaded or for beams subjected to both concentrated and distributed loads it is advisable to draw the shear and bending moment diagrams to determine the positions and magnitudes of the shear and moments. This procedure is explained in Chapter 3. The following example illustrates the design of web reinforcement for a typical girder.

Example.   A reinforced concrete girder has an effective depth of 26.5 in. and a width of 12 in. Its total depth is 29 in. The clear span between faces of supports is 18 ft, and the loading consists of two concentrated loads of 30 kips each brought to the third points of the span by the floor beams (Figs. 14-5a and b). If the allowable unit shear stress of the concrete $v_c = 60$ psi and $f_v = 20,000$ psi, design the required web reinforcement.

Solution: (1) The girder has a cross section of 12 × 29 in.; therefore its weight per linear foot is

$$\frac{12 \times 29}{144} \times 150 = 362 \text{ lb}$$

and the total weight of the girder is $362 \times 18 = 6516$ lb. The total load on the girder is $30,000 + 30,000 + 6516 = 66,516$ lb. Since the loading is symmetrical, $R_1 = R_2 = 66516 \div 2 = 33,258$ lb.

(2) Computing the shear at significant sections,

$$V_{(\text{at } R_1)} = 33,258 \text{ lb}$$

$$V_{(x=6-)} = 33,258 - (6 \times 362) = 31,086 \text{ lb}$$

$$V_{(x=6+)} = 33,258 - [(6 \times 362) + 30,000] = 1086 \text{ lb}$$

$$V_{(x=9)} = 33,258 - [(9 \times 362) + 30,000] = 0$$

With these shear values, the diagram is constructed as explained in Chapter 3; it is shown in Fig. 14-5c.

The maximum vertical shear is 33,258 lb; it occurs at the face of the supports; V_d, the value of the vertical shear at d distance from the face of the support, is

$$V_d = V - \left(\frac{d}{12} \times w\right) \quad \text{or} \quad V_d = 33,258 - \left(\frac{26.5}{12} \times 362\right)$$

and

$$V_d = 32,458 \text{ lb}$$

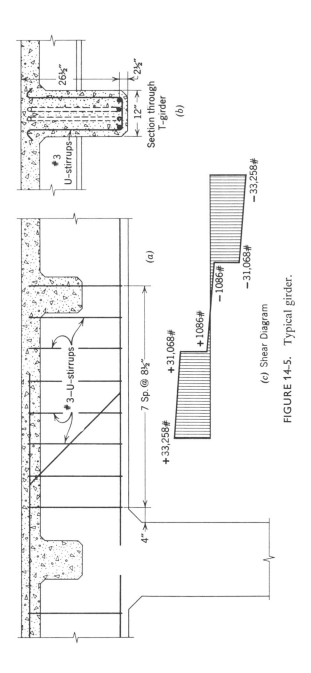

Section through
T-girder

(b)

#3
U-stirrups

#3–U-stirrups

+33,258#

7 Sp. @ 8½"

4"

(a)

26½"

2½"

12"

+31,068#

+1086#

−1086#

−31,068#

−33,258#

(c) Shear Diagram

FIGURE 14-5. Typical girder.

To find the unit shear stress at distance d from the face of support, we use Formula (6) from Art. 13-12.

$$v = \frac{V_d}{bd} \quad \text{or} \quad v = \frac{32,458}{12 \times 26.5} \quad \text{and} \quad v = 102 \text{ psi}$$

This exceeds the limiting value of 60 psi for v_c; therefore stirrups are required.

(3) The vertical shear just to the left of the first concentrated load is $V_{(x=6-)} = 31,068$ lb. Then

$$v = \frac{31,068}{12 \times 26.5} \quad \text{and} \quad v = 98 \text{ psi}$$

This indicates that stirrups are required from the face of the support up to the first concentrated load.

Now let us consider the shear immediately to the right of the first concentrated load. At this point

$$V_{(x=6+)} = 1068 \text{ lb} \quad \text{and} \quad v = \frac{1068}{12 \times 26.5} \quad \text{or} \quad v = 3.5 \text{ psi}$$

This stress is less than 60 psi, the allowable unit shearing stress for the concrete; hence no stirrups are required in the beam between the two concentrated loads.

$$V_c = v_c bd \quad \text{or} \quad V_c = 60 \times 12 \times 26.5 \quad \text{and} \quad V_c = 19,080 \text{ lb}$$

$$V' = V_d - V_c \quad \text{or} \quad V' = 32,458 - 19,080 \quad \text{and} \quad V' = 13,378 \text{ lb}$$

Assuming that #3 bars will be used for stirrups, $A_v = 2 \times 0.11 = 0.22$ sq in. and the maximum stirrup spacing, given by Formula (9) of Art. 13-12, is

$$s = \frac{A_v f_v d}{V'}$$

or

$$s = \frac{0.22 \times 20,000 \times 26.5}{13,378} \quad \text{and} \quad s = 8.7 \text{ in.}$$

Eight stirrups are used at each end of the girder as indicated in Fig. 14-5.

14-7 T-Beams

When a floor slab and its supporting beams are poured at the same time, the result is a monolithic construction in which a portion of the slab on each side of the beam serves as the flange of a T-beam. The part of the beam that projects below the slab is called the *web* or *stem* of the T-beam. This type of beam affords an economical form of construction and is commonly used; it is shown in Fig. 14-6a. For a simple beam the flange is in the compression zone, and there is

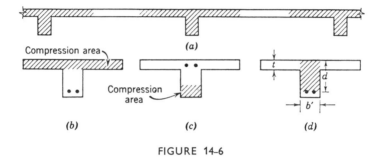

FIGURE 14-6

ample concrete to resist the compressive stresses, as shown in Fig. 14-6b. However, in a continuous beam there are negative bending moments over the supports, the flange here is in the tension zone, and the compressive stresses are in the web. See Fig. 14-6c. It is important to remember that only the area formed by the width of the web b' and the effective depth d are to be considered in computing the resistance to shear and to bending stresses over the supports. This is the hatched area $b'd$ shown in Fig. 14-6d. It is customary, when conditions permit, to have b', the width of the web, one half to one third of d, the effective depth.

The effective flange width to be used in the design of symmetrical T-beams shall not exceed one fourth the span length of the beam, and its overhanging width on either side of the web shall not exceed 8 times the thickness of the slab nor one half the clear distance to the next beam.

To determine the depth of a continuous T-beam, the negative bending moment over the support is first computed and the effective depth of the area $b'd$ is found by use of Formula (1), Art. 13-10:

$$d = \sqrt{\frac{M}{Rb}}$$

After d has been established, the area $b'd$ is investigated for shear. For rectangular beams

$$v = \frac{V}{bd} \qquad \text{Formula (6), Art. 13-12}$$

and for T-beams it becomes

$$b'd = \frac{V}{v}$$

This formula is used to establish the depth of T-beams if the depth is determined by the resistance of the concrete to shear.

The area of the negative moment reinforcement over the supports is found by the use of Formula (3), Art. 13-10:

$$A_s = \frac{M}{f_s jd}$$

In designing a floor construction consisting of a slab and beams, the slab is designed first; thus t, its thickness, is established before the T-beam is designed. See Fig. 14-6d. To find A_s for the positive bending moment between the supports we use the formula

$$A_s = \frac{M}{f_s[d - (t/2)]} \qquad \text{Formula (12)}$$

in which A_s = total area of the positive moment reinforcement in square inches,

M = maximum positive bending moment in inch-pounds,

f_s = allowable unit tensile stress in the steel in pounds per square inch,

d = effective depth of the T-beam in inches,

t = thickness of the flange of the T-beam in inches.

14-8   Design of a T-Beam

The various steps to be taken in the design of a T-beam are illustrated in the following example.

Example.   Design a fully continuous T-beam with a span of 20 ft
0 in. between faces of supporting girders. The floor slab, constituting
the flange of the T-beam, is 6 in. thick and the live load on the floor
is 100 psf. The T-beams are spaced 11 ft 0 in. on centers. Design the
beam in accordance with the following specification data:

$$f_c' = 3000 \text{ psi} \qquad\qquad f_y = 40,000 \text{ psi}$$
$$f_c = 1350 \text{ psi} \qquad\qquad f_s = 20,000 \text{ psi}$$
$$v_c = \text{limited to } 60 \text{ psi} \qquad f_v = 20,000 \text{ psi}$$
$$v = \text{limited to } 274 \text{ psi} \qquad n = 9$$

Design coefficients for these data are found in Table 13-4:

$$R = 225.6 \qquad k = 0.3831 \qquad j = 0.8723 \qquad p = 0.01293$$

Solution: (1) We will first determine the loading. Taking the weight
of reinforced concrete as 150 lb per cu ft, the 6-in. floor slab will
weigh 75 psf. This, with the live load of 100 psf, makes the total load
on the slab 175 psf. Because the beams are 11 ft 0 in. on centers,
$11 \times 175 = 1925$ lb *per lin ft* on the beam, not including the weight
of the web. This must be estimated; assume that it is 250 lb per lin
ft. Then $1925 + 250 = 2175$, say 2200 lb. This is *w*, the total live
and dead loads per linear foot on the T-beam. Therefore $W = 2200$
$\times 20 = 44,000$ lb, the total load on the T-beam. Since the loading
is symmetrical, the maximum vertical shear at the face of a support
is

$$V = R_1 = R_2 = 44,000 \div 2 = 22,000 \text{ lb}$$

(2) Bending moment: This is a fully continuous beam so the
maximum positive and negative design moments will have the same
magnitude (Fig. 14-1c). Therefore,

$$M = \frac{WL}{12} = \frac{44,000 \times 20 \times 12}{12} = 880,000 \text{ in-lb}$$

(3) Depth of beam: As stated in Art. 14-7, only the area $b'd$ is
effective at the support. Assuming 12 in. for the width of the web,

$$d = \sqrt{\frac{M}{Rb}} \qquad \text{or} \qquad d = \sqrt{\frac{880,000}{225.6 \times 12}} \qquad \text{and} \qquad d = 18 \text{ in.}$$

which is the effective depth.[3] Adding 2.5 in. for one half a bar diameter, plus fireproofing, $18 + 2.5 = 20.5$ in., say 21 in. for the total depth. The effective depth d then becomes $21 - 2.5 = 18.5$ in.

(4) Shear check: The area $b'd$ is now investigated to see that the maximum allowable value of $v = 274$ psi is not exceeded. The value of the unit shear stress is computed at distance d from the face of the support. Then

$$V_d = 22,000 - \left(\frac{18.5}{12} \times 2200 \right) \quad \text{or} \quad V_d = 18,600 \text{ lb}$$

and

$$v = \frac{V_d}{b'd} = \frac{18,600}{12 \times 18.5} = 84 \text{ psi}$$

This exceeds 60 psi but is less than 274 psi; consequently the dimensions of 12×18.5 in. for $b'd$ are adequate for both flexure and shear, but stirrups are required.

At this point a check of the stem weight assumed in Step (1) (250 lb per lin ft) should be made. Since the total depth of this beam is 21 in., including the slab thickness, the depth of the stem is $21 - 6 = 15$ in. Then

$$\frac{12 \times 15}{144} \times 150 = 188 \text{ lb per lin ft}$$

so the allowance originally made was amply large.

(5) Tensile reinforcement: The area of steel required at the center of the span is found from the formula

$$A_s = \frac{M}{f_s[d - (t/2)]}$$

This is a form of Formula (3), Art. 13-10, in which $d - (t/2)$ is substituted for jd. We know that the maximum positive moment is

---

[3] It should be noted that b in this formula is actually b' when applied to the stem or web width of T-beams; b is reserved for width of compression face, which becomes the flange width at midspan (Fig. 14-8). The 1971 ACI Code notation replaces b' with b_w.

880,000 in-lb; hence

$$A_s = \frac{880,000}{20,000[18.5 - (\text{\%})]} = 2.84 \text{ sq in.}$$

which is the required area of longitudinal reinforcement for the positive bending moment.

For the negative bending moment we use

$$A_s = \frac{M}{f_s j d} \qquad \text{Formula (3), Art. 13-10}$$

Then

$$A_s = \frac{880,000}{20,000 \times 0.8723 \times 18.5} \qquad \text{and} \qquad A_s = 2.72 \text{ sq in.}$$

Referring to Table 13-3, we see that 4-#8 bars have a cross-sectional area of (4 × 0.79), or 3.16 sq in. Because the beam is 12 in. wide, we can place the 4-#8 bars in a single layer. This layer consists of two bent and two straight bars, which will result in 3.16 sq in. for both the positive and negative bending moments. They are shown in Fig. 14-7.

(6) Web reinforcement: Since stirrups are required, we proceed in accordance with the procedure developed earlier:

$$V_c = v_c b'd \qquad \text{or} \qquad V_c = 60 \times 12 \times 18.5 \qquad \text{and} \qquad V_c = 13,300 \text{ lb}$$
$$V' = V_d - V_c \quad \text{or} \qquad V' = 18,600 - 13,300 \qquad \text{and} \qquad V' = 5300 \text{ lb}$$

Assuming that the stirrups are made of #3 bars, A_v, the cross-sectional area of one U-stirrup, is 2 × 0.11, or 0.22 sq in. Then

$$s = \frac{A_v f_v d}{V'} \qquad \text{or} \qquad s = \frac{0.22 \times 20,000 \times 18.5}{5300} = 15.4 \text{ in.}$$

However, the maximum stirrup spacing is $d/2$, or $18.5/2 = 9.25$ in., and we accept a stirrup spacing of 9 in.

To find the length of beam in which stirrups will be placed,

$$a = \frac{L}{2} \times \frac{V'}{V}$$

or

$$a = \frac{20}{2} \times \frac{5300}{22,000} \qquad \text{and} \qquad a = 2.4 \text{ ft or 29 in}$$

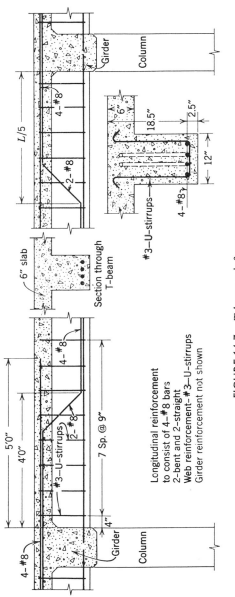

FIGURE 14-7. T-beam reinforcement.

Then $d + a + d = 18.5 + 29 + 18.5 = 66$ in., the total length at each end of the beam in which stirrups will be placed at 9-in. spacing See Fig. 14-7.

(7) Bond stress check: Using the flexural bond procedure of Art. 13-15, and noting that the sum of the perimeters of 4-#8 bars is $4 \times 3.142 = 12.57$ in.,

$$u = \frac{V}{\Sigma_0 jd} = \frac{22,000}{12.57 \times 0.872 \times 18.5} = 108 \text{ psi}$$

Referring to Table 13-5, we find the allowable bond stress for a #8 top bar to be 186 psi; therefore the bond stress is acceptable. The position and length of bars, determined in accordance with the provisions of Fig. 14-1c, are shown in Fig. 14-7.

14-9 Compressive Stress in T-Beam Flange

For average conditions the flange area is invariably large enough to resist the compression stresses that result from the positive bending moment at midspan. To show that this is true consider the following discussion. It is not a required step in the design of a T-beam.

Both Figs. 14-8a and b represent T-beam sections; the neutral axis is indicated for each section. In Fig. 14-8a the neutral axis lies in the slab, the flange area, whereas in Fig. 14-8b it falls below the slab. The areas above the neutral axis, the hatched areas, resist compressive stresses for positive moments. In Fig. 14-8b the area of the web below the slab subjected to compressive stresses is ignored in computations. If the dimensions of the parts of a T-beam are known, the position of the neutral surface may be accurately computed.

In Art. 14-7 specification requirements are given that limit the width of slab to be considered as the width of flange area of the T-beam. In the foregoing example we found that $b' = 12$ in., $t = 6$ in., and $d = 18.5$ in. Then b, the limiting width of flange, is

$$\frac{L}{4} = (20 \times 12) \div 4 = 60 \text{ in.}$$

or

$$b' + (2 \times 8 \times t) = 12 + (2 \times 8 \times 6) = 108 \text{ in.}$$

or

$$(\text{distance to adjacent beam}) \div 2 = (11 \times 12) \div 2 = 66 \text{ in.}$$

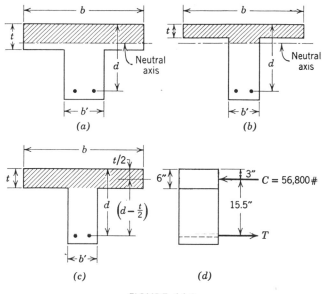

FIGURE 14-8

Of the three, 60 in. is the smallest magnitude and 60 in. is the width of flange to be used in computing the compressive stress.

In Art. 4-1 we found that the bending moment and the resisting moment for a beam in equilibrium are equal in magnitude. For the T-beam in the above example the bending moment was computed to be 880,000 in-lb. Referring to Figs. 14-8c and d (the latter being a side elevation of a portion of the beam), let C be the resultant of the compressive stresses in the flange; then $C \times 15.5$ is the resisting moment. Equating resisting moment to bending moment, $C \times 15.5 =$ 880,000 and $C = 56,800$ lb.

Although it is obviously inaccurate, let us assume temporarily that the compressive stresses are equally distributed on bt, the flange area, and that f_c psi is the uniformly distributed compressive stress. Then $C = f_c bt$, or $56,800 = f_c \times 60 \times 6$ and $f_c = 157$ psi. These computations indicate that the compressive stresses in the flange are well within the allowable, for, by data, f_c, the allowable compressive stress, is 1350 psi.

These computations are not justified because the exact position of the neutral surface was not found and also because we know that the compressive stresses are in direct proportion to their distances from the neutral surface. Nevertheless, we have shown by computations that the compressive stresses for the positive moment are well within the allowable and thus the foregoing procedure for establishing the effective depth is acceptable.

Problem 14-9-A.   A fully continuous T-beam has a span of 18 ft 0 in. between faces of supports. The floor slab is 5 in. thick and the live load is 150 psf. If the adjacent T-beams are 10 ft on centers, design the T-beam in accordance with the specification data given for the example of Art. 14-8.

14-10   Compression Reinforcement in T-Beam Stem

It was stated in Art. 14-7 that when a T-beam is continuous over a support, the negative bending moment produces tension in the flange and compression in the stem at the support (Fig. 14-6c). The design procedure followed in Arts. 14-7 and 14-8 provides a sufficient cross-section area of stem $b'd$, so that the allowable compressive stress in the concrete is not exceeded. However, if it is desired to use a beam of lesser depth, compression reinforcement may be provided by extending the main bottom bars from one span through the support and into the adjacent span a distance sufficient to develop the required compressive stress. The design of beams reinforced for compression is beyond the scope of this text, but it should be noted that *all* compression reinforcement in beams and girders must be enclosed by appropriate ties or stirrups whether or not required by shear considerations.

15

Reinforced Concrete Floor Systems

||||||||||||||||||||||||||||||||||||||||||||||||||||||||

15-1 Introduction

There are many different reinforced concrete floor systems, both cast-in-place and precast. The cast-in-place systems generally can be classified under four types:

1. One-way solid slab and beam
2. Two-way solid slab and beam
3. One-way ribbed or joist slab
4. Two-way flat plate, flat slab, or waffle slab

Each system has its distinct advantages, depending on the spacing of columns, magnitude of the loads, length of spans, and cost of construction. The floor area of the building and the purpose for which the building is to be used determine, in general, the positions and spacings of the columns. Whenever possible, the columns should be placed on common center lines; this is to simplify the framing. For the same reason it is desirable to repeat bays of the same dimensions.

15-2 One-Way Solid Slabs

One of the most commonly used floor systems consists of a solid slab supported by two parallel beams. The beams are usually supported by girders and the girders frame into columns. In this type of slab the tensile reinforcement runs in only one direction, from beam to beam. For this reason it is called a *one-way solid slab*. The number of beams in a panel or bay depends on their span, the column spacing, and the magnitude of the floor loads. The beams are spaced uniformly and usually frame into the girders at the center, third, or quarter points. The formwork for this type of floor is readily constructed. The one-way slab is economical for medium and heavy live loads for relatively short spans, 6 ft to 12 ft. Long spans for one-way slabs result in comparatively large dead loads. An example of the *beam and girder floor* is shown in Fig. 15-1; note that the floor consists of a one-way slab in which the tensile reinforcement runs from beam to beam.

To design a one-way slab a strip 12 in. wide, indicated by the hatched area in Fig. 15-1, is considered. This strip is designed as a beam whose width is 12 in. and on which is a uniformly distributed load. As shown in this figure, the strip is actually a fully continuous

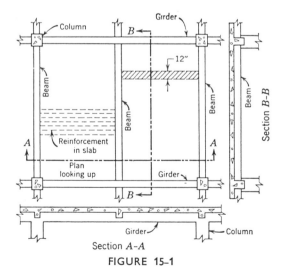

FIGURE 15-1

beam, the slab being a very wide shallow beam. As with any rectangular beam, the effective depth and tensile reinforcement are computed as explained in Art. 14-4. A minimum slab thickness is required to prevent excessive deflection. For one-way slabs of ordinary weight concrete (145 lb per cu ft) and steel reinforcement having $f_y = 40,000$ psi or 60,000 psi, the minimum thickness t may be selected from Table 15-1. The values in this table are valid only for

TABLE 15-1. Minimum Thickness of One-Way Slabs or Beams Unless Deflections Are Computed*

Member	End conditions	Minimum t †	
		$f_y = 40$ ksi	$f_y = 60$ ksi
Solid one-	Simple support	$L/25$	$L/20$
way slabs	One end continuous	$L/30$	$L/24$
	Both ends continuous	$L/35$	$L/28$
Beams or	Simple support	$L/20$	$L/16$
ribbed one-	One end continuous	$L/23$	$L/18.5$
way slabs	Both ends continuous	$L/26$	$L/21$

* Data abstracted from *Notes on ACI 318-71 Building Code Requirements* with permission of the American Concrete Institute.

† Valid only for members not supporting or attached to partitions or other construction likely to be damaged by large deflections.

members not supporting or attached to partitions or other construction likely to be damaged by large deflections. The span length L and thickness t must, of course, be in the same units.

15-3 Shrinkage and Temperature Reinforcement

In addition to tensile reinforcement in one-way floor and roof slabs, reinforcement placed at right angles to the main tensile steel must be provided for shrinkage and temperature stresses. For slabs in which deformed bars are used with specified yield strengths of less than 60,000 psi, the minimum ratio of this reinforcement to the gross concrete area is 0.002. This is the usual condition. In no case should these temperature bars be placed farther apart than 5 times the slab thickness nor more than 18 in.

Example.   What should be the spacing of #3 temperature bars used in a floor slab 5 in. thick?

Solution: Consider a strip of slab 12 in. wide and 5 in. thick. The strip has a gross cross-sectional area of 5 × 12 = 60 sq in., and the required area of temperature reinforcement is 0.002 × 60 = 0.12 sq in. The required area for a strip 1 *inch* wide is 0.12 ÷ 12 = 0.01 sq in. A #3 bar has a cross-sectional area of 0.11 sq in., so the number of inches of width to give 0.11 sq in. is 0.11 ÷ 0.01 = 11 in. The temperature reinforcement, therefore, consists of #3 bars spaced 11 in. on centers.

Table 15-2 eliminates the necessity of making such computations. After determining the steel area required per foot of width of slab (in this case 0.12 sq in.), a glance at the table shows that this area is provided by #3 bars spaced 11 in. on centers.

Problem 15-3-A*.   What size and spacing of temperature bars should be used for a 6½-in. floor slab? For a slab 4 in. thick?

15-4   Tensile Reinforcement in Slabs

In designing a one-way slab A_s, the area of main tensile steel required in a strip of slab 12 in. wide is computed. Either of the following formulas may be used:

$$A_s = pbd \qquad \text{Formula (2), Art. 13-10}$$

$$A_s = \frac{M}{f_s jd} \qquad \text{Formula (3), Art. 13-10}$$

In floor and roof solid slabs the maximum spacing of bars comprising the main tensile reinforcement is 3 times the slab thickness but not more than 18 in. Once A_s has been determined, the size and spacing of bars may be selected by use of Table 15-2.

Example.   In the design of a one-way solid floor slab, the required area of tensile reinforcement for a strip of slab 12 in. wide has been computed as $A_s = 0.36$ sq in. Determine appropriate bar sizes and spacings.

TABLE 15-2.   Areas of Bars in Reinforced Concrete Slabs per Foot of Width

Spacing (in.)	Areas of bars in square inches									
	#2*	#3	#4	#5	#6	#7	#8	#9	#10	#11
3	0.20	0.44	0.79	1.23	1.77	2.41	3.14	4.00		
3½	0.17	0.38	0.67	1.05	1.51	2.06	2.69	3.43	4.36	
4	0.15	0.33	0.59	0.92	1.33	1.80	2.36	3.00	3.81	4.68
4½	0.13	0.29	0.52	0.82	1.18	1.60	2.09	2.67	3.39	4.16
5	0.12	0.26	0.47	0.74	1.06	1.44	1.88	2.40	3.05	3.74
5½	0.11	0.24	0.43	0.67	0.96	1.31	1.71	2.18	2.77	3.40
6	0.10	0.22	0.39	0.61	0.88	1.20	1.57	2.00	2.54	3.12
6½	0.09	0.20	0.36	0.57	0.82	1.11	1.45	1.85	2.35	2.88
7	0.08	0.19	0.34	0.53	0.76	1.03	1.35	1.71	2.18	2.67
7½	0.08	0.18	0.31	0.49	0.71	0.96	1.26	1.60	2.03	2.50
8	0.07	0.17	0.29	0.46	0.66	0.90	1.18	1.50	1.91	2.34
8½	0.07	0.16	0.28	0.43	0.62	0.85	1.11	1.41	1.79	2.20
9	0.07	0.15	0.26	0.41	0.59	0.80	1.05	1.33	1.69	2.08
9½	0.06	0.14	0.25	0.39	0.56	0.76	0.99	1.26	1.60	1.97
10	0.06	0.13	0.24	0.37	0.53	0.72	0.94	1.20	1.52	1.87
11	0.05	0.12	0.21	0.33	0.48	0.66	0.86	1.09	1.39	1.70
12	0.05	0.11	0.20	0.31	0.44	0.60	0.79	1.00	1.27	1.56

* This bar (0.25 in. diameter) is a plain bar; all others listed are deformed.

Solution: Referring to Table 15-2, we note that either #4 bars spaced 6½ in. on centers or #5 bars with a 10-in. spacing may be used.

Problem 15-4-A   The required area of tensile reinforcement in a 12 in. width of a one-way solid slab is 0.43 sq in. Determine the spacing if #4 bars are used; if #5 bars are used.

15-5   Design of a One-Way Solid Slab

In general, the design procedure for a one-way slab follows that for a rectangular beam given in Art. 14-4, the rectangular section consisting of an imaginary strip of slab 1 ft wide. The bending moment for this strip is determined as though it were acting independently, and the effective depth and area of tensile steel computed. The spacing of bars in this 12-in. strip is the spacing that will be used throughout the full width of the slab.

It is not practicable to use web reinforcement in one-way slabs, and consequently the unit shearing stress v must be kept within v_c. If conditions result in shear stresses that exceed v_c, the depth of the slab must be increased.

Although simply supported single spans are frequently encountered, the majority of slabs used in building construction are continous or semicontinuous. In our discussion of one-way slab design, we will use the bending moment and shear factors given in Fig. 14-1 and Art. 14-3; for average conditions they are adequate. When estimating the weight of a slab, it is customary to allow 0.4 in. to 0.5 in. of thickness for each foot of span.

Example.   A one-way slab with fully continuous end conditions has a span of 11 ft 0 in. between faces of supporting beams. This slab is to support a live load of 190 psf, a floor finish of 20 psf, and a suspended metal lath and plaster ceiling. Design the slab in accordance with the following specification data:

$$f'_c = 3000 \text{ psi} \qquad f_y = 40{,}000 \text{ psi}$$
$$f_c = 1350 \text{ psi} \qquad f_s = 20{,}000 \text{ psi}$$
$$v_c = \text{limited to 60 psi} \qquad n = 9$$

Solution: (1) To determine the design load, we first estimate the weight of the slab. Assuming a thickness of 0.4 in. for each foot of span, $0.4 \times 11 = 4.4$, say 4.5 in., and the estimated weight is $(4.5 \div 12) \times 150 = 56$ psf. Then

$$
\begin{array}{rl}
\text{live load} = & 190 \\
\text{floor finish} = & 20 \\
\text{suspended ceiling} = & \underline{10} \\
\text{total superimposed load} = & \overline{220} \text{ psf} \\
\text{estimated slab weight} = & \underline{56} \\
\textit{design load} = & \overline{276} \text{ psf}
\end{array}
$$

Since the span is 11 ft, the total load on the 1-ft wide strip is $W = 276 \times 11 = 3036$ lb, and the end shear is $V = 3036 \div 2 = 1518$ lb.

(2) Depth of slab: Since the span is fully continuous (Fig. 14-2b), the maximum bending moment is

$$M = \frac{WL}{12} = \frac{3036 \times 11 \times 12}{12} = 33,400 \text{ in-lb}$$

and the required effective depth is

$$d = \sqrt{\frac{M}{Rb}} = \sqrt{\frac{33,400}{225.6 \times 12}} = 3.5 \text{ in.}$$

Allowing 0.25 in. for one half a bar diameter plus 0.75 in. for fireproofing below the bars (Art. 13-7), $3.5 + 0.25 + 0.75 = 4.5$ in., the total thickness of the slab. Checking this value against the minimum thickness requirements given in Table 15-1, $(11 \times 12) \div 35 = 3.77$ in.; hence our thickness of 4.5 in. is satisfactory.

(3) Tensile reinforcement: The area of tensile steel required in the 12-in. wide strip of slab is

$$A_s = \frac{M}{f_s jd} = \frac{33,400}{20,000 \times 0.8723 \times 3.5} = 0.55 \text{ sq in.}$$

Referring to Table 15-2, we find that #5 bars at a $6\frac{1}{2}$-in. spacing give us this area.

(4) Shearing stress: Computing the vertical shear at distance d from the face of support,

$$V_d = 1518 - \left(\frac{3.5}{12} \times 276\right) = 1438 \text{ lb}$$

and the unit shearing stress at this section is

$$v = \frac{V_d}{bd} \quad \text{or} \quad v = \frac{1438}{12 \times 3.5} \quad \text{and} \quad v = 34 \text{ psi}$$

This value is less than the limit of 60 psi for v_c, so the slab meets the requirements for shear.

(5) Bond stress check: From Table 13-3 we find that the perimeter of a #5 bar is 1.963 in. Because the bar spacing is 6.5 in., the number of bars in each 12-in. width of slab is $12 \div 6.5 = 1.85$. Then

$$u = \frac{V}{\Sigma_0 jd} = \frac{1518}{(1.85 \times 1.963) \times 0.8723 \times 3.5} = 136 \text{ psi}$$

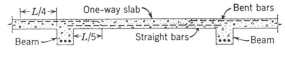

FIGURE 15-2

This is acceptable since it does not exceed the value of 298 psi for #5 top bars given as the maximum in Table 13-5.

(6) Temperature reinforcement: The required area in a strip of slab 12 in. wide is $4.5 \times 12 \times 0.002 = 0.108$, say 0.11 sq in. Referring to Table 15-2, #3 bars spaced 12 in. on centers satisfy this requirement.

(7) Arrangement of reinforcement: Because the slab is fully continuous, alternate bars may be bent up at the fifth points of the span and extended over the supports to the fourth points of the adjacent spans. The remaining bars are straight, placed in the bottom of the slab, and extended at least 6 in. into the supports or made continuous. In this example the clear fireproofing distance is $\frac{3}{4}$ in.; some codes require a full inch. See Fig. 15-2, in which the reinforcement for the adjacent spans is not shown. The #3 temperature bars are run at right angles to the main tensile reinforcement.

Problem 15-5-A*.   It is proposed to use a one-way solid slab 4 in. thick on a simply supported span of 8 ft. The live load is 100 psf and the tile flooring with its mortar setting bed weighs 25 psf. Specification data:

$$f'_c = 3000 \text{ psi} \qquad f_y = 40{,}000 \text{ psi}$$
$$f_c = 1350 \qquad f_s = 20{,}000 \text{ psi}$$
$$v_c = \text{limited to } 60 \text{ psi} \qquad n = 9$$

Determine (a) whether the slab is thick enough, and (b) the required size and spacing of bars for tensile reinforcement and temperature reinforcement, using the 4-in. thickness.

Problem 15-5-B.   Design a one-way fully continuous solid slab for a span of 12 ft 0 in. The live load is 150 psf, there is a suspended metal lath and plaster ceiling, and the floor finish weighs 30 psf. Specification data:

$$f'_c = 3000 \text{ psi} \qquad f_y = 60{,}000 \text{ psi}$$
$$f_c = 1350 \text{ psi} \qquad f_s = 24{,}000 \text{ psi}$$
$$v_c = \text{limited to } 60 \text{ psi} \qquad n = 9$$

15-6   Ribbed Slabs: Concrete Joists

One of the most economical systems of floor construction for relatively long spans and light loads consists of metal pan fillers between concrete joists. Standard pan forms produce inside dimensions of 20 in. or 30 in. in width, and are available in depths from 6 in. to 20 in. The pans are placed far enough apart to form joists or ribs 4 in. to 7 in. wide at the lowest point. A common condition consists of pans 20 in. wide, placed 25 in. on centers, making a rib 5 in. wide at the bottom, as shown in Fig. 15-3a. The top slab varies in thickness from 2 in. to $4\frac{1}{2}$ in.

Another type of ribbed slab can be made by employing concrete block or clay tile fillers in place of the metal pans. Concrete filler blocks are available in widths of 12 in., 16 in., and 24 in. and in depths of 6 in. to 14 in. Clay tile fillers are usually 12×12 in. in plan, and have depths ranging from 6 in. to 15 in. The tile filler blocks are frequently placed 16 in. on centers, thus making the joists 4 in. wide. See Fig. 15-3c.

Web reinforcement is not used in ribbed slabs. In metal pan systems the shear stress is resisted solely by the concrete. If, when investigating maximum shear stress, v is found to exceed v_c, a greater

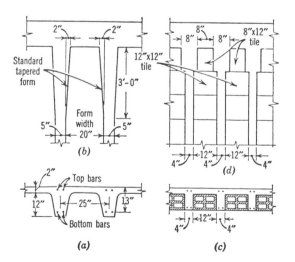

FIGURE 15-3

joist area at the supports is provided by using standard metal pans (sometimes called cores) with the sides tapered in plan, as shown in Fig. 15-3b. The degree of taper generally is such that the width of the joist is increased 4 in. in a 3-ft length. When permanent burned clay or concrete block fillers are used, the ACI Code permits the vertical shells of the fillers in contact with the concrete to be included in shear computations if the unit compressive strength of the block is at least equal to the specified strength of the concrete. Where a greater width of joist is required for shear at the supports, a tile filler 8 in. wide instead of 12 in. may be used as shown in Fig. 15-3d.

It will be noted from Fig. 15-3 that a ribbed slab basically consists of a number of small T-beams side by side. The tensile reinforcement is usually provided by two bars. Temperature reinforcement may be wire mesh or #2 bars (¼-in. diameter) run at right angles to the joists. All of the terms ribbed slab, concrete joists, and joist slab are commonly used to denote this type of floor system.

15-7 Design of a One-Way Joist Slab

Because a joist slab is composed of a number of small adjacent T-beams, its design may be accomplished by following the procedure for T-beams presented in Art. 14-8.

Example. Design a joist-slab system using metal pans for an interior clear span of 24 ft 0 in., the end conditions being fully continuous. The live load on the floor is 60 psf, the floor finish weighs 30 psf, and there will be a suspended metal lath and plaster ceiling. Specification data:

$$f'_c = 3000 \text{ psi} \qquad f_y = 40,000 \text{ psi}$$
$$f_c = 1350 \text{ psi} \qquad f_s = 20,000 \text{ psi}$$
$$v_c = \text{limited to } 66 \text{ psi} \qquad n = 9$$
$$(see \ Table \ 13\text{-}1)$$

Solution: (1) There are, of course, several combinations of pan depth and slab thickness (topping) that might be used. Let us assume a trial combination of pans 12 in. deep with a slab thickness of 2 in.; this is commonly called a "12 + 2" system, and its overall thickness is 14 in. Referring to Table 15-3, which gives weights in pounds per

TABLE 15-3.   Weights of Concrete Joist Slabs in Pounds per Square Foot

Metal pan filler, joists 5 in. wide, 25 in. on centers

Depth of pan (in.)	Thickness of slab (in.)		
	2	2½	3
6	42	48	55
8	48	54	61
10	54	60	67
12	61	67	74
14	68	75	81

Clay tile filler, joists 4 in. wide, 16 in. on centers

Depth of tile (in.)	Thickness of slab (in.)		
	2	2½	3
4	50	58	65
6	63	70	75
8	75	80	87
10	85	93	98
12	97	103	110

square foot for several combinations, we find that our 12 + 2 combination weights 61 psf. Establishing the design load,

$$
\begin{aligned}
\text{live load} &= 60 \\
\text{floor finish} &= 30 \\
\text{suspended ceiling} &= 10 \\
\text{joist slab} &= \underline{61} \\
\textit{design load} &= \overline{161 \text{ psf}}
\end{aligned}
$$

Because the joists will be placed 25 in. on centers (20-in. wide pans), the number of square feet per linear foot of joist is $25 \div 12 = 2.08$. Thus $161 \times 2.08 = 335$ lb, the load per linear foot of joist. Consequently, the total uniformly distributed load on each rib is

$W = 335 \times 24 = 8040$ lb, and the maximum vertical shear at the support is $V = W \div 2 = 4020$ lb.

(2) Bending moment: Because each joist constitutes a fully continuous beam, the maximum positive and negative design moments are

$$M = \frac{WL}{12} = \frac{8040 \times 24 \times 12}{12} = 192{,}960 \text{ in-lb}$$

(3) Depth of beam: Allowing 0.25 for one half a bar diameter plus 0.75 for fireproofing, the effective depth of our assumed $12 + 2$ combination is $14 - (0.25 + 0.75) = 13$ in. The stem or web width of this T-section is 5 in., making the area $b'd = 5 \times 13$ in. (Fig. 14-6d). Checking to determine whether the 13-in. value is adequate, the required d over the supports is

$$d = \sqrt{\frac{M}{Rb}} = \sqrt{\frac{192{,}960}{225.6 \times 5}} = 13 \text{ in.}$$

This agrees with the assumed d so the bending stresses at the supports are acceptable.

(4) Shear check: The vertical shear at distance d from the face of support is

$$V_d = 4020 - \left(\frac{13}{12} \times 335\right) \qquad \text{and} \qquad V_d = 3640 \text{ lb}$$

and the unit shearing stress is

$$v = \frac{V_d}{b'd} = \frac{3640}{5 \times 13} \qquad \text{and} \qquad v = 56 \text{ psi}$$

By data we known that the allowable unit shearing stress is 66 psi so the area $b'd$ is satisfactory for shear. If the shear stress had been in excess of 66 psi, tapered metal end pans could have been used and the area $b'd$ thus enlarged. This would have lowered the unit shearing stress.

(5) Tensile reinforcement: The area of steel required at the center of the span for positive bending moment is

$$A_s = \frac{M}{f_s(d - t/2)} = \frac{192{,}960}{20{,}000 \, (13 - 2/2)} = 0.8 \text{ sq in.}$$

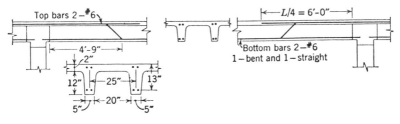

FIGURE 15-4

and for negative moment at the supports is

$$A_s = \frac{M}{f_s jd} = \frac{192,960}{20,000 \times 0.8723 \times 13} = 0.85 \text{ sq in.}$$

We accept, therefore, 2-#6 bars (2 × 0.44 = 0.88 sq in.) for each bending moment, pending check of the bond stress.

(6) Bond stress check: Using the flexural bond procedure of Art. 13-15, and noting that the sum of the perimeters of 2-#6 bars is 2 × 2.356 = 4.712 in.,

$$u = \frac{V}{\Sigma_0 jd} = \frac{4020}{(2 \times 2.356) \times 0.8723 \times 13} = 75 \text{ psi}$$

Because this stress is less than 248 psi given for top bars in Table 13-5, the bond stress is not excessive and the 2-#6 bars are adopted. The position and length of bars is shown in Fig. 15-4. In each joist, one of the 2-#6 bars is straight and the other is bent. The bent bars are turned up at the fifth points of the span and carried over the supporting beam to the quarter points of the adjacent spans. The straight bars extend into the supports a distance of at least 6 in.

Problem 15-7-A.   The end conditions of an interior panel of a ribbed floor slab are fully continuous and the clear span of the joists is 18 ft 0 in. If the live load on the floor is 70 psf and the floor finish weighs 25 psf, design the joist–slab system using metal pan fillers. Specification data is the same as that given for the foregoing example.

16

Reinforced Concrete
Columns

||||||||||||||||||||||||||||||||||||||||||||||||||||||||||

16-1 Introduction

The practicing structural designer customarily makes use of tables
for determining the size of columns employed in building construction. These tables are compiled in accordance with the governing
building code requirements. The provisions relating to the design
of columns in the 1971 ACI Code are quite different from those of
the working stress design method in the 1963 Code. The alternate
design method of ACI 318-71 requires that the service load capacity
of columns shall be taken as 40% of that computed in accordance
with ultimate strength design procedures.[1] Because of the complexity
of conditions relating to the design of columns, the following discussions are limited to the working stress design method for two of
the most common typical cases.

---

[1] The relation between service loads and ultimate loads is explained in Chapter
18.

354

A large percentage of the concrete columns encountered in building construction are known as *short columns*, and their length should not be greater than about 10 times their least cross-sectional dimension. Principal columns in buildings should have a minimum diameter of 10 in. or, if rectangular, a minimum thickness of 8 in., with a gross area not less than 96 sq in. Posts not continuous from story to story may have a minimum thickness of 6 in. if they constitute auxiliary supports rather than main members.

16-2 Tied Columns

A tied column has reinforcement consisting of longitudinal bars and separate lateral ties, as shown in Fig. 16-1a. The area of vertical reinforcement should not be less than 0.01 or more than 0.08 times the gross cross-sectional area. The minimum size of bar is #5, and the minimum number of bars in tied columns is four.

Lateral ties for tied columns should be at least ¼ in. in diameter and spaced apart not more than 16 bar diameters, 48 tie diameters, or the least dimension of the column. When there are more than

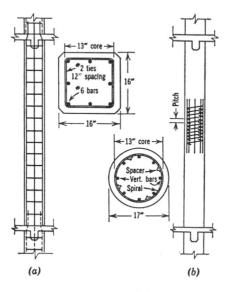

(a)   (b)

FIGURE 16-1

four vertical bars, additional ties should be provided so that every longitudinal bar will be held firmly in its designed position.

The reinforcement for tied columns should be protected by a covering of concrete, cast monolithically with the core, for which the thickness should not be less than $1\frac{1}{2}$ in.

The maximum permissible axial load on *short* tied columns is given by the formula

$$P = 0.85[A_g(0.25f_c' + f_s p_g)]$$

in which P = maximum permissible axial load in pounds,

A_g = gross area of the column in square inches,

f_c' = ultimate compressive strength of concrete in pounds per square inch,

f_s = allowable compressive stress in vertical reinforcement to be taken at 40% of the minimum values of the yield-point strength but not to exceed 30,000 psi,

p_g = ratio of the cross-sectional area of vertical reinforcement A_{st} to the gross column area A_g,

A_{st} = total area of longitudinal reinforcement in square inches.

16-3 Design of a Tied Column

As noted above, practical design of reinforced concrete columns is accomplished by the use of safe load tables.[2] Without these tables it is necessary to take trial sections and compute the allowable loads they will support. Because the cross-sectional area of the vertical steel may vary from 0.01 to 0.08 of the gross column area and also because concrete and reinforcement of different strengths may be used, it is obvious that several different combinations may satisfy the requirements. To illustrate the use of the foregoing formula let us compute the safe load on a specific column.

Example.   A short reinforced concrete tied column has a 16 × 16 in. cross section. The vertical reinforcement consists of 8-#6 bars for

[2] See Chapter 10 of *Simplified Design of Reinforced Concrete*, Third Edition, by Harry Parker (New York: Wiley, 1968).

which $f_y = 50,000$ psi and $f'_c = 3000$ psi. Compute the allowable axial load the tied column will support.

Solution: (1) The gross area of the column is $16 \times 16 = 256$ sq in. Because the yield point strength of the reinforcement is 50,000 psi, $f_s = 0.40 \times 50,000 = 20,000$ psi. The total area of the longitudinal reinforcement, A_{st}, is $8 \times 0.44 = 3.52$ sq in.; therefore

$$p_g = \frac{A_{st}}{A_g} = \frac{3.52}{256} \quad \text{and} \quad p_g = 0.0137$$

(2) Computing the allowable axial load,

$$P = 0.85[A_g(0.25f'_c + f_s p_g)]$$
$$= 0.85 \times 256[(0.25 \times 3000) + (20,000 \times 0.0137)]$$

and

$$P = 223,200 \text{ lb}$$

(3) Assuming that the lateral ties are made of #2 bars ($\frac{1}{4}$-in. diameter), the tie spacing is the smaller of

$$16 \text{ bar diameters, } 16 \times 0.75 = 12 \text{ in.}$$

or

$$48 \text{ tie diameters, } 48 \times 0.25 = 12 \text{ in.}$$

or

$$\text{the least column dimension} = 16 \text{ in.}$$

The 12-in. dimension controls; hence the ties for the column are made of #2 bars spaced 12 in. on centers. The vertical bars are placed not less than $1\frac{1}{2}$ in. from the face of the column.

Problem 16-3-A*. Compute the allowable axial load and the tie spacing for a short 14×14 in. tied column with 8-#5 bars for the vertical reinforcement; $f'_c = 3000$ psi and $f_y = 50,000$ psi.

Problem 16-3-B. A short tied column 12×12 in. in cross section is reinforced with 4-#6 bars for which $f_y = 50,000$ psi. If $f'_c = 3000$ psi, what allowable axial load will the column support? What is the spacing of the $\frac{1}{4}$-in. lateral ties?

16-4 Spiral Columns

A spiral column is a concrete column reinforced with vertical bars and a closely spaced continuously wound hooping, as indicated in

Fig. 16-1b. The purpose of the spiral reinforcement is to provide restraint against the tendency to crush, and also to support laterally the vertical reinforcing bars as well as the concrete core. Although spiral reinforcement is limited almost entirely to columns of circular cross section, it may be employed in rectangular cross sections as well.

The vertical reinforcement should not be less than 0.01 or more than 0.08 times the gross cross-sectional area. The minimum number of bars should be six and the minimum bar size #5.

The maximum permissible axial load for spiral columns is found by the formula

$$P = A_g(0.25f'_c + f_s p_g)$$

in which P = maximum permissible axial load in pounds,

A_g = gross area of the column in square inches,

f'_c = ultimate compressive strength of concrete in pounds per square inch,

f_s = allowable compressive stress in vertical reinforcement to be taken at 40% of the minimum values of the yield-point strength but not to exceed 30,000 psi,

p_g = ratio of the cross-sectional area of vertical reinforcement A_{st} to the gross column area A_g,

A_{st} = total area of longitudinal reinforcement in square inches.

The ratio of the spiral reinforcement p_s should not be less than the value given by the formula

$$p_s = 0.45\left(\frac{A_g}{A_c} - 1\right)\frac{f'_c}{f_y}$$

in which p_s = ratio of volume of spiral reinforcement to the volume of the concrete core (out to out of spirals),

A_g = gross area of the column in square inches,

A_c = area of the column core in square inches,

f'_c = ultimate compressive unit stress of concrete,

f_y = yield-point stress of the reinforcement.

The spiral reinforcement should consist of evenly spaced continuous spirals held firmly in place by at least three vertical spacer bars. The center-to-center spacing of the spirals should not exceed one sixth of the core diameter and the clear spacing between spirals should not exceed 3 in. nor be less than $1\frac{3}{8}$ in. or $1\frac{1}{2}$ times the maximum size of the coarse aggregate.

The column reinforcement should be protected everywhere by a covering of concrete cast monolithically with the core, for which the thickness must not be less than $1\frac{1}{2}$ in.

16-5   Design of a Spiral Column

In practice, the design of a spiral column is accomplished by the use of tables. If tables are not available, a trial section is assumed and its allowable load is computed by the use of formulas.

Example.   A short spiral column has an overall diameter of 17 in. with 2 in. of protective concrete covering for the vertical reinforcing bars. The vertical reinforcement consists of 9-#6 bars with an allowable f_s of 20,000 psi. The spiral reinforcement consists of cold-drawn wire for which $f_y = 60,000$ psi and $f'_c = 3000$ psi. Compute the allowable axial load and also the required spiral reinforcement.

Solution: (1) The area of 1-#6 bar is 0.44 sq in. (Table 13-3), making $A_{st} = 9 \times 0.44 = 3.96$ sq in. Since the gross diameter of the column is 17 in., the gross area is $D^2 \times 0.7854$, or $A_g = 17 \times 17 \times 0.7854 = 227$ sq in. Therefore

$$p_g = \frac{3.96}{227} = 0.01745$$

(2) Computing the allowable axial load,

$$P = A_g(0.25f'_c + f_s p_g)$$
$$= 227[(0.25 \times 3000) + (20,000 \times 0.01745)]$$

and

$$P = 249,300 \text{ lb}$$

(3) To determine the required spiral reinforcement we use the following formula;

$$p_s = 0.45\left(\frac{A_g}{A_c} - 1\right)\frac{f'_c}{f_y}$$

Because there is a 2-in. protective covering, the diameter of the core is 17 — 4 or 13 in. The area of the core A_c is 13 × 13 × 0.7854, or 132.7 sq in.

Substituting in the foregoing formula,

$$p_s = 0.45\left(\frac{227}{132.7} - 1\right)\frac{3000}{60,000} = 0.01597$$

the ratio of the volume of spiral reinforcement to the volume of the concrete core of the column.

Since the area of the core is 132.7 sq in., a 1 in. height of core contains 132.7 cu in. Therefore, each 1-in. height of core requires a volume of spiral steel equal to p_s × 132.7 = 0.01597 × 132.7 = 2.12 cu in.

Let us assume that the spiral will be made of wire of ⅜-in. diameter. The area of a ⅜-in. diameter is 0.11 sq in. (Table 13-3). The circumference of the 13-in. diameter core is 13 × 3.1416, or 40.84 in. Thus the volume of one complete spiral is 40.84 × 0.11, or 4.5 cu in.

Now the vertical length of column to give 4.5 cu in. of ⅜-in. spiral is 4.5 ÷ 2.12 = 2.1 in., the required pitch. Thus the spiral reinforcement consists of ⅜-in. cold-drawn wire with a 2-in pitch. This computation does not have to be made in practice because tables giving size and pitch of spirals for various diameter columns and values of f_y and f_c' are available.

Problem 16-5-A*.   An 18-in. diameter column has vertical reinforcement consisting of 10-#6 bars. The vertical bars have 2 in. of protective concrete covering. Compute the allowable axial load and the required spiral reinforcement. Specification data: $f_c' = 3000$ psi, $f_s = 20,000$ psi, the spiral reinforcement consists of ⅜-in. cold-drawn wire for which $f_y = 60,000$ psi.

17

Column Footings

||||||||||||||||||||||||||||||||||||||||||||||||||||||

17-1 Introduction

The primary purpose of a footing is to spread the loads so that the allowable bearing capacity of the foundation bed is not exceeded. In cities where experience and tests have established the allowable strengths of various foundation soils local building codes may be consulted to determine the bearing capacities for use in design. In the absence of such information, or for conditions in which the nature of the soil is unknown, borings or load tests should be made. For sizable structures borings at the site should always be made, and their results interpreted by a qualified soils engineer.

Footings may be classified as wall footings and column footings. In the former type the load is brought to the foundation bed as a uniform load per linear foot of wall; in the latter it is concentrated at the base of a concrete column or at the base plate of a structural steel column (Fig. 6-9). Sometimes two or more columns are supported on a single foundation which is called a combined footing. However, the independent footing supporting a single column is the most common type, and the one to be considered in this chapter.

361

17-2 Independent Column Footings

The great majority of independent or isolated column footings are square in plan, with reinforcement consisting of two sets of bars at right angles to each other. This is known as two-way reinforcement. The column may be placed directly on the footing block or it may be supported by a pedestal, the pedestal being supported by the footing block.

The area of the footing block is found by dividing the column load, plus the estimated weight of the footing block, by the allowable bearing capacity of the foundation bed. The weight of the footing varies from 4% to 10% of the load on the column. To design a footing it is common to assume an effective depth based on Formula (1), Art. 13-10. This assumed depth is then used to determine the values of v and u. If they are found to be excessive, a greater depth is taken, which, of course, will reduce the shear and bond stresses.

The ACI Code (1971 and 1963) requires that the bending moment be taken at the face of the column (or pedestal) at a section that extends completely across the footing. For the square footing shown in Fig. 17-1a the maximum bending moment is computed by considering the load on the hatched area cl. Then, if w is the upward pressure of the fundation bed in pounds per square foot and the dimensions a, c, and l are in feet,

$$M = c \times l \times w \times \frac{c}{2} = \frac{wlc^2}{2}\ \text{ft-lb} \qquad \text{or} \qquad M = 6wlc^2\ \text{in-lb}$$

the maximum bending moment. The dimension $c = (l - a) \div 2$. After the effective depth of the footing has been assumed, based on

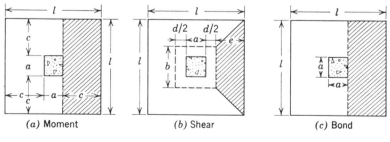

(a) Moment (b) Shear (c) Bond

FIGURE 17-1

the formula $d = \sqrt{M/Rb}$, the shear and bond stresses are investigated to see that d is acceptable.

The critical section for shear is taken perpendicular to the plane of the slab and is located at a distance $d/2$ out from the periphery of the column base. Using the letters shown in Fig. 17-1b, a is the side of the column or base plate, $b = a + d$, and

$$e = \frac{l}{2} - \frac{a}{2} - \frac{d}{2}$$

Then the hatched area used in determining the unit shearing stress is

$$\frac{l+b}{2} \times e$$

and, if w is the upward pressure from the foundation bed,

$$V = w \times \frac{l+b}{2} \times e$$

The unit shearing stress is computed by the formula

$$v = \frac{V}{b_0 d}$$

in which b_0 is the periphery of the critical section.

To compute the bond stress the critical section for V is taken at the face of the column. The footing area used in determining V, the vertical shear, is the hatched area shown in Fig. 17-1c. Consequently,

$$V = \frac{l(l-a)}{2} \times w$$

For reinforcing bars in footings deformed bars are invariably used. The ends of the bars should not be less than 3 in. nor more than 6 in. from the face of the footing. There should be not less than 3 in. from the reinforcing bars to the bottom of the footing. For footings on soil the thickness of concrete above the reinforcement at the edge of the footing should not be less than 6 in.

17-3   Design of an Independent Column Footing

The determination of column footing dimensions and reinforcement is not subject to precise analysis and several different solutions to the same problem may be acceptable. The following example illustrates a working stress design procedure that may be employed in the design of an independent column footing with two-way reinforcement.

Example.   A 16-in. square column exerts a load of 240 kips on a two-way block footing. The foundation bed on which the footing is placed has a working bearing capacity of 4000 psf. Design the footing in accordance with the following specification data:

$$f'_c = 3000 \text{ psi} \qquad \text{Grade 40 bars}$$
$$f_c = 1350 \text{ psi} \qquad f_s = 20,000 \text{ psi}$$
$$v_c = \text{limited to 110 psi} \qquad n = 9$$
$$(see \ Table \ 13\text{-}1)$$

Formula coefficients for these stresses are found in Table 13-4:

$$R = 225.6 \qquad k = 0.3831 \qquad j = 0.8723 \qquad p = 0.01293$$

Solution: (1) To estimate the footing weight let us assume that the weight is 7% of the column load. Then $240,000 \times 0.07 = 16,800$ lb. Hence $240,000 + 16,800 = 256,800$ lb, the total load on the foundation bed. The bearing capacity of the foundation bed is 4000 psf; therefore $256,800 \div 4000 = 64.2$, say 64 sq ft. Accept a block footing 8×8 ft square. Thus $l = 8$ ft. See Fig. 17-2.

(2) The weight of the footing does not cause it to bend; hence $240,000 \div 64 = 3750$ psf which is the value of w, the upward pressure of the foundation bed that produces bending, shear, and bond stresses. From Art. 17-2, the maximum bending moment is

$$M = 6wlc^2 = 6 \times 3750 \times 8 \times 3.33 \times 3.33 = 1,990,000 \text{ in-lb}$$

In this equation 3.33 ft $= c$, which is found by the formula $(l - a) \div 2$ or $(8 - 1.33) \div 2$.

(3) The required minimum effective depth is

$$d = \sqrt{\frac{M}{Rb}} = \sqrt{\frac{1,990,000}{225.6 \times 96}} = 9.6 \text{ in.}$$

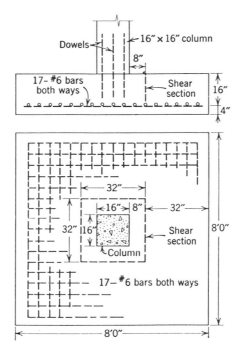

FIGURE 17-2

In this equation $b = 8$ ft or 96 in. A depth of 9.6 in. is probably too small to satisfy requirements for shear, so let us arbitrarily increase the depth to $d = 16$ in. Subsequent computations for shear and bond stresses will determine whether the assumed 16-in. effective depth is sufficiently large.

(4) The next step is to compute v, the actual unit shear stress, to see whether the assumed d is adequate. From Fig. 17-1b,

$$e = \frac{l}{2} - \frac{a}{2} - \frac{d}{2} \quad \text{or} \quad e = \frac{96}{2} - \frac{16}{2} - \frac{16}{2} \quad \text{and} \quad e = 32 \text{ in.}$$

Also, $b = a + d$ or $b = 16 + 16$ and $b = 32$ in. See Fig. 17-2. Then the hatched area shown in Fig. 17-1b is

$$\frac{b + l}{2} \times e \quad \text{or} \quad \frac{32 + 96}{2} \times 32 = 2048 \text{ sq in.}$$

and

$$\frac{2048}{144} = 14.2 \text{ sq ft}$$

Since w, the upward pressure on the footing, is 3750 psf, $V = 14.2 \times 3750$ and $V = 53,400$ lb. Hence

$$v = \frac{V}{b_0 d} \quad \text{or} \quad v = \frac{53,400}{32 \times 16} \quad \text{and} \quad v = 104 \text{ psi}$$

Because this value is less than the allowable 110 psi, the assumed 16-in. effective depth is acceptable for shear.

(5) Determining the area of the tensile reinforcement,

$$A_s = \frac{M}{f_s j d} \quad \text{or} \quad A_s = \frac{1,990,000}{20,000 \times 0.8723 \times 16} = 7.12 \text{ sq in.}$$

Accept 17-#6 bars; $17 \times 0.44 = 7.48$ sq in.

(6) Investigating the bond stress,

$$V = \frac{l(l - a)}{2} \times w$$

or

$$V = \frac{8 \times (8 - 1.33)}{2} \times 3750 = 100,000 \text{ lb}$$

Reference to Table 13-3 shows that the perimeter of a #6 bar is 2.356 in. Then

$$u = \frac{V}{\Sigma_0 j d} \quad \text{or} \quad u = \frac{100,000}{(17 \times 2.356) \times 0.8723 \times 16} = 178 \text{ psi}$$

In this equation 17 is the number of #6 bars. Since 178 psi is less than the allowable 351 psi (Table 13-5), the 17-#6 bars are adopted.

(7) From these computations we accept a square footing 8×8 ft in plan with an effective depth of 16 in. By adding 4 in. for protection, the total thickness of the footing becomes $16 + 4 = 20$ in. The tensile reinforcement consists of two bands of 17-#6 bars each, running at right angles to each other. This information is recorded in Fig. 17-2.

At 150 lb per cu ft the weight of the footing block is

$$8 \times 8 \times 1.67 \times 150 = 16,000 \text{ lb}$$

Therefore the estimated weight of 16,800 lb made in Step (1) of the computations was ample.

Problem 17-3-A. Design a square independent column footing to take a column load of 219 kips. The column is 14 in. square, and the allowable bearing capacity of the foundation bed is 3000 psf. Specification data is the same as that for the foregoing example.

18

Ultimate Strength Design

||||||||||||||||||||||||||||||||||||||||||||||||||||||||||

18-1 Introduction

The straight-line distribution of compressive stress (Fig. 13-1) is valid at working stress levels because the stresses developed under load vary approximately with the distance from the neutral axis, in accordance with elastic theory. However, shrinkage and cracking of the concrete, together with the phenomenon of creep under sustained loading, complicate the stress distribution. Over time, stresses computed in reinforced concrete members on the basis of elastic theory are not realistic. Generally speaking, the serviceability of the working stress design method is maintained by the differentials provided between the allowable compressive stress f_c and the specified compressive strength of the concrete f'_c, and between the allowable tensile stress f_s and the yield strength of the steel reinforcement f_y. These differentials are, in effect, factors of safety (Art. 1-12).

Laboratory investigations over several years have revealed that stresses in both concrete and steel at *ultimate load* can be determined with greater precision than at working or *service load*. This condition has led to the development of ultimate strength design which is becoming the predominant design method for important building structures. As noted earlier (Art. 13-2), the 1971 ACI Code is built primarily around ultimate strength design, referred to in the code as *strength design*. Extended consideration of this method is beyond the scope of this book, but the following brief discussion should serve as an introduction to the considerations on which it is based.

18-2   Loads and Load Factors

The live and dead loads we have dealt with thus far in the book are called *service loads*; taken together, they represent a best estimate of the actual load a structural member may be called upon to support. When the ultimate strength design method[1] is used, the service loads must be increased by a specified *load factor* in order to provide a factor of safety. These load factors are different for live load, dead load, and wind or earthquake loading. Section 9.3 of ACI 318-71 specifies several combinations of load factors to be considered but we will limit our attention to the basic relationship of Code equation (9-1):

$$U = 1.4D + 1.7L$$

in which U represents the *design* load, *design* shear, or *design* moment, as the case may be; D represents the service dead load; and L represents the service live load.

18-3   Capacity Reduction Factors

In addition to the use of load factors, the theoretical capacity of a structural member is reduced by a capacity reduction factor, called the ϕ factor. The ϕ factors provide for variations in materials, workmanship, construction dimensions, etc., and also take into account the structural importance of a member and the adequacy of the theory on which its strength calculations are based. The basic equations for strength in bending, shear, and column action are multiplied by the appropriate ϕ factor to give a magnitude less than the theoretical value. Capacity reduction factors prescribed by the

---

[1] Called strength design method hereafter to conform with ACI 318–71.

1971 Code include the following:

Bending calculations = 0.90
Spiral columns = 0.75
Tied columns = 0.70
Shear = 0.85
Bearing on concrete = 0.70
Bending in plain concrete = 0.65

It will be noted that columns have lower factors than beams.

18-4   Bending Stresses in Rectangular Beams

Figure 18-1a is an abridgment of Fig. 13-1 showing the straight-line compressive stress distribution assumed in working stress design; f_c represents the maximum allowable working value of the extreme fiber stress. Figure 18-1b illustrates an assumed parabolic stress distribution when the value of the extreme fiber stress has reached f'_c, the specified compressive strength of the concrete (i.e., the "ultimate strength"). Figure 18-1c shows the equivalent rectangular concrete stress distribution permitted by the ACI Code for use in the strength design method. The rectangular "stress block" is based on the assumption that a concrete stress of $0.85f'_c$ is uniformly distributed over the compression zone with dimensions equal to the beam width b and the distance a which locates a line parallel to and above the neutral axis. The value of a is determined from the expression $a = \beta_1 \times c$, where β_1 (beta one) is a factor that varies with the compressive strength of the concrete, and c is the distance from the extreme

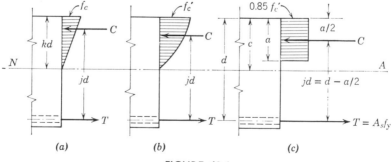

(a)   (b)   (c)

FIGURE 18-1

fiber to the neutral axis. For concrete having f'_c equal to or less than 4000 psi, the code gives $a = 0.85c$.

Under the rectangular stress block assumption, the magnitude of the resultant of the compressive stresses is

$$C = 0.85f'_c \times b \times a$$

and it acts at a distance of $a/2$ from the top of the beam. The arm of the resisting couple jd then becomes $d - (a/2)$ and the theoretical resisting moment as governed by the concrete is

$$M_t = C\left(d - \frac{a}{2}\right) = 0.85f'_c ba \times \left(d - \frac{a}{2}\right)$$

and the theoretical moment strength as governed by the reinforcement is

$$M_t = T\left(d - \frac{a}{2}\right) = A_s f_y\left(d - \frac{a}{2}\right)$$

If *balanced* conditions exist, i.e., the concrete reaches its full compressive strength when the steel reinforcement reaches its yield strength, the two above equations are equal to each other and

$$0.85f'_c ba = A_s f_y \qquad \text{or} \qquad a = \frac{A_s f_y}{0.85f'_c b} = \frac{pf_y d}{0.85f'_c}$$

Although balanced beams are generally uneconomical, the concept is useful in developing strength design procedures.

An alternate form for the theoretical moment strength formula is

$$M_t = bd^2 f'_c \omega(1 - 0.59\omega)$$

where ω (omega) equals pf_y/f'_c, and p (rho) is the reinforcement ratio A_s/bd. To find the usable design load moment, the theoretical moment strength must be multiplied by the appropriate capacity reduction factor, or

$$M_u = \phi M_t$$

To assure a ductile failure by yielding of the steel reinforcement rather than permitting the sudden brittle failure characteristic of concrete under excessive compression, the ACI Code limits the reinforcement ratio in beams to 0.75 of the ratio that would produce balanced conditions. The balancing ratio ρ_b may be found from the formula

$$\rho_b = \frac{0.85 f'_c \beta_1}{f_y} \times \frac{87,000}{87,000 + f_y}$$

The use of this formula, as well as that for M_t stated above, is greatly facilitated by available tables that give values of ρ_b and ω for different values of f'_c and f_y.

18-5   Design of Rectangular Beams for Bending

The general procedure in the design of rectangular beams for flexure, using the strength design method, is illustrated in the following example.

Example.   The service load bending moments on a rectangular beam 10 in. wide are 58 kip-ft for dead load and 38 kip-ft for live load. If $f'_c = 4000$ psi and $f_y = 60,000$ psi, determine the depth of the beam and the tensile reinforcement required.
Solution: (1) The design load moment is determined first

$$U = 1.4D + 1.7L$$
$$M_u = 1.4M_d + 1.7M_t$$
$$= (1.4 \times 58) + (1.7 \times 38) = 146 \text{ kip-ft}$$

(2) Applying the capacity reduction factor, $\phi = 0.90$, the required theoretical moment strength is

$$M_t = \frac{M_u}{\phi} = \frac{146}{0.90} = 162 \text{ kip-ft} \quad \text{or} \quad 1944 \text{ kip-in.}$$

(3) Computing the reinforcement ratio for a balanced condition,

$$\rho_b = \frac{0.85 f'_c \beta_1}{f_y} \times \frac{87,000}{87,000 + f_y}$$
$$= \frac{0.85 \times 4000 \times 0.85}{60,000} \times \frac{87,000}{87,000 + 60,000}$$
$$= 0.0482 \times 0.591 = 0.0285$$

Reducing this ratio in accordance with ACI Code requirements,

$$\rho_{max} = 0.75\rho_b = 0.75 \times 0.0285 = 0.0214$$

(4) To determine the effective depth we use the formula

$$M_t = bd^2 f'_c \omega(1 - 0.59\omega) \quad \text{or} \quad d^2 = \frac{M_t}{bf'_c\omega(1 - 0.59\omega)}$$

For the conditions of this example,

$$\omega = \frac{\rho f_y}{f'_c} = \frac{0.0214 \times 60,000}{4000} = 0.321$$

and

$$\omega(1 - 0.59\omega) = 0.321 \times (1 - 0.1894) = 0.2602$$

Substituting (with f'_c in ksi since M_t is in kip-inches),

$$d^2 = \frac{M_t}{bf'_c\omega(1 - 0.59\omega)} = \frac{1944}{10 \times 4 \times 0.2602} = 186.5$$

and

$$d = \sqrt{186.5} = 13.6 \text{ in.}$$

(5) Calculating the required area of steel,

$$A_s = \rho bd = 0.0214 \times 10 \times 13.6 = 2.91 \text{ sq in.}$$

This area can be supplied by 4-#8 bars ($A_s = 3.16$ sq in.) or 3-#9 bars ($A_s = 3.00$ sq in.). Selecting the latter combination and adding to the effective depth one half bar diameter, an allowance for #4 stirrups, plus 2 in for concrete protection, the overall depth becomes $13.6 + 0.564 + 0.5 + 2.0 = 16.66$ in. Use 17 in.

(6) In actual design, the adopted cross section and reinforcing steel would have to be checked for minimum reinforcement ratio and for deflection controls. In addition, computations for development length of the tensile steel and for web reinforcements (stirrups) would, of course, have to be made.

18-6 Scope of Ultimate Strength Design

The example in the preceding article demonstrated the principal steps in design for flexure only. The strength design method, however, has applications far beyond beam design. It possesses particular significance in the design of columns, and in its application to rigid frames where all members are subjected to combined bending and direct stress. The reader who wishes to study strength design and its applications further is referred to the following publications:

Reinforced Concrete Fundamentals, Third Edition, by Phil M. Ferguson (New York: Wiley, 1973).

Reinforced Concrete Design, Second Edition, by Chu-Kia Wang and Charles G. Salmon (New York: Intext, 1973).

V

ROOF TRUSSES

|||||||||||||||||||||||||||||||||||||||||||||||||||||

Stresses in Trusses

||||||||||||||||||||||||||||||||||||||||||||||||||||||

19-1 Introduction

A truss is a framed structure composed of straight members so arranged and connected at their ends that the stresses are either axial compression or tension. Basically a truss consists of a number of triangles framed together

A *bay* is that portion of a roof structure bounded by two adjacent trusses; the spacing of the trusses on centers is the width of the bay. A *purlin* is a beam spanning from truss to truss that brings to the trusses the loads due to snow, wind, and weight of the roof construction. The portion of a truss that occurs between two adjacent joints of the upper chord is called a *panel*. The load brought to an upper-chord joint or *panel point* is, therefore, the roof design load in pounds per square foot multiplied by the panel length times the bay width; this is called a *panel load*. Figure 19-1a gives the names of other elements of a typical roof truss.

The height or *rise* of a truss divided by the span is called the *pitch*; the rise divided by half the span is the *slope*. Unfortunately these two terms are often used interchangeably. A less ambiguous way of expressing the slope is to give the amount of rise per foot of span. A

377

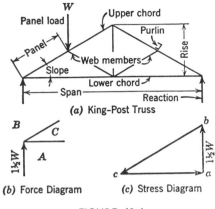

(a) King-Post Truss

(b) Force Diagram   (c) Stress Diagram

FIGURE 19-1

roof that rises 6 in. in a horizontal distance of 12 in. has a slope of "6 in 12." Reference to Table 19-1 should clarify this terminology.

19-2   Force Polygon

In designing a roof truss, the designer must first determine the magnitude and character of the stress in each member. By character is meant the kind of stress, tension or compression. This may be accomplished by graphical methods.

With the exception of wind loads, trusses are generally symmetrically loaded. In the truss shown in Fig. 19-1a, for example, there would be three equal vertical panel loads, W lb each; and, because the truss is symmetrical, each upward force or reaction at the supports would equal $\frac{1}{2} \times 3W = 1\frac{1}{2}W$. The lower left-hand end, or heel joint, of the truss is represented diagrammatically in Fig. 19-1b. The forces, read clockwise about the joint, are three in number: AB, the reaction an upward force of $1\frac{1}{2}W$ lb; BC, the upper chord; and CA,

TABLE 19-1.   Roof Pitches and Slopes

Pitch	$\frac{1}{8}$	$\frac{1}{6}$	$\frac{1}{5}$	$\frac{1}{4}$	1/3.46	$\frac{1}{3}$	$\frac{1}{2}$
Degrees . .	14° 3′	18° 26′	21° 48′	26° 34′	30° 0′	33° 40′	45° 0′
Slope	3 in 12	4 in 12	4.8 in 12	6 in 12	6.92 in 12	8 in 12	12 in 12

the lower chord, both of which are of unknown magnitude. These forces are concurrent, and because, by data, they are in equilibrium a stress diagram corresponding to the forces must close. Therefore draw an upward force ab to some convenient scale representing the reaction $1\frac{1}{2}W$ (Fig. 19-1c). From point b draw a line parallel to BC, and from point a draw a line parallel to CA. The intersection of these two lines determines point c. The magnitudes of the stresses in the members BC and CA are found by scaling their lengths in the stress diagram (Fig. 19-1c), at the same scale at which ab was drawn.

To determine the character of the stresses at this joint, we refer to Figs. 19-1b and c. Consider first the member BC about the point ABC. Because the forces were read in a clockwise direction, note the sequence of the letters: B first, then C. In the stress diagram we find that bc reads downward to the left. In the force diagram, if BC reads downward to the left, this is *toward* the reference point ABC; hence the member BC is in compression. Next consider the member CA about the point ABC. In the stress diagram we find that ca reads from left to right. If we consider the member CA in the force diagram as reading from left to right, we read *away* from the reference point ABC; therefore the member CA is in tension.

The length of a member in a truss bears no direct relation to the magnitude of its stress. The magnitude of the stress is determined by the length of the line in the stress diagram corresponding to the truss member.

The stress diagram (Fig. 19-1c) is the force polygon for the three concurrent forces at the heel of the truss. The stress diagram for the entire truss would consist of combined force polygons of the forces at all the different joints of the truss. Figure 19-2a is a four-panel Fink truss; the complete stress diagram for vertical loads is shown in Fig. 19-2b.

19-3 Stress Diagrams

The panel load for the vertical loading on the truss shown in Fig. 19-2a is 4 kips. The two end loads are 2 kips each, making a total vertical load of $4 + 4 + 4 + 2 + 2 = 16$ kips. Because the truss is symmetrical, each upward force at the support (reaction) is $16 \div 2 = 8$ kips.

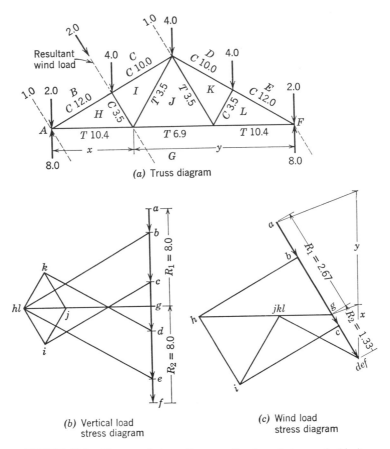

(a) Truss diagram

(b) Vertical load
stress diagram

(c) Wind load
stress diagram

FIGURE 19-2.   Truss and stress diagrams (loads and stresses in kips).

The panel loads and reactions being known, *the first step in con-structing a stress diagram is to draw a force polygon of the external forces.* These forces are *AB*, *BC*, *CD*, *DE*, *EF*, *FG*, and *GA*, and the magnitudes of all are known. Therefore, at a suitable scale, draw *ab* (Fig. 19-2*b*), a downward force equal to 2 kips. The next external force is *BC*, and from point *b*, just determined, draw *bc* downward, equivalent to 4 kips. Continue with *CD*, *DE*, and *EF*. This completes the downward forces. The line just drawn is called the load line, the length being equivalent to 16 kips. The next external force is *FG*, an upward force of 8 kips. This determines the position of point *g*; and *GA*, the remaining external force, completes the polygon of the external forces. Because the loads and reactions are vertical, the force polygon of the external forces is a vertical line.

In conjunction with the polygon just drawn, draw a force polygon for the forces *AB*, *BH*, *HG*, and *GA* about the point *ABHG*. From *b*, draw a line parallel to *BH*; and from *g*, draw a line parallel to *HG*. Their intersection determines point *h*. Next consider the members about the joint *BCIH*. From *c*, draw a line parallel to *CI*; and through *h*, draw a line parallel to *IH*, their intersection determining point *i*. The next joint is *HIJG*. Through *i*, draw a line parallel to *IJ*; and through *g*, draw a line parallel to *JG*, thus establishing point *j*. In the same manner, take the joints *CDKJI* and *DELK*. This completes the stress diagram. See Fig. 19-2*b*. The magnitudes of the stresses in the members are found by scaling the lengths of the lines in the stress diagram just completed. The character of the stresses is found as explained in Art. 19-2. When recording the character of the stresses, a minus sign ($-$) usually denotes compression, and a plus sign ($+$) denotes tension. However, in some books, these designations are reversed. In order to avoid confusion, the system employed in this text uses the symbol (*C*) for compressive stress and (*T*) for tensile stress. The important point to remember is that a member in compression tends to be made shorter and resists this shortening by *pushing against* the joints at its ends. A tension member, on the other hand, tends to become longer and resists the lengthening by *pulling away* from its end joints. For the vertical loads the character and magnitude of the stresses are shown on the truss diagram of Fig. 19-2*a*.

19-4 Wind Load Stress Diagram

The stresses in members of a roof truss due to wind loads are found by constructing a stress diagram as previously described. It is assumed that the wind exerts a pressure perpendicular to the roof surface. In the truss diagram (Fig. 19-2*a*), the wind, indicated by dotted lines, is shown coming from the left, the total load being $1 + 2 + 1 = 4$ kips. To draw a stress diagram for the wind loads, proceed as before. Construct the force polygon of the external forces, namely

AB, *BC*, *C–DEF*, *DEF–G*, and *GA*, the latter two being the wind-load reactions. It will be noted that, because the wind comes from the left, there are no forces *DE* and *EF*; consequently, the letters *D*, *E*, and *F* represent a single point in the stress diagram. Draw *ab*, *bc*, and *c–def*. See Fig 19-2c. It may be assumed that the reactions due to wind loads are parallel to the direction of the wind; and because the wind comes from the left, the left-hand reaction will have a greater magnitude than that on the right for this particular truss. For the purpose of finding the magnitudes of the reactions, we may consider that all the wind loads are concentrated in a single line of action at *BC*. The resultant wind load continued divides the lower chord into two parts, x and y, and the magnitudes of the reactions are to each other as the division x is to y. The line *a–def*, representing the total wind load, is therefore divided in the same proportion as the two divisions of the lower chord, x and y. To accomplish this, erect a line from the point *def* in a length equivalent to the length of the lower chord and divide it into sections x and y. From the upper extremity of the line just drawn, draw a line to point *a* and a parallel line from the point separating x and y to the load line. This determines point *g* and, consequently, the reactions R_1 and R_2. The force polygon of the external forces is now completed, and the stress diagram is then drawn as previously described. It will be found that the letters *j*, *k*, and *l* fall at one point, indicating that there are no stresses in *JK* and *KL* when the wind comes from the left. For design purposes, however, the stresses in *JK* and *KL* are taken the same as for *JI* and *IH*, respectively, since the direction of the wind may be reversed.

19-5   Roof Loads

The dead load is always present on the roof trusses. It consists of the roof covering, such as slate and felt, and the sheathing, rafters, purlins, and weight of the truss. Tables 5-8 and 19-2 give approximate weights of some of the materials commonly used. The actual weight of a steel truss cannot be determined accurately in advance, but it may be estimated by means of Table 19-3. The approximate weight of the truss having been computed, it is distributed among the panel loads on the upper chord. After the truss has been designed, its actual weight may be computed and compared with the

TABLE 19-2. Approximate Weights of Roofing Material

Roofing material	Approximate weight per square foot (lb)
Corrugated galvanized iron	
20 gage	2.0
18 gage	3.0
Nailing concrete per inch of thickness	8.0
Sheathing	
White Pine, Spruce, and Hemlock, 1″ thick	3.0
Southern Yellow Pine, 1″ thick	3.5
Shingles	
Wood	3.0
Asbestos	5.0
Slag roof, 4-ply felt	5.0
Slate	
$\frac{3}{16}″$ thick	7.0
$\frac{3}{8}″$ thick	12–14
Tile, clay	
Plain	14.0
Spanish	10.0
Tin plate, IX thickness	0.75

TABLE 19-3. Approximate Weight of Steel Trusses in Pounds per Square Foot of Roof Surface

Span	Slope of roof			
Feet	45°	30°	25°	Flat
Up to 40	5	6	7	8
40–50	6	7	7	8
50–60	7	8	9	10
60–70	7	8	9	10
70–80	8	9	10	11

TABLE 19-4. Snow Loads for Roof Trusses in Pounds per Square Foot of Roof Surface

	Slope of roof				
Locality	45°	30°	25°	20°	Flat
Northwestern and New England states	15	20	30	35	40
Western and Central states	10	15	25	30	35
Pacific and Southern states	0	5	10	10	10

estimated weight. If the two are not in reasonable agreement, another weight should be assumed and the truss redesigned. Except for unusual cases, this step is seldom necessary.

Because the snow load is vertical, it is generally added to the dead load for the purpose of drawing the vertical load stress diagram. The magnitude of the snow load depends on the pitch of the roof and the locality. Table 19-4 lists snow loads for various pitches and locations and may be used when no specific loads are given by local building codes. For accuracy, a stress diagram may be drawn for snow loads alone, the stresses in the truss members being directly proportional to those for dead loads.

TABLE 19-5. Wind Pressure on Roof Surfaces

Slope of roof	Normal pressure
Degrees	Pounds per square foot
15	14.0
20	18.0
25	22.0
30	26.0
35	30.0
40	33.0
45	36.0
50	38.0

The wind pressure, when coming at right angles to the side of a building, probably never exceeds 40 psf. Since it is customary to consider that the wind acts in a direction perpendicular to the roof surface, the magnitude of the wind pressure decreases as the pitch becomes smaller. Table 19-5 lists wind pressures for roofs of various pitches based on a wind pressure of 40 psf on a vertical surface.

Obviously a condition of dead load, maximum snow, and maximum wind load cannot occur simultaneously. The conditions causing maximum stresses most likely to occur are the following combinations:

1. Dead load and maximum snow
2. Dead load, maximum wind, and minimum snow
3. Dead load, minimum wind, and maximum snow

For accuracy we may draw separate stress diagrams for dead, snow, and wind loads, tabulating the stresses and using for the design load for the various truss members the greatest stress to be found for any of the foregoing combinations. It is customary to consider one half the maximum snow and wind loads as the minimum values.

19-6 Equivalent Vertical Loading

A convenient and practical method of computing stresses in truss members is to use an equivalent vertical load for the combined snow and wind loads. For the usual simple roof trusses of moderate span supported on masonry walls, the maximum stresses in the members determined from the combinations stated in Art. 19-5 are substantially the same as for a uniform vertical load acting over the entire roof surface. The method of equivalent vertical loading requires only one stress diagram to be drawn and will answer for any probable combination of dead, wind, and snow loads when the proper equivalent vertical live load is chosen. Many building codes specify a minimum vertical live load that must be used for buildings within their jurisdiction; but in the absence of mandatory provisions, selection of the proper equivalent load depends on the judgment and experience of the designer. Table 19-6 may be used as a guide in this connection, but local knowledge of snowfall and wind velocity may make it prudent to increase these values in individual cases.

TABLE 19-6. Equivalent Vertical Loads for Combined Snow and Wind Loads in Pounds per Square Foot of Roof Surface

Locality	Slope of roof				
	45°	30°	25°	20°	Flat
Northwestern and New England states	28	25	24	35	40
Western and Central states	28	25	24	30	35
Southern and Pacific states	28	25	24	22	20

20

Design of a Steel Truss

||||||||||||||||||||||||||||||||||||||||||||||||||||||||||

20-1 General

A steel Fink truss is used in this chapter to demonstrate the design of roof trusses. The procedure consists of four principal steps: determination of loads and stresses, design of compression members, design of tension members, and design of joints. Each of these operations is explained in subsequent articles related to the data given below.

   Data: Fink trusses with eight panels (Fig. 20-1*a*) will be used on a span of 60 ft and at a spacing of 18 ft on centers. The roof construction consists of ⅜-in. slate laid on a 4-in. slab of nailing concrete which spans between the purlins. The building is located in one of the Central states, and the upper chord has a slope of 30°. Truss members will be fabricated from A36 steel, and allowable unit stresses are controlled by the AISC Specification.

20-2 Loads and Stresses

The design will be carried out using the equivalent vertical loading for combined snow and wind loads (Art. 19-6). Table 19-6 shows a

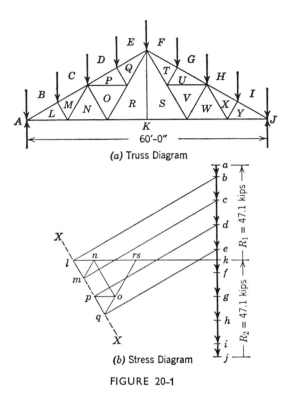

(a) Truss Diagram

(b) Stress Diagram

FIGURE 20-1

value of 25 psf for our location and slope of roof. The weights of the roofing and concrete deck are found by use of Table 19-2. We will first establish the panel loads and then determine the magnitude and character of the stress in each member by the graphical method explained in Art. 19-3.

(1) The design load for the roof is

$$\begin{aligned}
\tfrac{3}{8}\text{-in. slate roofing} &= 12 \\
4\text{-in. nailing concrete} &= 32 \\
\text{snow and wind equivalent} &= \underline{25} \\
\text{total} &= 69 \text{ psf}
\end{aligned}$$

The load brought to each panel point is equal to 69 multiplied by the panel length times the bay width, as explained in Art. 19-1. The panel length may be measured from the truss diagram (Fig. 20-1a) drawn to scale, and in this case is 8.6 ft. The panel area is therefore 8.6 × 18 = 155 sq ft and the panel load, not including the weight of the truss, is 69 × 155 = 10,700 lb or 10.7 kips.

Since the approximate weight of the truss (Table 19-3) is 8 lb per sq ft of roof surface, and we may take seven panel areas as being directly supported by one truss, the estimated truss weight is 8 × 7 × 155 = 8680 lb. If we assume this weight allowance to be distributed equally among the top chord panel points, the portion going to each is 8680 ÷ 8 = 1080 lb or 1.08 kips. The vertical load at each panel point is then 10.7 + 1.08 = 11.78 kips.

The total load on the truss is 8 × 11.78 = 94.2 kips and each reaction is 94.2 ÷ 2 = 47.1 kips.

(2) Because the truss is symmetrically loaded, it is necessary to draw only half of the stress diagram. First the force polygon of the external forces is drawn, as explained in Art. 19-3. Here, again, since the loads and reactions are vertical, the force polygon of external forces is a vertical line. See Fig. 20-1*b*. Then, beginning with the joint *ABLK* and following with *BCML* and *LMNK*, the points *l*, *m*, and *n* are readily found in the stress diagram. The joint *CDPONM*, however, presents difficulties, for there are two letters to be found; this is also true with respect to joint *NORK*.

The eight-panel Fink truss of this type requires a special solution, one of the simplest methods being the following. It is known that each successive member of the upper chord, beginning with the one nearest the support, has the same rate of decrease in stress. In this instance the stresses in *BL* and *CM* have been determined, and by referring to the stress diagram the decrease in stress is observed. The stresses in *DP* and *EQ* will be reduced in the same ratio, and their magnitudes are found by drawing the dotted line *X–X* through points *l* and *m* in the stress diagram. The intersection of a line drawn through *d*, parallel to *DP*, and the line *X–X* determines the point *p;* and a line through *e* parallel to *EQ* intersects *X–X* at *q*. Determining the points *n*, *o*, and *r* now presents no difficulties. This completes one half the stress diagram and is sufficient for our purpose, for the truss is symmetrically loaded and the stresses in the members in each half of the truss are alike. It should be noted that no stress exists in the member *RS*, but a minimum size steel angle is used to prevent the possible deflection of this portion of the lower chord.

Table 20-1 is a tabulation of the stresses found in the Fink truss. The first column designates the member, the second the character

TABLE 20-1. Stresses in Fink Truss

Truss member	Character of stress	Magnitude of stress (kips)	Section
BL	C	82.0	2 L 4 × 3½ × 7⁄16
CM	C	76.4	2 L 4 × 3½ × 7⁄16
DP	C	70.7	2 L 4 × 3½ × 7⁄16
EQ	C	64.8	2 L 4 × 3½ × 7⁄16
KL	T	71.3	2 L 3½ × 2½ × 3⁄8
KN	T	61.2	2 L 3½ × 2½ × 3⁄8
KR	T	40.7	2 L 2½ × 2 × 5⁄16
LM	C	10.2	2 L 2½ × 2 × 5⁄16
PQ	C	10.2	2 L 2½ × 2 × 5⁄16
NO	C	20.4	2 L 3½ × 2½ × 5⁄16
MN	T	10.2	2 L 2½ × 2 × 5⁄16
OP	T	10.2	2 L 2½ × 2 × 5⁄16
OR	T	20.4	2 L 2½ × 2 × 5⁄16
QR	T	30.6	2 L 2½ × 2 × 5⁄16
RS	0	0	1 L 2½ × 2 × 5⁄16

of the stress, and the third the magnitude of the stress in the various members; the fourth column will be considered later. These stresses are found by measuring the length of the lines in the stress diagram at the same scale used in laying off the force polygon of external forces.

20-3 Compression Members

Steel trusses with riveted joints, and some with welded joints, have members made up of two angles separated at the connections by $\frac{3}{8}$-in. or $\frac{1}{2}$-in. gusset plates. Generally, the most efficient section consists of two angles having unequal legs, the long legs placed back to back. The compression members are struts and are designed as columns of given length and axial loads. The designer must use whatever column formula is required by the building code, as described in Art. 6-5. Table 6-3 gives allowable axial loads for sections formed of two angles with a $\frac{3}{8}$-in. space between backs. This table was computed in accordance with the column formulas given in Art. 6-5, and the proper section may be selected by referring to the table without computations. It should be remembered that the limiting value of l/r for main compression members is 200 and sometimes the size of a column is determined by this ratio rather than by the magnitude of the load. In Table 6-3, which gives allowable loads for double-angle struts, the loads below the heavy horizontal lines are for columns whose slenderness ratio is greater than 120. For instance, with respect to the X–X axis, a strut made up of two $4 \times 3 \times \frac{5}{16}$ in. angles, long legs back to back and $\frac{3}{8}$ in. apart, has a least radius of gyration of 1.27 in. and the slenderness ratio of the strut, if it is 14 ft long, is $l/r = (14 \times 12)/1.27 = 132$. If the length is 12 ft, $l/r = (12 \times 12)/1.27 = 113$. Refer to Table 6-3 and note that the heavy horizontal line occurs between these two lengths.

If the length of the upper chord is not too great, it is economical to use the same cross section for the entire length. The member BL has a compressive stress of 82 kips. The remaining members of the upper chord have smaller stresses so the stress in BL will determine the size of the entire upper chord. It is suggested that Art. 6-8 relating to double-angle struts be reviewed before proceeding with the example presented below.

Example. Determine the size of the compression members in the truss under consideration, using $\frac{3}{8}$-in. gusset plates and A36 steel. *Solution:* (1) From Table 20-1 we find that the stress in member BL is 82 kips. Using Table 6-3 as explained in Art. 6-8, we note that 84 kips is the allowable load for a strut 8 ft long composed of 2 angles $4 \times 3\frac{1}{2} \times \frac{3}{8}$ with the long legs $\frac{3}{8}$ in. back to back. However, BL has a length of 8.6 ft and consequently the allowable load on this section will be less than 84 kips.

Investigating the section by the method shown in Art. 6-8, or interpolating directly in Table 6-3, reveals that the allowable load on the 8.6-ft length is approximately 80 kips which is less than the 82 kips required. Therefore, select 2 angles $4 \times 3\frac{1}{2} \times \frac{7}{16}$; this section will support 97 kips on 8 ft so it will be amply strong for 82 kips on the 8.6-ft length. It will be used for all of the top chord.

(2) The member NO has a design load of 20.4 kips and its length is 10 ft. We find in Table 6-3 that a member composed of 2 angles $3\frac{1}{2} \times 2\frac{1}{2} \times \frac{5}{16}$ is acceptable. Its allowable load is 42 kips; slenderness ratio is the governing factor for this member.

(3) The design loads for the compression members LM and PQ are each 10.2 kips; their effective lengths are 5 ft. The minimum section commonly used in trusses is composed of 2 angles $2\frac{1}{2} \times 2 \times \frac{5}{16}$. Reference to Table 6-3 shows that a section composed of these two angles is acceptable for both members. All of the selected sections are recorded in the fourth column of the Table 20–1; corresponding members in the right half of the truss are, of course, similar.

20-4 Tension Members

If a member in tension fails, the rupture will occur at the section where the area is smallest. This reduced cross section will occur at a rivet hole, as explained in Art. 7-8. In trusses of this type $\frac{3}{4}$-in. and $\frac{7}{8}$-in. rivets are generally used. In computing the net, or effective area, the rivet hole diameter is considered to be $\frac{1}{8}$ in. greater than the diameter of the rivet. It should be noted that the length of a member in tension is not a factor in determining its cross section. Table 7-5 gives the effective net areas for *single* angles with one rivet hole deducted. Because two angles are generally used for truss members, the proper section is readily found by dividing one half the design load by the allowable tensile stress and selecting the required section directly from the table.

Example.   Design the tension members of the truss under consideration, using $\frac{3}{4}$-in. rivets and A36 steel for which the allowable unit tensile stress is 22 ksi (Table 5-2).
Solution: (1) First, let us consider the member KL, in which there is a tensile stress of 71.3 kips. The member is composed of two angles and the proper section is determined by the use of Table 7-5. Because the load is 71.3 kips, the load to be resisted by each angle is 71.3 ÷ 2 = 35.7 kips. The net area required for each angle is 35.7 ÷ 22 = 1.62 sq in. Referring to Table 7-5, we see that one angle $3\frac{1}{2} \times 2\frac{1}{2} \times \frac{3}{8}$, punched for a $\frac{3}{4}$-in. rivet, has a net area of 1.78 sq in.; hence we accept for KL a section consisting of 2 angles $3\frac{1}{2} \times 2\frac{1}{2} \times \frac{3}{8}$.

(2) The member *KN* resists a tensile load of only 61.2 kips so the same section will be used as for *KL*, making the bottom chord continuous from the heel joint (*ABLK*) to joint *NORK*.

(3) The design load on member *KR* is 40.7 kips. Each angle will resist half of this or 20.4 kips. Therefore the required area of each angle is 20.4 ÷ 22 = 0.928 sq in. Table 7-5 shows that one angle $2\frac{1}{2} \times 2 \times \frac{5}{16}$, punched for a $\frac{3}{4}$-in. rivet, affords a net area of 1.04 sq in.; therefore 2 angles $2\frac{1}{2} \times 2 \times \frac{5}{16}$ are accepted for member *KR*.

(4) The remaining members in tension have relatively small stresses so the minimum section of 2 angles $2\frac{1}{2} \times 2 \times \frac{5}{16}$ will be used for members *MN*, *OP*, *OR*, and *QR*. A single angle $2\frac{1}{2} \times 2 \times \frac{5}{16}$ will be used for member *RS* even though the stress diagram indicates zero stress; its purpose is to counteract the tendency of the lower chord to sag.

Problem 20-4-A. An eight-panel Fink truss has a span of 63 ft 0 in. The total vertical load on the truss is 104 kips; thus the panel loads are 13 kips each, except those over the supports which are 6.5 kips each. Using the letter notation shown in Fig. 20-1a, tabulate the character and magnitude of the stresses in the various members and also, assuming the steel to be A36, the structural steel sections to be used. The allowable tensile stress is 22 ksi and connections will be made with $\frac{3}{4}$-in. rivets.

20-5 Riveted Joints

The design of riveted connections is discussed in Arts. 7-4 and 7-5, and the number of rivets to be used at various joints of the truss is determined by the methods described. In general, the required number of rivets at the end of a member is the design load on the member divided by the controlling value of one rivet. To determine the number of rivets, we must know the design load, thickness of the gusset plate and angles, size of the rivets, and allowable unit stresses.

Regardless of the magnitude of the stress, it is customary to use at least two rivets at the ends of each member. The maximum spacing on centers, called the pitch of rivets, is usually 6 in., and occasionally this requirement, in conjunction with the length of the gusset plate, may determine the number of rivets. The minimum pitch of rivets is

3 diameters of a rivet. It is essential that the two angles which make up a truss member act as a unit, and, to accomplish this, *stitch rivets* are placed in the member between the joints. For compression members they should not be placed more than 2 ft on centers and for tension members not more than 3 ft 6 in.

When a member is continuous at a joint, as, for example, *BL* and *CM* at the joint *BCML* (Fig. 20-1), the rivets are not required to transmit the entire chord stress on either side of the joint. The load transmitted by the rivets is the difference in magnitude of the two loads, in this instance $82.0 - 76.4 = 5.6$ kips, and this is the load that determines the number of rivets required.

In theory, the lines of action of truss members meet in a point at the joints, and these lines should coincide with the center of gravity axes of the members. To avoid eccentricity the lines of rivets should be placed on these lines, but because of the practical difficulty in driving rivets close to the outstanding legs the rivets are placed on the gage lines. Standard gage dimensions are given in Table 7-1. In practice, the lines of action are made to coincide with the gage lines of the sections.

Example 1.   Compute the number of rivets in the members *BL* and *LK* about the joint *ABLK* (Fig. 20-1a). Gusset plates are $\frac{3}{8}$ in. thick and rivets are $\frac{3}{4}$ in. in diameter. The allowable unit stresses are to be in accordance with the AISC Specification: shear $= 15.0$ ksi and bearing $= 48.6$ ksi.

Solution: (1) The stress in member *BL* is 82.0 kips and the angles are $\frac{7}{16}$ in. thick (Table 20-1). Referring to Table 7-3, we find the following allowable working values for $\frac{3}{4}$-in. rivets:

$$\text{double shear} = 13.25 \text{ kips}$$

$$\tfrac{3}{8}\text{-in. bearing} = 13.7 \text{ kips}$$

Therefore, the controlling value of one rivet is 13.25 kips and the number required is $82.0 \div 13.25 = 6+$. Use seven rivets.

(2) Similarly, 13.25 kips is the controlling value of one rivet for the member *LK*, and the number required is $71.3 \div 13.25 = 5+$. Use six rivets.

Example 2. Compute the number of rivets required in the members about the joint *BCML*.

Solution: (1) As noted earlier, the load in the upper chord angles to be transmitted across this joint is 82.0 − 76.4 = 5.6 kips. Since the controlling value of one rivet is 13.25 kips, the number required (56.0 ÷ 13.25) is less than one. For practical reasons two rivets are used, the rivets being placed on the gage line.

(2) The stress in member *ML* is 10.2 kips and the angles are $\frac{5}{16}$ in. thick. Again, the controlling value of one rivet is 13.25 kips and the number required (10.2 ÷ 13.25) is less than one. Use two rivets.

The number of rivets at the other joints is determined in a similar manner and recorded on a design sketch such as the one shown in Fig. 20-2. No fewer than two rivets are used at the end of a member. For symmetry, or because of the size of the gusset plate, the number of rivets is frequently greater than that demanded by the stress. Not shown in Fig. 20-2 are the intermittent rivets and fillers (frequently called stitch rivets) required in double-angle compression members.

20-6 Welded Joints

Welds may be substituted for rivets in gusset-type connections. When this is done, the design procedure is similar to that employed in Example 2 of Art. 8-5 except that two angles are involved, each carrying half the total stress in the member. Figure 8-6b would then show angles welded to both sides of the plate. However, in order to realize the full effectiveness of welding, the chord members may be made of structural tees (produced by splitting the web of a wide flange shape or I-beam), and the double-angle members welded directly to the stem of the tee section. Figure 8-9k indicates the general arrangement of this type of construction and Fig. 20-3 shows a detail of a typical joint. (See the listing in Art. 5-3 for the method of designating structural tees.)

20-7 Shop Drawings

The drawings used to fabricate structural steel work are known as *shop drawings*. These drawings give complete details, spacing of rivets or length and arrangement of welds, dimensions of members and

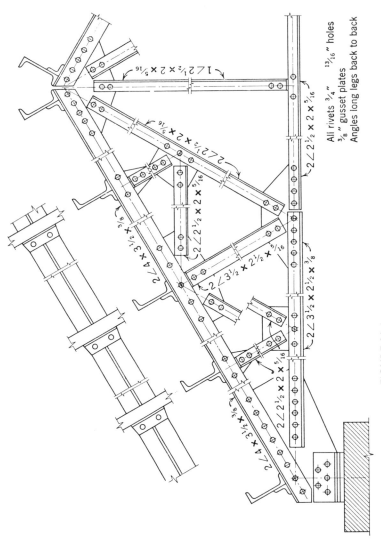

FIGURE 20-2. Design sketch for riveted truss.

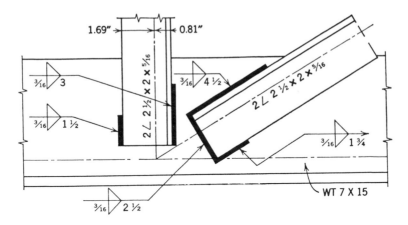

FIGURE 20-3. Welded truss joint.

gusset plates, and any other information required to fabricate the truss or other structural element. Shop drawings are made by the fabricating company from general information supplied on a design drawing or sketch (Fig. 20-2) by the designer. They are submitted to the architect or engineer for approval before the manufacturing process is begun.

Answers to Selected Problems

||||||||||||||||||||||||||||||||||||||||||||||||||||||

The answers given below are for those problems marked with an asterisk (*) in the text. In general they are carried to three significant figures, except in certain cases where extension of the numerical answer seemed desirable as an aid in interpreting the result.

Chapter 1

1-7-A 3.33 sq in.

1-7-C $\frac{7}{8}$-in. diameter, required area = 0.6013 sq in.

1-7-F 196 kips

1-15-A 19,350 lb

1-15-C 29,549 ksi (say 29,500 ksi or 29,500,000 psi)

Chapter 2

2-6-B $R_1 = 3$ kips $R_2 = 3$ kips

2-6-E $R_1 = 14.4$ kips $R_2 = 9.6$ kips

2-6-G $R_1 = 10,100$ lb $R_2 = 11,900$ lb

2-7-B $R_1 = 4.4$ kips $R_2 = 5.6$ kips

2-7-D $R_1 = 4430$ lb $R_2 = 7570$ lb

2-7-H $R_1 = 5625$ lb $R_2 = 4375$ lb

2-8-C $R_1 = 6090$ lb $R_2 = 2610$ lb
2-8-D $R_1 = 2.46$ kips $R_2 = 3.14$ kips
2-8-I $R_1 = 6750$ lb $R_2 = 5250$ lb

Chapter 3

3-3-B	Max. $V = 1830$ lb	$V = 0$ at $x = 9$ ft
3-3-C	Max. $V = 9050$ lb	$V = 0$ at $x = 11.2$ ft
3-3-G	Max. $V = 4333$ lb	$V = 0$ at $x = 9$ ft
3-6-A	Max. $V = 5250$ lb	Max. $M = 18,380$ ft-lb
3-6-B	Max. $V = 1114$ lb	Max. $M = 4285$ ft-lb
3-6-D	Max. $V = 13.7$ kips	Max. $M = 65.4$ kip-ft
3-6-J	Max. $V = 6700$ lb	Max. $M = 28,300$ ft-lb
3-7-A	Max. $M = 44$ kip-ft	
3-7-C	Max. $M = 22,500$ ft-lb	
3-7-G	Max. $M = 15$ kip-ft	
3-8-B	Max. $V = 1500$ lb	Max. $M = 9500$ ft-lb
3-8-D	Max. $V = 2700$ lb	Max. $M = 12,750$ ft-lb
3-12-A	Max. $V = 3P/2$	Max. $M = PL/2$
3-12-B	Max. $V = 2P$	Max. $M = 3PL/5 = PL/1.67$

Chapter 4
4-4-A   4.4 in.
4-4-D   3.58 in.
4-5-B   205.3 in.^{4}
4-5-F   682.3 in.^{4}
4-8-A   399.8 in.^{4}
4-8-B   295.5 in.^{4}
4-9-C   5.31 in.

Chapter 5
5-10-B   W 10 × 25
5-10-D   W 21 × 55
5-10-G   W 10 × 21
5-10-J   W 14 × 30
5-11-A $f_v = 3.36$ ksi < 14.5 ksi
5-11-F $f_v = 2.56$ ksi < 14.5 ksi
5-14-A   W 14 × 30
5-14-C   W 12 × 27, $D = 0.23$ in.
5-14-E   W 14 × 30
5-15-C   W 10 × 21
5-15-E   W 12 × 27

5-17-A W 10 × 21
5-17-E W 16 × 36
5-19-A W 8 × 20
5-20-D W 14 × 30
5-20-F 15.3 kips
5-23-A 12 × 15 × 1⅛
5-24-A No, allowable load = 81.4 kips

Chapter 6
6-6-A 244 kips
6-6-D 378 kips
6-7-B W 12 × 65 or W 10 × 77 if space is limited
6-8-A 2L 4 × 3½ × ⁵⁄₁₆
6-8-B 159 kips
6-13-A 12 × 14 × 1 in.

Chapter 7
7-5-A 9.02 kips
7-5-F Four rivets
7-8-A 2L 3 × 2 × ⁵⁄₁₆

Chapter 8
8-7-B 100 kips

Chapter 10
10-4-A 4 × 10, required $S = 49.8$ in.^{3}
10-4-B 4 × 16, required $S = 123.4$ in.^{3}
10-5-A $v = 83.1$ psi <85 psi (acceptable)
10-5-B Yes, $v = 75$ psi < 80 psi
10-6-A $D = 0.307$ in., allowable = 0.53 in.
10-6-D $D = 0.4$ in., allowable = 0.53 in.
10-7-B 12 × 16
10-10-A 2 × 12, 16 in. on centers; minimum required $F_b = 890$ psi; minimum required $E = 1,000,000$ psi
10-11-A 156 psf
10-12-B 2 × 8, 16 in. on centers (Eastern Hemlock No. 2, among others)

Chapter 11
11-4-A 139,000 lb
11-5-A 10 × 10 Douglas Fir, dense select structural

11-5-B Yes, it will support 59 kips
11-6-A 5080 lb

Chapter 13
13-10-A Effective depth = 20 in.; total depth = 22.5 in.; A_s required
= 3.11 sq in.; use 4-#8 bars
13-14-A b = 10 in., d = 20.8 in., A_s = 3-#9 bars, #4 U-stirrups, s =
8 in.; length of beam requiring stirrups = 54 in.
13-15-A 19 in.
13-16-A Embedment provided = 24 + 3 = 27 in.; required l_d (Table
13-6) = 17.6 in.; by $(M_t/2V) + l_a$ test, l_d may not exceed
28.2 in.; yes, development length provided is adequate

Chapter 15
15-3-A #3 bars with 8½-in. spacing; #3 bars with 13-in spacing
15-5-A (a) 4-in. thickness is satisfactory; minimum t for deflection =
3.84 in.; d for bending = 2.5 in.; t = 2.5 + 0.25 + 0.75 =
3.5 in. < 4 in. (b) Using d = 4.0 − 1.0 = 3 in., A_s = 0.321
sq in.; use #5 bars with 11-in. spacing; temperature bars =
#3 bars at 13½-in. spacing.

Chapter 16
16-3-A 167,000 lb; ¼-in. ties at 10-in. spacing
16-5-A 279,000 lb; ⅜-in. spiral at 2-in. pitch

Index

Abbreviations, x
Aggregates, concrete, 289
Air-entrained concrete, 294
Allowable loads, for steel beams, 113, 118, 125
for steel columns, 156, 160
for wood columns, 264
for wood plank floors, 252
Allowable stress design, 148, 208
Allowable stresses, for bolts, 104
for bond, 311
for concrete, 292
for connectors, 98
for rivets, 184
for structural lumber, 220
for structural steel, 98
for welds, 196
Allowable unit stress, 13
American Concrete Institute, 14
Building Code Requirements, 290
American Institute of Steel Construction, 14

American Institute of Timber Construction, 261, 274
American Society for Testing and Materials, 13
American Standard channels, 72
American Standard I-beams, 70
Angles, equal legs, 74
net sections, 191
unequal legs, 74
in tension, 189
Answers to selected problems, 398
Arc welding, 195
Areas of bars, 295
Axial load, 166

Bars, areas of, 295
perimeters of, 295
in slabs, 345
Base plates, steel column, 173
Bay, floor, 132
roof, 377
Beam connections, steel, 192
Beam diagrams, typical loadings, 60
Beam formula, 8, 64

Beams, 23
bending stresses in, 8, 61
continuous, 202, 318
with continuous action, 202
deflection of, 107, 231
glued laminated, 261
laterally unsupported, 119
with light loads, 124
overhanging, 47
reinforced concrete, 318
safe loads for, 113, 118, 125, 252
steel, 91
types of, 23
unbraced length of, 120
wood, 219, 233
Bearing plates, steel beams, 138
Bearing pressure on masonry, 139
Bearing-type connections, 182
Bending, 8
design for, 102, 227, 297
theory of, 61
Bending factors, 168
Bending moment, 40, 320
negative, 47
Bending moment diagrams, 41
Bending moment formulas, 57, 60
Bending stresses, 8, 61
in rectangular concrete beams, 370
Bolted connections, 179
Bolts, 187

allowable stresses for, 184
Bond stress, 310
Building Code Requirements for Reinforced Concrete (ACI), 292

Cantilever beams, 23, 53
Capacity reduction factors, 369
Cement, 293
Center of moments, 18
Centroid, 65
Channels, used as beams, 118
Column base plates, 173
Column footing, 361
Column formulas, 153, 263
Column live load reduction, 172
Columns, allowable loads, 156, 160
allowable stresses, 155
bending factors, 168
design of steel, 159
design of wood, 265, 271
eccentrically loaded, steel, 171
effective length of, 152
end conditions, 153
reinforced concrete, 354
spaced wood, 271
spiral concrete, 357
steel, 149
steel pipe, 166
structural tubing, 166, 169
tied concrete, 356
wood, 262
Column sections, steel, 150
Compact section criteria, 101

Compact sections, 100
Compression, 5, 6
Compression flange, un-
    braced length of, 102
Compression members,
    trusses, 390
Concentrated loads, 24
Concentric load, 166
Concrete, 289
  air-entrained, 294
  allowable stresses, 292
  fireproofing, 128
  floor systems, 341
  joists, 349
  modulus of elasticity,
    297
  proportioning mixes, 290
  steel reinforcement, 294
  strength of, 290
  water-cement ratio, 291
  workability, 291
Concrete joists, 349
Concrete joist slab, 350
Connections, beam, 192
  bearing type, 182
  bolted, 179
  free-end, 193
  friction type, 182
  moment-resisting, 193
  riveted, 179
  welded, 194
Continuous beams, 24,
    318
Crippling of beam webs,
    141

Dead load, 132
Decimal equivalents,
    front endpaper
Deflection, steel beams,
    107
  wood beams, 231

Deflection coefficients,
    109
Deflection formulas, 111
Deformation, 9
Development length of re-
    inforcement, 313
  of tension bars, 316
Diagonal tension, 301
Direct stress, 4
Direct stress formula, 15
Distributed loads, 24, 27
Double-angle struts, 162,
    164
Double shear in rivets,
    182
Dresses sizes, lumber, 220

Eccentrically loaded col-
    umns, 171
Edge distance for connec-
    tors, 283
Effective column length,
    152
Elastic limit, 9
Electric arc welding, 195
Enclosed bearing, rivets,
    182
End distance for connec-
    tors, 281
Equilibrium, 19
  laws of, 20
Equivalent tabular load,
    118
Equivalent vertical load-
    ing, 385

Factor of safety, 11
Fasteners, 179
Field welding, 200
Fillet weld, 196
Fink truss, 387
Fireproofing,weightof,129

Flexure, design for, 102
Flexure formula, 8, 62
 application of, 83
Flexure formulas, rein-
     forced concrete beams,
     297
Floor bay, 132
Floor framing, steel, 131
 beam design, 132
 girder design, 135
Floor joists, span ta-
     bles, 239
Floor systems, reinforced
     concrete, 341
Floors, plank, 250
Footings, reinforced con-
     crete, 361
Force, 3
Force polygon, 378
Formula coefficients,
     concrete, 299
Foundations, grillage,
     177
Framing connections,
     steel, 192
Friction-type connections,
     182

Gage lines, 180
Girders, 23
 concrete, 330
 steel, 131, 135
Glued laminated beams,
     261
 columns, 274
Grillage foundations, 177

Hardwoods, 220
High-strength bolts, 187
Hooke's law, 9, 10
Horizontal shear, 34
 wood beams, 229

I-beams, 70, 92
Inflection point, 50, 322
Investigation, 11

Joints, riveted, 181
 welded, 195
Joists, concrete, 349
 open web, 142
 wood, 219
Joist tables, wood, 239

Kip, 4, 98

Laterally unsupported
     beams, 119
Lateral support of beams,
     101, 119
Laws of equilibrium, 20
Length of span, 320
Lever arm, 18
Light loads on beams, 124
Live load, 132
Live load reduction in
     columns, 172
Live loads, minimum, 134
Load factors, 214, 369
Loads, allowable, for
     steel beams, 113, 118,
     125
 for steel columns, 156,
     160
 for wood columns, 264
 for wood plank floors,
     252
 kinds of, 24
 concentrated, 57
 distributed, 27, 58
 eccentric, 171
 equivalent tabular, 118
 equivalent vertical, 385
 reduction in, 172
 roof, 382

typical, 59
Lumber sizes, 76, 220
Lumber, structural, 219
  allowable stresses for,
  220

Materials, weights of,
  133
Mechanics, 3
Miscellaneous shapes, 94
Modulus of elasticity,
  11, 12, 297
Moment, 18
  bending, 40
  resisting, 61
  statical, 65
Moment arm, 18
Moment diagram, 41
Moment of a force, 18
Moment of inertia, 64, 79
  of rectangular sections,
  80
  transferring of, 85
Moments, center of, 18
Movable partitions, 132

National Design Specifi-
  cation, Stress-Grade
  Lumber, 221
National Forest Products
  Association, 14, 227,
  279
Negative bending moment, 47
Net sections, 189
Neutral axis, 62
Neutral surface, 61
Nomenclature, AISC, 95
Noncompact shapes, 100
Notation, reinforced con-
  crete, 296

One-way concrete slabs,

design of, 345
  minimum thickness of, 343
  shrinkage and temperature
  reinforcement, 343
  tension reinforcement,
  344
Open web steel joists,
  142, 144
Overhanging beams, 24, 29,
  47

Panel point, 377
Partitions, movable, 132
Pedestal, on footing, 362
Perimeter of bars, 295
Permanent set, 10
Pipe columns, 166
Pitch, of rivets, 180
  of roofs, 377
Plank floors, 250, 252
Planks, 219
Plastic design, steel,
  148, 208
  theory of, 208
Plastic hinge, 209
Plastic moment, 209
Plastic range, 208
Plastic section modulus,
  211
Plate girders, 147
Plug welds, 204
Point of inflection, 50
Portland cement, 293
Posts and timbers, 219
Properties of sections,
  61, 64
Purlins, 377

Radius of gyration, 64, 87
Rafter roofs, 255
Rafter span tables, 255,
  257

Reactions, 4, 21, 25
Reduction in loads, 172
Reinforced concrete, 287
  bond stress, 310
  design methods, 289
  notation, 296
  stresses in, 289
  ultimate strength design, 290
  working stress design, 289
Reinforced concrete beams, 318
  design bending moments, 320
  design of rectangular, 322
  development length of reinforcement, 313
  diagonal tension in, 301
  flexure design formulas, 297
  shear reinforcement, 301
  span conditions, 318
  T-beams, 332
  web reinforcement, 301
Reinforced concrete columns, 354
  short columns, 355
  spiral columns, 357
  tied columns, 355
Reinforced concrete floor systems, 341
  concrete joists, 349
  one-way joist slab, 350
  one-way solid slabs, 342, 345
  ribbed slabs, 349
Reinforcing bars, 295
  development length of, 316
Relation between shear
  and bending moment, 42
Resisting moment, 61
Restrained beam, 24
Ribbed slabs, 349
Riveted connections, 179
Riveted joints, roof trusses, 393
  structural action in, 181
  types of, 181
Riveting, 179
  edge distance, 180
  gage lines, 180
Rivets, allowable stresses for, 184
  bearing in, 182
  double shear in, 182
  enclosed bearing, 182
  pitch of, 180
  single bearing, 182
  single shear, 182
  working values for, 183
Rolled steel shapes, 92
  designations of, 94
Roof loads, live and dead, 382
Roof trusses, 375
  compression members, 390
  design of, 387
  equivalent vertical loading, 385
  force polygon, 378
  riveted joints, 393
  snow load, 384
  stress diagram, 379
  tension members, 392
  weight of, 383
  welded joints, 395
  wind load, 381

S shapes, 70
Safe loads, based on

deflection, 125
Safe load tables, channels, 118
open web steel joists, 144
plank floors, 252
solid wood columns, 266, 270
W and S shapes, 113, 114
Section modulus, 64, 82
plastic, 211
Service loads, 369
Shape factor, 213
Shear, 5, 6
diagrams, 37
horizontal, 34
investigation for, 105
vertical, 34
Shear and moment formulas, 59
Shear stress formula, 16
Shop drawings, 395
Shop welding, 200
Shrinkage reinforcement, 343
Simple beam, 23
Single bearing in rivets, 182
Single shear in rivets, 182
Slabs, one-way, 342
design of, 345
Slenderness ratio, 150, 152
wood columns, 262
Slope of roofs, 377
Slot welds, 197, 204
Snow load, 384
Softwoods, 220
Solid wood columns, allowable loads, 264
design of, 265

Spaced wood columns, 271
Spiral concrete columns, 357
Spiral reinforcement, 358
Split ring connectors, 275
Standard I-beams, 70
Statical moment, 65
Statics, 3
Steel, structural, 91
allowable stresses for, 98
grades of, 91
Steel beams, 91
fireproofing of, 128
rolled shapes, 92
web crippling, 141
Steel columns, 159
Steel grillage, 177
Steel pipe columns, 166
Steel truss, design of, 387
Stiffness, 12
Stirrups, 301
Strength design, 369
Strength of concrete, 290
Strength of materials, 3
Strength, ultimate, 10
Stress, allowable unit, 13
direct, 4
kinds of, 5
tensile, 4
unit, 5
unit compressive, 6
Stress diagrams, 379
Stresses, allowable for connectors, 98
allowable for steel, 99
in trusses, 377
in welds, 196
Stress-strain diagram, 209
plastic range of, 209

Stringers, 219
Structural design methods, 148
Structural lumber, 219
  allowable stresses, 220, 222
  moisture content of, 221
  properties of, 76
Structural mechanics, 3
Structural shapes, 92
  designations of, 94
Structural Steel for Buildings-AISC Specification, 93
Structural steel, grades of, 91, 93
Structural tubing columns, 166, 169
Struts, 149
  double angle, 162, 164
Suggestions for study, ix
Symbols used in steel design, 95

Tables, list of, back endpapers
T-beams, 332
  compression in flange, 338
  compression in stem, 340
  design of, 333
Temperature reinforcement, 343
Tensile force, 4
Tensile stress, 4
Tension, 5
Tension members, roof trusses, 392
Tension reinforcement in slabs, 344
Theory of bending, 61
Tied concrete columns, 355

Timber connectors, 275
  allowable loads, 282
  edge distance, 283
  end distance, 281
  species groups, 277
  strength of joints, 277
Timber Engineering Company, 277
Timbers and posts, 219
Transfer-of-axis equation, 85
Types of beams, 23
Types 1 and 2 construction, steel, 193
Typical loadings, 59

Ultimate strength, 10
Ultimate strength design, 368
  scope of, 373
Unbraced length of compression flange, 102
Unfinished bolts, 187
Uniformly distributed load, 24
Unit compressive stress, 6
Unit stress, 5
U.S. Forest Products Laboratory, 277

Vertical shear, 34

W shapes, 6
Water-cement ratio, 291
Web reinforcement, 301
  formulas for, 302
  of girders, 328
  notation for, 302
  under uniform loading, 303
Weights of materials, 133

Welded connections, 194
  details of, 205
Welded joints, design of,
  200
  in roof trusses, 395
  types of, 195
Welding, base metal, 197
  electric arc, 195
  electrodes for, 197
  field, 200
  shop, 200
Welds, continuous action
  of, 202
  stresses in, 196
  types of, 195, 196
Western Wood Products
  Association, 270
Wide flange sections, 66
Wind load, 381
  stress diagram, 381
Wood, allowable stresses
  in, 220

Wood beams, 219
  deflection, 231
  design procedure, 233
  end bearing, 238
  horizontal shear, 229
  limiting span length,
    231
Wood columns, 262
  design formulas, 263
  glued laminated, 274
  slenderness ratio, 262
  spaced columns, 271
  types of, 262
Wood construction, 217
Wood floor joists, span
  tables, 239
Wood joists, design of,
  246
Wood species, 220

Yield point, 9

List of Tables

		Page
1-1	Selected Allowable Stresses	14
3-1	Beam Diagrams and Formulas	60
4-1	Selected Wide Flange Sections, Properties for Designing	66
4-2	American Standard I-Beams, Properties for Designing	70
4-3	American Standard Channels, Properties for Designing	72
4-4	Selected Angles, Equal Legs—Properties for Designing	74
4-5	Selected Angles, Unequal Legs—Properties for Designing	75
4-6	Properties of Structural Lumber	76
5-1	Structural Steels for Buildings—1969 AISC Specification	93
5-2	Allowable Unit Stresses for Structural Steel	99
5-3	Deflection Coefficients for Uniformly Distributed Loads	110
5-4	Allowable Uniform Loads for Selected W and S Shapes Used as Beams.	114
5-5	Allowable Uniform Loads for Selected Channels Used as Beams	120
5-6	Beam Safe Load Table Based on Allowable Deflection	126
5-7	Approximate Weight of Concrete Fireproofing for Beams	129
5-8	Weights of Building Materials	133
5-9	Minimum Live Loads	134
5-10	Allowable Bearing Pressure on Masonry Walls	139
5-11	Standard Load Table for Open Web Steel Joists, J-Series	144
6-1	Allowable Unit Stresses for Columns of A36 Steel	155
6-2	Allowable Axial Loads on Columns—Selected W Shapes	160
6-3	Double-Angle Struts, Allowable Concentric Loads	164
6-4	Typical Column Safe Load Table, Standard Steel Pipe	167
6-5	Typical Column Safe Load Table, Square Structural Tubing	169
6-6	Live Load Reduction Computations	174
7-1	Usual Gage Dimensions for Angles	180
7-2	Allowable Tension and Shear Unit Stresses for Rivets and Bolts	184
7-3	Allowable Loads for Rivets and Threaded Fasteners in Bearing-Type Connections	185
7-4	High-Strength Bolt Tension	189
7-5	Effective Net Areas of Angles with One Hole Deducted	191
8-1	Allowable Working Strength of Fillet Welds	197
8-2	Relation between Material Thickness and Minimum Size of Fillet Welds	198
10-1	Allowable Unit Stresses for Structural Lumber	222
10-2	Maximum Spans for Floor Joists—Live Load 40 psf	242
10-3	Maximum Spans for Floor Joists—Live Load 30psf	244
10-4	Safe Uniform Loads on Plank Floors	252